INDIAN COELOMYCETES

INDIAN COELOMYCETES

J. Muthumary
Professor
Centre for Advanced Studies in Botany
University of Madras
Chennai

MJP PUBLISHERS

CHENNAI NEW DELHI TIRUNELVELI

Cataloguing-in-Publication Data

Muthu Mary, J (1951 -).
Indian Coelomycetes / by J. Muthumary.-
Chennai: MJP Publishers, **2023**
xviii, 348 p.; 24 cm.
Includes, References, Index.
ISBN 978-81-8094-162-7 (pb.)
1. Coelomycetes-India I. Title
579 : 550954 MUT MJP 145

ISBN 978-81-8094-162-7

Printed and bound in India

MJP PUBLISHERS
New No. 5, Muthu Kalathy Street
Triplicane
Chennai 600 005

Publisher : J.C. Pillai
Managing Editor : C. Sajeesh Kumar
Project Editor : P. Parvath Radha
Acquisitions Editor : C. Janarthanan
Editorial Team : B. Ramalakshmi, V.R. Padma, R. Hemalatha, S. Jeevasruthi, B. Annalakshmi, N. Yamuna Devi, C. Devi
CIP Data : Prof. K. Hariharan, Librarian
RKM Vivekananda College, Chennai.

To

My Mentor

Professor C.V. Subramanian

FOREWORD

We are in the age of Science where descriptive science currently receives low priority, especially biological science. There is much concern about conserving our biota, though we have only incomplete inventories of these. The low priority for descriptive biology in our scheme of education, exploration and study has inevitably led to lack of taxonomists to do the inventories. There are many groups of plants and animals and fungi that await exploration and description. The tropics are full of beautiful fungi awaiting study. Among the fungi, many groups are poorly explored and the Coelomycetes are no exception.

Any study of Coelomycetes therefore must be welcomed. Professor Muthumary of the Centre of Advanced Study, University of Madras has been studying Coelomycetes for over three decades. Apart from a descriptive account of Septoria species from southern India, she is known for her work on a range of Coelomycetes, including several endophytes, also from southern India. The present work titled *Indian Coelomycetes,* as she says modestly in the Preface, represents a summation of her work and that of her students. She devotes the introductory pages to methods of collection, techniques of culturing and microscopic study of Coelomycetes. She provides descriptions supported by illustrations (line drawings and photomicrographs) of over 140 species belonging to about 80 genera. That the entire work is based on the author's own collections gives it an aura of authenticity. The host-fungus index, and the indices of hosts and fungi, will be useful to users of the compendium.

I believe such descriptive work fills a lacuna in our efforts towards inventories that are relevant in mapping fungal diversity of the tropics. I therefore welcome this volume.

C. V. Subramanian
Former Director
Center for Advanced Studies in Botany
University of Madras
Chennai

PREFACE

This book is written to help mycologists in the study of Coelomycetes in as simple a way as possible. It is an introductory identification book designed for those interested in Coelomycetes. The Coelomycetes are a very different and difficult group of fungi to classify and identify. In India there are no books available for identification of Coelomycetes. This book is meant to solve the problem in identifying Indian Coelomycetes. Coelomycetes are from many substrates like leaf, twig and flowers of plants. Also they are known from soil and other substrates. Tropical and temperate, Antarctic and Arctic zones are abound with these fungi. Plants which are of little economic significance are also affected by these fungi. Leaf lesions, stem cankers, blights and anthracnoses are some of the diseases caused by these fungi. Therefore it is important that these organisms are identified correctly.

The taxonomy of this group had met lot of criticism from mycologists because of the anomalies in structure of the conidiomata. Undue importance was given to the conidiomatal structure in the past inspite of the conidiomata being important for classification of this group. There are really no generic distinction between acervular and sporodochial fungi in culture or with moniliaceous, dematiaceous and synnematous fungi. Mason (1937) interpreted all such conidiomata as different forms of a basic fungal stroma. In the present treatment, the conidiomatal types of various fruit bodies in the traditional method are followed for classification.

In the book I have made an attempt to give descriptions, illustrations and photomicrographs for about 78 genera in 142 species. My research on this group of fungi over several years has resulted in this book. As will be seen in this book it is readily apparent there still remain an enormous work is to be done in our country. The book is a summation of the research on Coelomycetes I have investigated or come across by me and by my students. I have provided descriptions for 142 Coelomycetes collected from Tamilnadu, India. Photomicrographs are given wherever necessary. The list of Coelomycetes given in the book provides an easy reference for the reader.

I hope the book will be useful for identification of Coelomycetes. The index in the book includes the list of Coelomycetes described, Host index, Fungus Index and Host-Fungus index. There is also a Bibliography section separately. The very important works on Coelomycetes is given at the end for the use of mycologists. Hope this book will help the readers to identify Coelomycetes properly and find out the references for Coelomycetes. I am pleased to introduce the readers to the "Indian Coelomycetes" for teaching and for learning mycology.

J. Muthumary

ACKNOWLEDGEMENTS

Thanks are due to my students Dr. S. Masilamani, Dr. R. Senthilkumaran and Dr. M. Murugan for their generous support in supplying details of some photomicrographs.

Special thanks are due to Mr. S. Kamalraj and Miss. S. Yogeswari for their help with the usage of computer in typing the manuscript and also in arranging the figures. It is an enormous work and I am thankful to MJP publishers for their cooperation in publishing the book.

J. Muthumary

CONTENTS

1

INTRODUCTION

IMPORTANCE OF FUNGI

Fungi constitute a major component in the microbial world. Mycology, the study of fungi, is important for students of life sciences. It is recognized as a major course for those who study microbiology for their career. Our everyday life is affected by fungi and they are present everywhere. They cause diseases in plants, animals and humans, produce industrially important products, decompose litter and play a very important role in the ecosystem. It is important to study their biology, physiology, taxonomy, genetics and biochemistry. In order to control the diseases caused by fungi, plant pathology with reference to fungi becomes essential. The study in relation to the secondary metabolites or other useful biocompounds produced by fungi is essential to exploit them for our own use. But sadly the fungi are being ignored or neglected in many studies. Majority of the fungi have never been screened for antibiotics or other useful biocompounds. The role of ecologists is very important to study fungi for their association with other organisms. Fungi depend upon other sources for their food, and are heterotrophic. They break down dead plant and animal material into simple compounds for the survival of other organisms. Mycorrhizal fungi have association with roots of higher plants and help the plants to absorb essential minerals from poor soils. With algae and cyanobacteria, mutually fungi form lichens. Both cultivated and wild mushrooms are delicacies for humankind. Baking and brewing industries use many types of yeasts for fermentation.

Fungi are known to produce pharmaceutically important compounds like antibiotics—penicillin and cephalosporin. *Tolypocladium* produces the immuno-suppressive anti-rejection transplant drug, cyclosporin. Various fungi produce steroids, hormones and even birth-control pills for commercial use. Mitosporic fungi like *Aspergillus*, produce citric acid commercially. This citric acid finds its use in cola preparation. Fermentation of cheese and cheese products use species of *Penicillium*. *Trichoderma* species produce enzymes which are used in softening stone-washed jeans. Recent researches show that anticancer drugs like taxol can be extracted from some of the endophytic fungi. Fungi are also detrimental to human life. Many diseases of animals and humans are caused by these fungi either directly or through their toxins. Contamination and spoilage of food is caused by many fungi. They destroy wood, timber, cloth, cotton, wool and any type of degradable material. In association with lichens, they cause cracks in rock. Various aspects of fungi like ecology, physiology, genetics, biochemistry and taxonomy can be used to exploit these organisms to be used by humankind. Progress in molecular biology and its application to the fungi is very much essential in the current context.

FUNGAL BIOTECHNOLOGY

Genetic manipulation in fungal biotechnology needs a set of tools to be used with fungi (Talbot, 2001) which includes transformation technology and a number of cytological, biochemical and genetic techniques. Genomic technologies, like yeast two-hybrid libraries and DNA microarrays can be employed with fungi (Phizicky and Fields, 1995; Yang and Speed, 2002; Kumar and Snyder, 2001). Fungal biotechnology laboratories construct vectors to be used to sequentially disrupt genes in transformation research (Royer *et al.* 1999). Fungal genome sequence is widely done to manipulate cloning of genes in fungi. The use of fungal transposons from filamentous fungi has been understood (Nicosia *et al.* 2001). Agrochemical and pharmaceutical fields need identifying essential genes. Generating conditional lethal mutants is essential during transformation. Selection of controllable promoters seem to be an attractive approach in recent studies (Hamer *et al.*, 2001). Biotechnological processes with fungi also include post-transcriptional gene silencing (PTGS). Other molecular techniques like EST collections, physical maps, micro-arrayed DNA or cDNA collections are found to be useful with fungi. Screening for desired phenotypes is possible with Restricted Enzyme Mediated Insertions (REMI) (Brown *et al.* 1998; Sweigurd and Ebbole, 2001). To enhance the biochemical activities of fungi, biotechnological applications are necessary. For strain improvement, mutagenesis and screening are used widely. Genome shuffling, as done with bacteria, can be employed for fungi also for strain improvement (Zhang *et al.*, 2002). Both industry and academia will be benefitted by the use of these biotechnological tools in fungi.

It is reported that the number of genes in filamentous fungi is roughly twice that reported for yeasts and similar to those found in *Drosophila.* Genome analysis in *Neurospora crassa* showed that many of the genes are not functionally evaluated. The function of the numerous genes in filamentous fungi is yet to be investigated. Many molecular secrets with reference to filamentous fungi are yet to be deciphered. The future of biotechnological research in filamentous fungi is very bright.

Fungi have attracted the attention of ecologists, biologists and biochemists in several ways over many years because of their diversity, complex life cycles and their usefulness to humankind. Many industries use fungi as sources of biocompounds in recent years. The correct identification of the fungi used in producing such compounds becomes very essential and therefore fungal taxonomy as an important and practical discipline is very much preferred. There are many schools of thought that traditional morphological applications to fungal taxonomy will fail to result in solved framework of these organisms. However, strong evolutionary basis with accurate information can be obtained even at species level by using new molecular appliances. Current record shows that there are approximately 80,000 species of fungi (Kirk *et al.*, 2001). Of them, about 60% of the fungi belong to Ascomycota and their asexual forms. But there are still many more fungi yet to be discovered (Hawksworth, 2001). The asexual stages of fungi occupy approximately 40% of the described species. Many of the industrially important fungi come under the asexual forms. Defining a fungal species includes different ways and methods. Generally all fungi are described based on morphological species concept. Morphological characters are used to distinguish species from other "morpho-species". In spite of being a subject of debate, this species concept is necessary for identification purposes. Based on this, the Coelomycetes collected from the Southern part of India are described. Transmission Electron Microscopic studies of the conidia and conidiogenous cells are given for *Silbospora cistina* (Cooke) Sutton and *Pestalotiopsis disseminata* (Thuem.) Stey. in the present work.

ENDOPHYTIC COELOMYCETES

Primary metabolism in organisms include processes for cell growth and function like protein synthesis, DNA replication and fatty acid biosynthesis. In contrast, secondary metabolism includes the biochemical pathways that are not essential for the growth and reproduction of the organisms. With reference to that, endophytic organisms have a unique ecological niche influencing plant diversity. The term endophyte was introduced by De Bary (1866) who defined endophytes as "Organisms that colonize internal plant tissues" (*endon*-within; *Phyton*-plant). Mycologists use the term fungal endophytes for fungi causing asymptomatic infections in aerial tissues of plants.

Histological investigations carried out on *Taxodium mucronatum* by the author suggested that the endophytic fungi were observed to grow both in the intercellular and intracellular spaces in the cortical tissue of the twig. The most common method to study endophytes is to isolate them from the host after surface sterilization. Endophytic fungi grow as elongated, convoluted, sparsely branched hyphae, running through intercellular spaces parallel to the long axis of leaves and stems (Clay, 1989). Among the endophytic Coelomycetes, *Bartalinia robillardoides* and *Camarosporium palliatum* are commonly isolated from plants like *Catharanthus roseus* and *Terminalia bellirica* (Bhuvaneswari and Muthumary, 2005). From several medicinal plants and at least a dozen tree species, *Botryodiplodia theobromae* has been reported as endophytes (Suryanarayanan *et al.* 2002). *Pestalotiopsis* spp. the most commonly isolated endophytic fungi of rainforest plants (Nag Raj, 1993) belong to Coelomycetes. In fact it is not rare to isolate one of these species from any tropical plant. In the author's laboratory, *Phyllosticta* spp. were recovered from several of the selected medicinal plants. Endophytic fungi are increasingly recognized as sources of novel bioactive compounds and secondary metabolites for biological control (Strobel, 2003). In addition to studying the distribution and ecology of fungal endophytes from medicinal plants, special attention should be given to screening them for potent metabolites. Symbiotic relationship of host and microbes produces new and interesting bioactive compounds which may find use in pharmaceutical industries. Some of the potentially important compounds isolated from endophytic fungi include cryptocin, cryptocandin, preussomerin, phomopsichalasin and torreyanic acid. Above all, endophytes have been proved to be an effective source of taxol, a cytotoxic drug used in the treatment of cancer (Wani *et al.*, 1971).

Taxol

Taxol, a highly functionalized diterpene, was first discovered in the bark of Western yew (*Taxus brevifolia*), a relatively rare and slow-growing tree by Wani *et al.*, (1971) and the yield is very low (0.01–0.03%). Efforts were therefore made by several research groups to address the supply problem of taxol in a variety of ways. Any fungus, for example, that can be manipulated to increase the production of taxol by some technique, can serve as an inexhaustible source of taxol, and the supply problem of this drug can be solved forever. After several years of effort, a novel endophytic fungus, *Taxomyces andreanae*, was discovered in *T. brevifolia* by Strobel *et al.*, (1993), which was able to produce taxol. The endophytes of medicinal plants provide a good source of compounds of biological activity and endophytes are an untapped reservoir of potentially novel, effective drugs (Stierle *et al.*, 1995). The investigations and the significance in the discovery of fungi that produce taxol indicate that there are abundant resources of fungi that produce taxol.

Most of them are endophytes of *Taxus* spp. Therefore, a study has reported the isolation of taxol-producing endophytic fungi from tropical medicinal plants in India for the first time. These newly isolated fungi are different from those endophytic fungi that produce taxol previously reported. Hence, from both the ecological and economic viewpoints, a microbial source will be the replacement for the yew.

2

STUDY OF COELOMYCETES

COLLECTION AND EXAMINATION OF COELOMYCETES

The specimen with Coelomycete fruit bodies are collected from the field using a hand lens. After transportation to the laboratory, the sample with the fructification is examined under a stereo-microscope to study the nature and position of the conidiomata in relation to the substratum. The sample with a few conidiomata is placed on the slide for whole mounts to study the ostiole and conidiomatal setae. Squash mounts are prepared with distilled water by applying slight pressure for the release of conidiogenous cells and conidia. Semi-permanent slides are prepared using either lactophenol or lactophenol with cotton blue. Occasionally 1% erythrosin in 10% NH_4OH was found suitable to study conidiogenous cells and conidia. The slides are sealed using DPX for future reference. No staining is required for brown-coloured conidia and distilled water mounts are used for best results in studying cellular or mucilaginous appendages. Lactophenol cotton blue and lactofuchsin are suitable for hyaline conidia and conidiogenous cells (Hawksworth *et al.*, 1983).

Sample Collection

Coelomycetes grow on dead or live stems, branches, twigs, leaves, bark and fruits. They normally cause necrotic lesions or cankers. When looking for Coelomycetes on these substrates, the presence of conidiomata should be confirmed by using a hand lens. The collection data should be recorded along with the name of the host as it is essential for identification of the species. The specimen should be air-dried to be stored in the herbarium. Proper microscopic examination of the material in the laboratory is essential because characters such as the type of conidiomata, the relationship of the conidiomata to the substrate, the conidiogenous cells, the tissues comprising the conidiomatal wall layers and the ostiole are important for the correct identification of the species. Apart from these characteristics, conidium development, conidial morphology, paraphyses and behaviour in artificial culture are also taken into consideration for proper identification.

Coelomycetes could be collected from any environmental location. Many angiosperms harbour fruitbodies of Coelomycetes on leaf, bark, twig, branch, fruits and seeds. Other members of the plant kingdom like lichens and algae also possess fungi. Coelomycetes occur either in living tissues or live as saprotrophs on decaying substrates. Close examination of the substrate invariably with the help of a magnifying glass or hand-held lens is very essential to confirm the presence of conidiomata. These fruitbodies or conidiomata are either superficial or fully or partly immersed in the substrate. Conidial masses of varying colours ranging from orange and pink to black may be visible under moist conditions, sometimes in the form of glistening conidial cirrhi. Very close and careful observation is required to detect the presence of conidiomata on the substrate.

A Coelomycete mycologist going for a field collection trip may need a hand lens, paper and plastic pouches, knife for trimming specimen from bark and wood, field notebook, scissors, big collection bag to accommodate fungal collection and a plant taxonomy and identification manual to identify the plants in the area of collection.

Each sample collected should have the name of the host plant on which the fungus grows, because invariably the determination of the species name is linked with the identity of the host. Enough sample material should be collected, because in case the collection is described as new to science, ample material is needed to serve as type specimen and for deposition in different international herbaria. In instances where the host plant is not identifiable in the field, inclusion of fruits and flowers along with the other plant parts may facilitate correct identification later. Moist samples may be dried properly before being brought to the laboratory to avoid the growth of unwanted moulds and spoilage of samples. The diseased plant material may be processed in the same way followed for preserving botanical herbarium specimen.

Laboratory Study

Laboratory study and microscopic examination of the sample is necessary for identification of Coelomycetes. Generally Coelomycetes have minute fructifications and possess very small conidiogenous cells and conidia. The position of the conidiomata in relation to their host is very important. For example, when growing on a leaf, whether the conidiomata are superficial, epidermal, subepidermal or cuticular is to be determined. This can be determined later in the laboratory after taking sections of the sample for correct identification. Air-drying is found to be the best method of preservation for samples of Coelomycetes.

Slide Preparation by Sectioning

Coelomycetes are special fungi where the nature of the fructification is essential for correct identification. For this purpose, sections of the fructification should be made either with free hand or with any other simple microtome. Usually a freezing microtome with a cryostat is used for taking vertical section of the fruit body. In case of fresh material, no soaking is necessary. If the sample is dry or lignified tissue, softening is necessary using ethanol, water, lactic acid or 2% KOH for 1–2 hours.

While sectioning the samples for microscopical observation, a small strip or a bit of the specimen with few conidiomata is placed on a microscopic slide and held by the finger firmly. The sample is sectioned vertically using a razor blade so that a series of sections are obtained on the slide. These sections will be approximately 10–15 μm thick. Thereafter the sections are transferred with the help of camel hair brush to distilled water in a watch glass. These sections are examined under a stereo-microscope to select the best section, which is used for temporary slide preparation. It is very important to use fresh edges of razor blades for each sectioning. The best section selected is mounted on to a microscopic slide with proper mountant similar to lactic acid or lactophenol-cottonblue and sealed with nailpolish or DPX. This will be a semipermanent preparation done with great care. To study conidiogenous cells and conidiogenesis, tease mounts of the conidioma are done on the slide and preserved similarly. When observing the sample with microscopes equipped with phase contrast optics, it is not necessary to stain conidia or conidiogenous cells. Distilled water gives very good results. The use of 3% erythrosine in 10% NH_4OH solution will be found useful to observe clearly the conidial cell wall and appendages. Sometimes plain distilled water with the sample in a phase contrast optics will be enough to observe the very small conidia and conidiogenous cells.

A modified Leifson's flagella staining method has been successfully applied to study appendages in Coelomycetes and other appendaged conidial forms (Punithalingam, 1981, 1984, a, b; Punithalingam and Woodhams, 1982, 1984), (Masilamani and Muthumary, 1994).

Flagellar staining For staining appendages, a modified Leifson's flagellar staining technique is used (Punithalingam and Woodhams, 1984). This technique is simple, quick and cheap.

The stain is a mixture of solution A, containing 1.2% basic fuchsin in 95% industrial methylated spirit, and solution B, containing 1.5% tannic acid and 0.75% sodium chloride in distilled water. Phenol (1/1000) is added to solution B to prevent mould growth during storage. The slides are prepared by placing conidiomata on microscope slides and squeezing them in a drop of water to obtain conidia. These are then spread to obtain a thin film on a glass slide, dried at room temperature and stained with 1 ml of staining solution. The stain is allowed to stand on the surface of the thin film for 2–5 s, rinsed with running tap water, dried at room temperature and mounted in Euparol (ASCO Laboratories, England).

Nuclear staining For staining nuclei in hyaline conidia, HCl–Giemsa technique of Hrushovetz (1956), adapted and modified by Punithalingam (1983) is used. Material is either fixed directly in Carnoy's fluid containing absolute alcohol and glacial acetic acid (3 : 1) or a mature conidioma is transferred to a microscope slide, teased and spread to produce a thin film of conidia, air-dried at room temperature, fixed in Carnoy's fluid for 15 min., rinsed thrice in 95% ethyl alcohol and stored in 70% ethyl alcohol for a minimum period of 24 h or stored for several days until staining. Before staining, the slides are immersed in 1N HCl for 5 min. at room temperature (23–25°C) followed by hydrolysis in 1N HCl at 58.5°–60°C for 8–9 min. The slides are rinsed thrice in distilled water each for 1 min., rinsed in phosphate (K_2HPO_4) buffer at pH 6.9 and stained in Giemsa stain (Gurrs improved R66 BDH Chemical Ltd., UK). The slides are again rinsed with phosphate buffer then in distilled water, air-dried at room temperature and mounted in Euparol.

For staining pigmented conidia, since the modified Giemsa–HCl method and or the aceto-orcein method did not give good results, the following technique was used. Conidia were spread on a glass slide in a minimum amount of water and air-dried at room temperature. Slides were fixed in Carnoy's solution for 15 min., and stored for 24 h in 70% ethyl alcohol. A longer period of storage of up to 10 days gave better results. The slides were then immersed in 1N HCl for 5 min., at room temperature. Hydrolysis was carried out for 10–20 min. at 60–68°C. Hydrolysis at higher temperatures beyond 60°C and for a longer period of 10–20 min. gave better results compared with the technique suggested by Punithalingam (1983). Slides were rinsed 5 times in distilled water at least for 1 min., each, followed by 4 rinses in phosphate buffer at pH 6.9. They were stained in Giemsa stain for 30 min., rinsed in phosphate buffer at pH 6.9 followed by rinsing in distilled water. Instead of directly mounting in Euparol the slides were counterstained. They were immersed in 60% glacial acetic acid (BDH Chem., U.K.) for 10 min. for decolorization of the Giemsa stain and immersed in 1% citric acid for 5 min. This improved staining. They were then treated with a few drops of aceto-orcein, gently heated and mounted in Euparol.

For pigmented and thick-walled conidia, a conidial wall cracking technique was used. A thin film of conidia from conidiomata was spread on a glass slide and air-dried at room temperature. A coverslip was placed on the material and tapped, followed by slight pressure. The routine Giemsa staining method which gave better results was used, since the walls of conidia cracked after tapping and application of pressure, and cytoplasm with nuclei came out from the cracked conidia which were stained and easily seen.

CULTURE OF COELOMYCETES

Isolation of Coelomycetes

General mycological methods of isolation of fungi are applicable for Coelomycetes also. Coelomycetes are very slow-growing when compared to other fungi especially *Aspergillus, Penicillium* and other microbes. Serial dilution may be used to isolate these fungi on a minimal medium such as tap water agar. Better sporulation is obtained when the cultures are subjected to 12 h nuv (near ultra violet) followed by 12 h dark (Leach, 1962). Plastic Petri dishes stimulate sporulation at 25°C. Sterilized wheat straw or dissected bits of the host leaf or stem stimulate the production of conidiomata when included in the agar media. Various media were observed to stimulate growth of several Coelomycetes and studies are yet to be explored on growth condition of these fungi. Oatmeal agar is reported to stimulate production of conidiomata and conidia in *Phoma* spp. (Dorenbosch, 1970; Boerema *et al.*, 1965, 1968, 1971, 1973, 1977). Gramincolous species grow on PDA, PCA and OMA (Sutton, 1965, 1968a–c). Punithalingam (1970a–c, 1974a) used PDA and oatmeal agar in studying various species of *Phyllosticta.* In studying various species of Coelomycetes like *Septoria, Phyllosticta, Botryodiplodia, Pestalotiopsis* and *Colletotrichum.* (Muthumary (1986a–c, 1993), Masilamani and Muthumary (1996) and Jayachandra and Muthumary (2001) used PDA and OMA. Riddel's (1950) slide culture method was found to be very useful in studying various stages of development of conidiomata (Muthumary and Vanaj 1986; Muthumary, Jayachandra and Preetha, 1993; Masilamani and Muthumary, 1994; Masilamani and Muthumary, 1996; Jayachandra and Muthumary, 1998; Jayachandra and Muthumary, 2001; Murugan and Muthumary, 2001; Murugan and Muthumary, 2003).

Generally the leaf or twig with the conidiomata are incubated in a moist chamber after surface sterilization with 0.01% mercuric chloride. After 2–3 days the conidial mass from the conidiomata is issued through the ostiole on to the leaf surface in tendrils. The conidia are collected aseptically in sterile water containing 5 mg/ml streptomycin. A few drops of the spore suspension is spread over on 2% tap water agar and incubated at 24°C for 6 h. Individual germinated conidia are transferred to oatmeal agar plates. Subcultures are also made from conidiomata produced in culture. Slide cultures are used to study the mode of conidial germination and the initial stages of the development of the conidiomata (Riddell, 1950).

With reference to culturing the Coelomycetes in artificial conditions and media, several anomalies were observed. Sutton and Alcorn (1984) suggested that new species of Coelomycetes should be described on the basis of data collected from fungi grown on synthetic media. However, many examples of anomalies in the fungus were observed when grown in artificial culture media when compared with the characters found in nature.

The author observed that many species of *Pestalotiopsis* produced pycnidial conidiomata in artificial culture media but only acervuli of these fungi are found on leaves or natural substrates. The same phenomenon had been reported in *Pestalotiopsis* (then *Pestalotia*) by Guba (1961). There are many examples found in literature in support of this argument. During the studies on *Discosia* species, they produce sporodochial and intermediate type of structures when grown in artificial media but typical applanate or distinct stromatic conidiomata are characteristically produced on the natural host. Nag Raj (1993) suggested that by growing the isolate on sterilized plant tissues like leaves, one can avoid the anomalies found in Coelomycetes. Literature is available on growth of these fungi without anomalies in cultures (Sutton, 1962, 1968c; Booth, 1971; Hawksworth 1974).

Microscopy

The general method of microscopic examination is the best for identification of Coelomycetes. It is less time-consuming and cost-effective when compared to either TEM or SEM methods. The preliminary examination of the sample may be done with the help of a simple binocular dissecting microscope. The microscopical features can be very clearly studied with a compound microscope equipped with phase and differential interference contrast (DIC) optics. The phase and DIC attachments will help to clearly observe conidiogenous cells and the flagella of the conidia along with any other special features of the conidia and conidiogenous cells. Generally unstained sections and squash mounts in plain distilled water serve the purpose very well.

A Carl Zeiss Axiotar microscope equipped with bright field, phase contrast and Nomarskii will be useful to observe the microscopical characters of the fungi. Photographs are taken with a camera attached to the microscope. For drawings, a camera lucida attached to the microscope is used. Nowadays a digital camera attached to the microscope is found to be very useful in documentation of the microscopical features of fungi.

Preservation of Coelomycetes

Investigations on the biochemistry and molecular biology of Coelomycetes have been out by many mycologists and biochemists in recent times. The *ex situ* conservation of Coelomycetes for a long period of time has become important to the scientific community. Generally, the Coelomycetes are slow-growing in artificial culture media and may be overgrown by fast growing fungi like *Aspergillus* and *Penicillium*. Serial dilution on tap water may prevent contamination when isolating Coelomycetes. Sutton suggested 12 h near ultraviolet (black light) irradiation/12 h dark (Leach, 1962) in plastic Petri dishes for better sporulation in Coelomycetes. Incorporation of sterilized cultural substrates such as wheat straw, cut leaves or stem with the basic agar may stimulate conidiomatal initial in culture. Generally, Coelomycetes are stored under mineral oil, as dried culture or on agar slant. However, long-term storage in mineral oil may lead to morphological and physiological alteration such as loss of sporulation capability, attenuation or loss of virulence in Coelomycetes. The oil depth used during preservation has been shown to affect the viability of fungal species. An oil layer of 0.2–0.5 cm prevents evaporation of water and dryness of the culture medium with sufficient oxygen whereas 1 cm oil creates anaerobic condition causing cellular damage (Pampianskaya, 1964). Organisms maintained in culture collection and subjected to frequent transfers may lose some of their original metabolic activity. Longevity of microorganism under mineral oil varies considerably depending upon the species, storage, temperature and culture media. The use of soil culture for long-term storage has been discouraged because of the risk of mutation and contamination by other organisms and Pampianskaya (1964) reported 100% viability and retention of pathogenicity in 24 strains of *Septoria* stored in soil up to one year. The endophytic species of Coelomycetes were reported to be viable for 2 years on agar slants. However, different storage regimes do not guarantee for the morphological, biochemical and genetic stability of endophytic fungi preserved in culture collections for more than a year.

The preservation of Coelomycetous fungi in culture collection demands attention regularly because they may easily lose their sporulating capability. Periodic monitoring is necessary to check the viability of Coelomycetes in culture. The suggested preservation methods for Coleomycetes include storage under mineral oil and sterilized water and agar slants with autoclaved plant tissue. The method recommended by Suryanarayanan and Kumaresan (1998) using agar slants within Eppendorf tubes topped by mineral oil may help in easy transportation and storage.

CULTURING COELOMYCETES

In view of the variability and diversity in the form of the Coelomycete conidiomata it is no easy task to accommodate them satisfactorily in a system of classification based on conidiomata. One finds acervular, stromatic and pycnidial conidiomata in the species of *Colletotrichum* grown in culture. Similarly, conidiogenous cells are formed directly from the mycelium in culture though such fungi are known to produce pycnidial conidiomata in nature (Subramanian and Reddy, 1974; Baxter, 1981: Baxter *et al.*, 1985). Therefore, as more and more data to this effect become available, the distinction between Hyphomycetes and Coelomycetes may be abandoned because of the presence of intermediary stages between Hyphomycetes, acervular, stromatic and cupulate conidiomata. Species of *Pestalotiopsis* have not been studied in detail with reference to the development of conidiomata. Investigation in the author's laboratory on the various stages of development of conidiomata in species of *Pestalotiopsis* revealed interesting observations. The fungus *Pestalotiopsis* produces conidia inside mature pycnidial conidiomata in culture. The central cells in the primordium showed signs of more schizogenous and lysigenous activity to form the central cavity. The remnants of the disorganized cells at the centre of the primordium were visible in many transverse sections of the pycnidial conidiomata. The remaining portions of the cells often gelatinize and fill up the cavity with the mucilaginous matrix. Further developmental stages showed the formation of the conidium simultaneously with the cavity formation. As the conidia mature, they are released from the conidiogenous cells and fill up the cavity. The formation of conidiogenous cells inside the cavity of the conidiomata is another interesting feature studied in the species of *Pestalotiopsis.* It was the pioneering work of Punithalingam (1979) who studied in detail, the process of conidiation in *Ascochyta* spp. in culture, which produced both temporary and permanent conidiogenous cells inside the pycnidium. Earlier workers who studied the development of pycnidia in culture examined only mature pycnidia but the development was not given due consideration. It was Punithalingam (1979) who made a thorough investigation on the various stages of conidiation inside a pycnidial conidioma. According to him, the cells lining the pycnidial cavity produce conidia initially only to get sloughed off as soon as their function is over. Once a layer of conidiogenous cells lining the pycnidial cavity is completely dislodged, a new layer of cells lining the cavity functions as conidiogenous cells. Thus, the processes of conidiation and dislodgement are repeated continuously. These are the two main processes by which the pycnidial cavity gradually increases in size and becomes filled with conidia and sloughed off conidiogenous cells. This is another way adopted by nature to produce greater number of conidia inside a pycnidium where the space is very limited. During the development, as the primordium increases in size, the central cavity is to be formed because the primordium is a compactly packed pseudoparenchymatous tissue. Nag Raj (1981) suggested that the slime, which originates through lysis occurring during cavity formation, play an important role in conidium dispersal. The presence of mucilaginous matrix during the formation of the pycnidial conidiomata was reported in certain other pycnidial Coelomycetes also (Masilamani and Muthumary, 1994, 1995, 1996). The pioneering work of Punithalingam (1979) who studied in detail the process of conidiation in *Ascochyta* species in culture revealed a similar process in this fungus.

The production of temporary and permanent conidiogenous cells in *Ascochyta* is construed as a general feature of *Ascochyta* species in culture (Punithalingam, 1979). Similarly, the dual conidiation process was also observed in species of *Pestalotiopsis* in the author's investigation. The dual conidiation process was observed in species of *Phyllosticta caryotae* Shen in culture. Punithalingam (1979) reported that some of the observations made with *Ascochyta* in culture were in conformity with the interpretations of Archer (1926) during his investigation on cavity formation and ostiole development in some members of Sphaeropsidales. This suggests that probably these fungi which produce pycnidial conidiomata in culture follow some specific regularity in their conidiomatal development. It is not possible to conclude that these stages of developmental processes occur during the formation of pycnidia in nature.

3

TAXONOMY OF COELOMYCETES

The life cycle of a fungus includes a sexual phase and an asexual phase. During the sexual phase the spores or propagules are formed from cells where meiosis has occurred. When meiosis is absent during the formation of propagules, the fungi are known as anamorphic fungi (Deuteromycetes or Fungi imperfecti). Members of Ascomycetes or Basidiomycetes produce propagules from cells where meiosis has occurred. They are known as teleomorphic fungi. Anamorphic fungi are mostly prevalent and are also economically important. The Hyphomycetes produce conidia on separate hyphae and not on any specialized conidiomata. Agonomycetes produce sterile mycelial forms or they may produce structures like chlamydospores or sclerotia. Coelomycetes produce conidia within specialized fruitbodies like pycnidial, acervular, cupulate or stromateic conidiomata. More than 1000 genera in about 7000 species are known under Coelomycetes. They occur as parasites on higher plants or as saprotrophs. Modern reassessments on Coelomycete taxonomy are many in recent times—Sutton in Ainsworth *et al.* (Eds.). *The fungi* 4A:513, 1973; Nagraj (in Cole and Kendrick (Eds.) *Biology of conidial fungi* 1: 43, 1981; Nagraj, Coelomycetous anamorphs with appendage-bearing conidia, 1993; Sutton, coelomycetes I–VII, Mycological papers 1961–1981; VI 1977, Generic names proposed for coelomycetes; Nagraj *et al.* Icones generum coelomycetum I–XIII, 1972–1982 Nagraj *et al.* CJB 49–67, Genera of Coelomycetes I–XXVIII; Mathur *The Coelomycetes of India*, 1979; Kendrick, *The Whole fungus* 1979; Sutton, *The Coelomycetes* 1980; Nagraj in Cole and Kendrick, *Biology of Conidial Fungi* 43–84, 1981, Punithalingam, 1970(a–c) and 1974(a–b).

Coelomycetes are anamorphic fungi in which conidia are produced within a cavity lined by fungal or host tissue or sometimes by these two tissues. These groups of fungi are known mainly from temperate and tropical regions, which are capable of surviving in a wide variety of ecological situations. They have been reported from leaf litter and other organic debris, soil, water, air and also as parasites on many plant parts causing diseases. Many species are pathogenic to economically important crops and ornamental plants causing lesions, blights, anthracnose, cankers and diebacks. Many Coelomycetes are also responsible for rots in fruits and seeds. Recently many of these fungi have been reported to be present as endophytes in a variety of plant species.

These groups of fungi were recognized as early as 1796. Great progress has been achieved by research efforts of many mycologists in these fungi [Sutton, 1961, 1962, 1963, 1967a and b, 1968(a–b), 1969, 1971, 1972a and b, 1973, 1974, 1975, 1977, 1980; Punithalingam, 1970(a–c), 1974(a–b), 1979, 1989(a–b); Nag Raj and Kendrick, 1971, 1972 a and b, 1978, 1981, 1986(a–b)]. Studies on Indian Coelomycetes were initiated in our country by C.V. Subramanian and K. Ramakrishnan in the University Botany Laboratory in the year 1953(b) on the nature of the spore-appendage in *Neottiospora* Desm. *Plagionema* Subram. & Ramakr. was described for a Coelomyceteous fungus with a basal, lateral and apical appendaged conidia, which was later reduced to synonymy with *Ciliochorella* Syd. (Subramanian and Ramakrishnan, 1953a). This was the first research paper published on Indian Coelomycetes (*Pers. Communication* – C.V. Subramanian). Subsequently papers were published on Indian Coelomycetes by Subramanian and Ramakrishnan (Subramanian and Ramakrishnan (1954, 1958) and Subramanian, 1961). In fact, the appendaged Coelomycetous genus *Monochaetina* was erected by C.V. Subramanian for an Indian fungus with a single apical appendage. And also *Basipilus*, another Coelomycetous generic name, was coined for a fungus producing spores with 1–2 basal appendages. Therefore, it was C.V. Subramanian and Ramakrishnan who studied Coelomycetous fungi for the first time in our country. Studies on Indian Coelomycetes had been pursued by Subramanian and Reddy (1974) and followed by Muthumary and co-workers (Masilamani and Muthumary 1994, 1995, 1996; Muthumary, 1986a–c, 1987a–c, 1988, 1993; Muthumary and Sutton, 1986; Muthumary and Vanaja, 1986; Muthumary *et al.*, 1986, 1993; Muthumary and Masilamani, 1989). The primary aim of this book is to present data relating to about 142 species in 78 genera of Coelomycetes collected mainly from various parts in Tamil Nadu and a few other in India and to provide an easy and practical approach to the identification of these fungi. Details on conidiomata, conidiogenous cells, conidium development and conidial morphology in each species is given.

TAXONOMIC CHARACTERS OF COELOMYCETES

In Coelomycetes the conidium is formed in specialized structures such as acervuli or pycnidia. It may not be possible to simply classify these conidiomata into two broad categories because of their morphological diversity. However, the five morphological categories—acervuli, pycnidia, stromata, cupulate and pycnothyrioid pycnidia—have been widely accepted.

Acervuli are subcuticular, inter-epidermal, sup-epidermal or sub-peridermal in the host and the enclosing fungal layer is absent. The conidiogenous cells arise from the upper layers of the fungal stroma lining the cavity of the acervulus. The overlying host tissues are ruptured irregularly by the developing mass of conidia. Often septate, brown, thick-walled setae are seen randomly distributed throughout the conidiomata while in some, they are marginal. In some species, the initially closed pycnidia finally become wide open and described as acervuli by mistake. Cupulate conidiomata have also been described as acervuli in many instances. Generally species of *Colletotrichum, Pestalotiopsis, Marssonina, Phloeospora, Colletogloeum, Stilbospora, Coryneum* and *Seimatosporium* are characterized by acervular conidiomata.

The cuplike forms are very characteristic in their conidiomata which are discrete, usually superficial and often adorned with sterile hyphae or setae that are hyaline and pigmented and often septate. Species of *Heteropatella, Diarimella, Stauronema, Dinemasporium* and *Pseudolachnea* are characterized by cupulate conidiomata.

The sporodochial, acervular and cupulate conidiomata are very difficult to be differentiated when they are grown in artificial culture media as they are closer to each other than with pycnidial forms.

The pycnidial conidiomata show a range of variability from simple flask like structures to complex stromatic forms. Simple pycnidium is a globose to flask-shaped, often unilocular structure with an ostiole for the release of conidia. The pycnidia may be immersed, erumpent or superficial. Some are provided with hyaline or pigmented setae which are often septate.

The pycnothyrial conidiomata are distinct from pycnidial or stromatic categories. They are generally flat, hemispherical or irregular. Often they are superficial or subcuticular. They are composed of an upper shield, sometimes supported by a central column of cells. The conidiogenous cells are seen only at the base of the conidioma. Sometimes there is a regular ostiole. Otherwise dehiscence may occur by a slit, irregular tears or longitudinal fissure.

Stromatic conidiomata are distinct from pycnidia and include a variety of pulvinate and clavate forms. Some stromatic forms incorporate host and fungal cells whereas in some the conidiomata are composed solely of fungal tissue. In multiloculate conidiomata, each locule may have a separate ostiole.

The wall tissues of the conidiomata have usually been described as pseudoparenchymatous and plectenchymatous. In a simple pycnidium, a single tissue type is present in the wall. In a stromatic conidioma, several tissue types may be involved. Often the pale brown cells of the wall become darker near the ostiole. In some pycnidia, the wall is several cells thick (*Phyllosticta)* and is composed of outer textura of dark brown, thick-walled cells and inner textura prismatica of hyaline, thin-walled cells. Very complex wall tissues are present in some genera, e.g. *Cornicularie1la.* In the more common *Pestalotiopsis,* the outer layers of the conidiomata, are thick-walled and pigmented whereas the inner layers are less pigmented and thin-walled.

Conidium Ontogeny

The significance of conidiogenesis in classification of conidial fungi gained recognition only after the publication of Hughes (1953a–b) system. The concepts of Hughes have been extended to many Coelomycetes in the following years and new genera of Coelomycetes were established based on conidium ontogeny. According to Sutton (1980), the application of conidium ontogeny to the taxonomy of Coelomycetes is somewhat more difficult because in Coelomycetes conidiogenous structures are much smaller and consequently more difficult to interpret.

Not much information is available regarding the electron microscopic details on conidium ontogeny in Coelomycetes. Similarly, time-lapse photomicrography, which is readily applied in the study of Hyphomycetes has not been used at all in the Coelomycetes. Therefore, basic problems of conidium ontogeny in many Coelomycetous fungi are yet to be solved.

The available data on the conidium ontogeny in Coelomycetes show that phialides and proliferating phialides are very common in many Coelomycetous forms followed by blastic-single and blastic-sympodial ontogeny. Blastic-annellidic and polyblastic ontogeny is less common especially in pycnidial forms.

Conidia

The conidia in Coelomycetous fungi exhibit diversity in form, from amerospores to helicospores, and the complex form of conidium is seen in *Geastrumia.* Sutton (1973) recognized two types of septa in Coelomycetes: the eusepta are present usually in common genera and the distosepta are present in genera like *Coryneum, Massariothea* and *Pestalotiopsis.* Concolorous pigmented conidia are common whereas the versicoloured conidium where the end cells remain hyaline and the median cells show variable degrees of pigmentation is present in species of *Pestalotiopsis.* In many taxa with hyaline conidia, the walls are usually smooth, and develop pigmentation and ornamentation of the walls when they mature. Often the conidium in many Coelomycetes bear appendages which are described variously as pedicels, setae and cilia. Appendages are common in Coelomycetes which are adapted to special ecological niches such as marine habitats, aquatic environment and also terrestrial conditions. According to Nag Raj (1993) nearly 40% of Coelomycetes possess appendages in their conidia.

The appendages may be cellular (*Bartalinia lateripes*) or extracellular ((*Phyllosticta, Urohendersonia*) (Nag Raj, 1993). Based on cytological studies Punithalingam (1989) listed five major types of appendages

1. enucleate mucilaginous
2. enucleate, tubular and filiform
3. enucleate, tubular
4. nucleate, tubular
5. nucleate with apical cell/appendages.

Both cellular and extracellular appendages vary in the developmental processes. Lot of information with reference to the cytochemistry of the extracellular appendages is very essential for a better understanding of these novel structures. A comprehensive treatment of appendaged Coelomycetes by Nag Raj (1993) provides very valuable information about these appendages.

4

DESCRIPTION OF COELOMYCETES

LIST OF COELOMYCETES

1. *Amerosporiopsis gaubae* Petrak
2. *Amerosporium concinnum* Petrak
3. *Aplosporella beaumontiana* Ahmad
4. *Ascochyta avenae (*Petrak*)* Sprague & Johnson
5. *Ascochyta festucae* Punithalingam
6. *Ascochyta phaseolorum Saccardo*
7. *Ascochyta pisi* Lib.
8. *Ascochyta sorghi* Sacc.
9. *Asteroma psidi* Prasad, Tyagi & Agnihotri
10. *Bartalinia robillardoides* Tassi
11. *Botryodiplodia theobromae* Pat.
12. *Camarosporellum nervisequum* (Tassi) Tassi
13. *Camarosporium quaternatum* (Hazsl.) Schulz
14. *Ceuthospora lauri* (Grev.) Grev.
15. *Chaetoconis polygoni* (Ell. & Ev.) Clements,
16. *Chaetodiplodia sp.*
17. *Chaetomella acutiseta* Sutton & Sarbhoy
18. *Chaetomella raphigera* Swift

19. *Chaetopatella indica* Vasant Rao & Verghese
20. *Chaetospermum carneum* Tassi.
21. *Chaetospermum chaetosporum* (Patquillard) Smith & Ramsbottem
22. *Ciliochorella mangiferae* Sydow
23. *Coleophoma crateriformis* (Dur. & Mont.) Hohn.
24. *Coleophoma cylindrospora* (Desm.) Hohn.
25. *Coleophoma empetri* (Rostr.) Petrak
26. *Colletogloeum feroniae* (T.S. & K. Ramakr.) Sutton
27. *Colletogloeum sissoo* (Syd.) Sutton
28. *Colletotrichum capsici* (Syd.) Butler & Bisby
29. *Colletotrichum circinans* (Berk.) Vogl.
30. *Colletotrichum dematium* (Pers. Ex Fr.) Grove
31. *Colletotrichum falcatum* Went.
32. *Colletotrichum gloeosporioides* (Penz.) Sacc.
33. *Colletotrichum lindemuthianum* (Sacc. & Magan.) Br. & Cav.
34. *Coniella castaneicola* (Ell. & Ev.) Sutton
35. *Coniella fragariae* (Oud.) Sutton
36. *Coniothyrium fuckelii* Sacc.
37. *Coniothyrium palmarum* Cda.
38. *Conostroma quercicola* Muthumary
39. *Coryneum indicum* Sutton & Rizwi
40. *Coryneum quercinum* Muthumary
41. *Coryneum modonium* (Sacc.)
42. *Cryptosporiopsis malicorticis* (Cordl.) Nannfeldt
43. *Cylindrogloeum trillii* (Ell. & Ev.) Arx.
44. *Cytospora chrysosperma* Pers.
45. *Cytosphaera mangiferae* Died.
46. *Diarimella setulosa* Sutton
47. *Dinemasporium aberrans* Sutton
48. *Dinemasporium strigosum* (Pers. Ex Fr.) Saccardo
49. *Diplodia sycina* Mont.

50. *Discogloeum veronicae* (Lib.) Petrak
51. *Discosia artocreas* (Tode ex Fries) Fr.
52. *Discosia strobilina* Lib.
53. *Discosia vagans de* Not.
54. *Dwiroopa sundara* Subramanian and Muthumary
55. *Dwiroopella ramya* Subramanian and Muthumary
56. *Entomosporium mespili* (Dc Ex Duby) Sacc.
57. *Fusicoccum indicum* Tilak & Rokde.
58. *Hainesia lythri* (Desm.) Hohn.
59. *Harknessia eucalypti* Cooke
60. *Harknessia podocarpi* Lindquist & Sutton
61. *Harknessia shearii* Petrak
62. *Harknessia thujina* Ell. & Ev.
63. *Hendersonia sp.*
64. *Heteropatella indica* Muthumary
65. *Hyalotiella transvalensis* Papendorf
66. *Kendrickomyces indicus* Sutton, Rao & Mhaskar
67. *Marssonina rosae* (Lib.) Died.
68. *Melanconium sp.*
69. *Microsphaeropsis sp.*
70. *Mirimyces pulcher* Nag Raj
71. *Monochaetia kansensis* (Ellis & Barhsolomew) Saccardo
72. *Monochaetia lutea* Swart and Griffiths
73. *Monostichella indica* Sutton
74. *Mycotribulus mirabilis* Nag Raj & Kendrick
75. *Paramassariothea falcata* Subram. & Muthumary
76. *Pestalotiopsis adusta* (Ell & Ev.) Steyaert
77. *Pestalotiopsis breviseta* (Sacc.) Stey.
78. *Pestalotiopsis cruenta* (Syd.) Stey.
79. *Pestalotiopsis disseminata* (Thuem.) Stey.
80. *Pestalotiopsis funerea* (Desm.) Stey.

81. *Pestalotiopsis guepinii* (Desm.) Stey.
82. *Pestalotiopsis maculans* (Corda) Nag Raj
83. *Pestalotiopsis palmarum* (Cooke) Stey.
84. *Pestalotiopsis sydowiana* Bres.
85. *Pestalotiopsis theae* (Sawada) Stey.
86. *Pestalotiopsis uvicola* (Spegazzini) Bissett
87. *Pestalozziella parva* Nag Raj
88. *Phaeocytostroma sacchari* (Ell. & Ev.) Sutton
89. *Phloeospora maculans* (Bereng.) Allesch.
90. *Phloeosporella leucosceptri* (Keissl.) Sutton
91. *Phoma glomerata* (Cda) Wollenw. & Hochapt.
92. *Phoma sp.*
93. *Phomopsis abdita* Sacc.
94. *Phomopsis archeri* Sutton
95. *Phomopsis dracaenae* Sahni
96. *Phomopsis eugeniae* Punithalingam
97. *Phomopsis lantanae* (Costa & Camara) Sutton
98. *Phomopsis mangiferae* Ahmad
99. *Phomopsis obscurans* (Ell. & Ev.) Sutton
100. *Phomopsis terminaliae* (P. Henn) Sutton
101. *Phomopsis tersa* (Sacc.) Sutton
102. *Phomopsis sp.*
103. *Phyllosticta arecae* Hohnel
104. *Phyllosticta artocarpina* Sydow & Butler
105. *Phyllosticta amaryllidicola* Van der Aa
106. *Phyllosticta citricarpa* (Mc Alp.) Van der Aa
107. *Phyllosticta codiaeicola* Died.
108. *Phyllosticta colocasiicola* Hohnel
109. *Phyllosticta cruenta* (Kunze ex Fr.) Kickx.
110. *Phyllosticta ficicola* Pat.
111. *Phyllosticta kigeliae* Died.

112. Phyllosticta state of Guignardia mangiferae Roy.
113. *Phyllosticta mimusopsis-elengi* Vasant Rao
114. *Phyllosticta murnadensis* (Ponnappa) Van der Aa
115. *Phyllosticta murrayicola* Van der Aa
116. *Phyllosticta musarum* (Cooke) Van der Aa
117. *Phyllosticta pervincae* Bissett & Darbyshire
118. *Phyllosticta plumbaginicola* Vasanth Rao
119. *Phyllosticta spinarum* (Diedicke) Nag Raj & Morelet
120. *Phyllosticta sp.*
121. *Pilidium acerinum* Kunze
122. *Pseudolachnea longiciliata* (Hino and Katumoto) Sutton
123. *Pyrenochaeta ilicis* Wilson
124. *Robillarda depazeoides* (W. & C.) Sacc.
125. *Robillarda sessilis* (Saccardo) Saccardo
126. *Satchmopsis brasiliensis* Sutton & Hodges
127. *Scyphospora phyllostachydis* Muthumary
128. *Seiridium anceps* (Sacc.) Sutton
129. *Septoria thespesiae* Ramakr. T.S. & K.
130. *Septocyta ruborum* (Lib.) Petrak
131. *Sphaerellopsis filum* (Biv-Bern. Ex Fr.) Sutton
132. *Sphaeropsis tumefaciens* Hedges
133. *Stagonospora brideliae* Thirum. & Narasimhan
134. *Stauronema cruciferum* (Ellis) Sydow
135. *Stilbospora cistina* (Cke) Sutton
136. *Suttoniella sp.*
137. *Tetranacrium gramineum* Hudson & Sutton
138. *Tiarosporella graminis* (Pirozynski & Shoemaker) Nag Raj
139. *Tiarosporella madreeya* (Subram. & Ramakr.) Nag Raj
140. *Truncatella laurocerasi* (Westd.) Stey.
141. *Urohendersonia pongamiae* Nag Raj & Ponnappa
142. *Waydora typica* (Rodway) Sutton

1. *Amerosporiopsis gaubae* Petrak

Source

Petrak, *Bot. Arch.*, **43**: 84 (1941); Sutton, B.C., *The Coelomycetes*, CMI, England, p. 373 (1980).

Description

Conidiomata pycnidial, globose, immersed, unilocular, thick-walled, 180–200 µm diam. Pycnidial wall made up of thick-walled, brown cells at the outside and thin-walled, hyaline cells at the inner side giving rise to conidiogenous cells. Ostiole central, circular. Conidiogenous cells lining the pycnidial cavity, phialidic, cylindrical, straight, hyaline, 10–12 × 2.5–3.0 µm. Conidia single-celled, hyaline, smooth, elliptical, apex obtuse, base acute to truncate, 5–8 × 2–2.5 µm.

Specimen collected

On dead leaves of *Cupressus* sp., Kodaikanal, Tamil Nadu, Nov. 1987; On dead leaflets of Benth., Tamil Nadu, Jul. 1988, collected by J. Muthumary. IC. NO: 112.

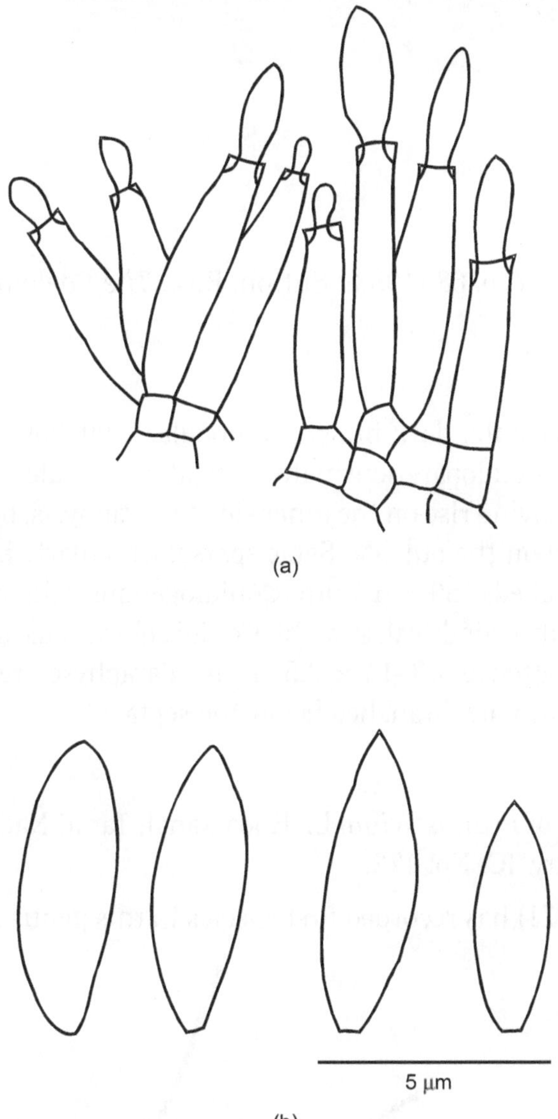

Figure 4.1 *Amerosporiopsis gaubae* (a) Developing conidia (b) Mature conidia

2. *Amerosporium concinnum* Petrak

Source

Petrak, *Sydowia,* **7**: p. 68 (1953); Sutton, B.C., *The Coelomycetes,* CMI, England, p. 619 (1980).

Description

Conidiomata separate, dark brown, unilocular, 300–500 μm diam., superficial, cupulate, wall pseudoparenchymatous, hyaline to pale brown, at the sides of elongated cells giving rise on the inner side to paraphysis, becoming thick-walled, and dark brown on the outside. Setae sparse, irregularly inserted, straight, dark brown, thick-walled, 150 × 6.5 μm. Conidiogenous cells lining the basal cavity, phialidic, smooth, cylindrical, straight. Conidia olivaceous, fusiform, aseptate, apex conical, base flattened, 13–14 × 2.5–3 μm. Paraphyses restricted to the lateral wall, incurved, septate, branched below the septa.

Specimen collected

On dead leaves of *Prunus avium* L., Kodaikanal, Tamil Nadu, Jul. 1989, collected by J. Muthumary. IC. No: 172.

Mathur (1971) has recorded first species in this genus from India.

Figure 4.2 V.S. of conidiomata (×200) of *Amerosporium concinnum*

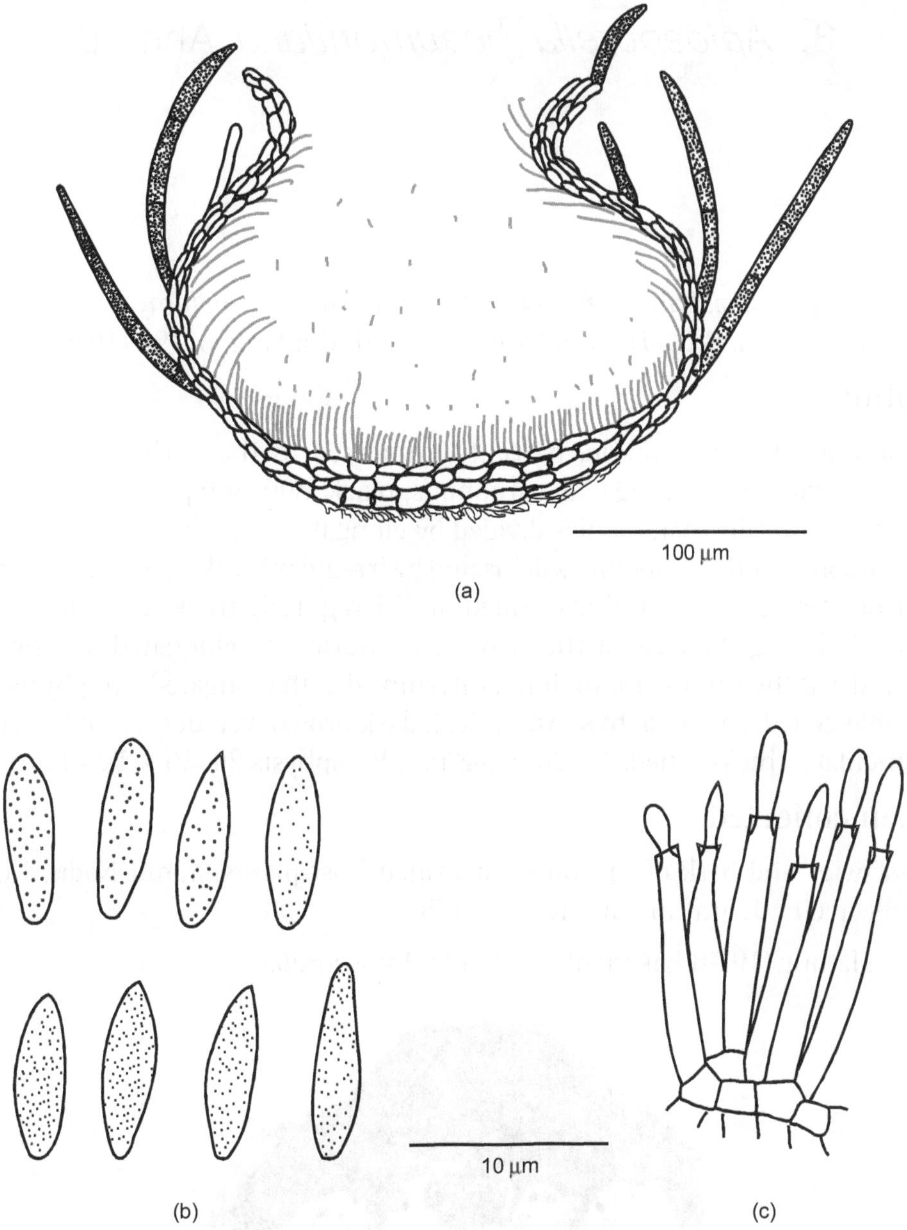

Figure 4.3 Amerosporium concinnum. (a) V.S. of conidioma, (b) Mature conidia, (c) Developing conidia

3. *Aplosporella beaumontiana* Ahmad

Source

Ahmad, *Biologia, Lahore,* **8**: 144 (1962); Mathur, *The Coelomycetes of India*, p. 16 (1979); Sutton, B.C., *The Coelomycetes*, CMI, England, p. 190 (1980).

Description

Conidiomata eustromatic, partially immersed in the periderm, erumpent to superficial, often aggregated, carbonous, black, 500–800 μm in diam. In vertical section, multilocular, locules divided by elongated, loosely arranged cells, with a common opening, sometimes dehiscing by irregular breaks at the tip of the stroma. Stromatic cells brown, thick-walled, wall irregularly thick. Conidiogenous cells mostly lining the base of the stroma, cylindrical to elongated, blastic, bearing conidia at the tip, 10–15 μm long, intermixed with elongated paraphysis. Conidia single-celled, spherical to subspherical, dark brown, verrucose, ridges irregularly reticulate, thick-walled, 15–20 × 4–8 μm. Paraphysis 20–40 × 1.5–2 μm.

Specimen collected

On twigs and bark of several unidentified host plants, Tamil Nadu, Nov. 1989. collected by J. Muthumary. IC. NO: 173.

Mathur (1979) has mentioned only *Aplosporella*.

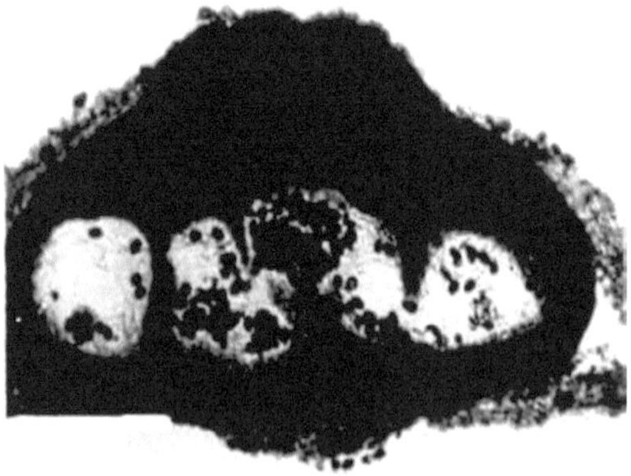

Figure 4.4 V.S. of conidiomata (×200) of *Aplosporella beaumontiana*

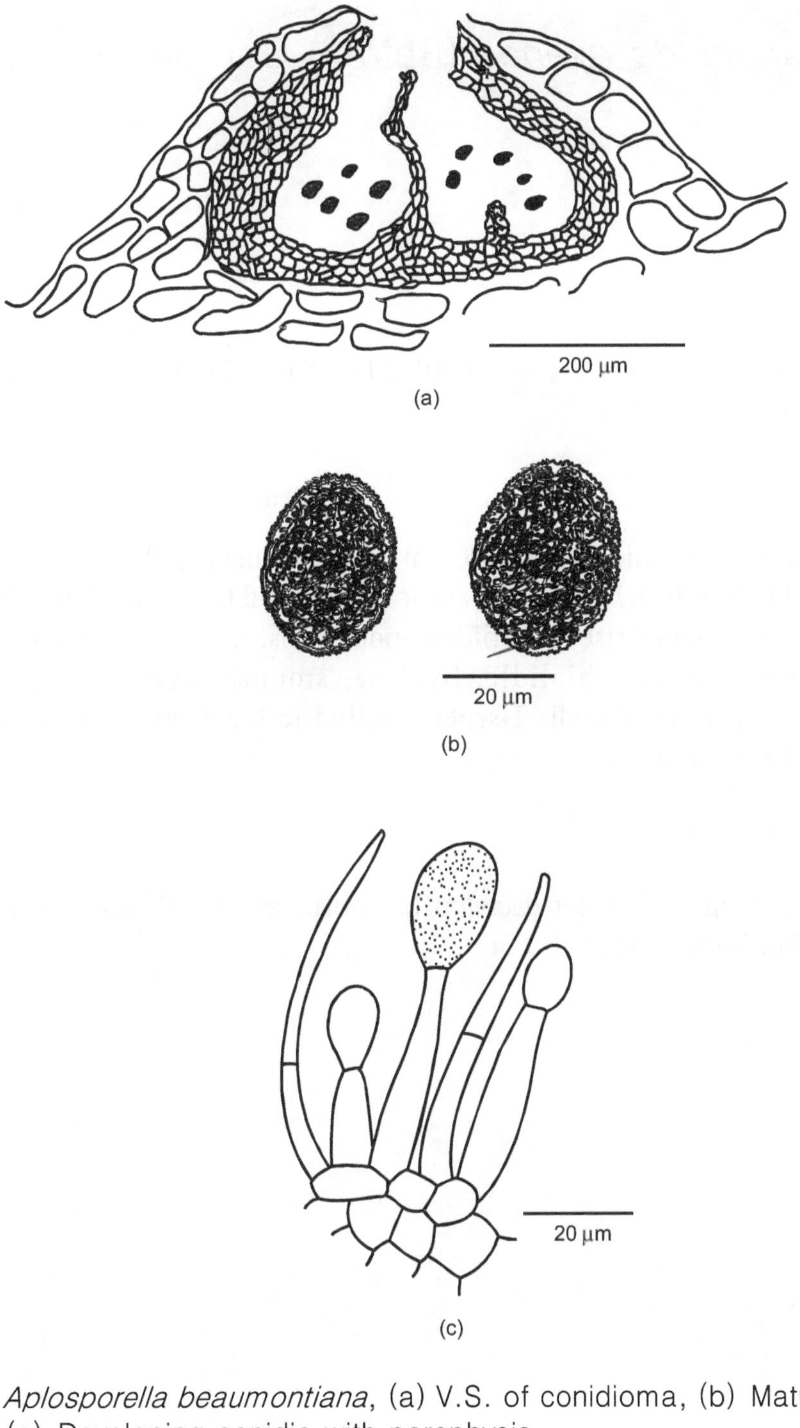

Figure 4.5 *Aplosporella beaumontiana*, (a) V.S. of conidioma, (b) Mature conidia, (c) Developing conidia with paraphysis

4. *Ascochyta avenae* (Petrak) Sprague & Johnson

Source

Sprague & Johnson, *Mycologia*, **40**: 297 (1948); Punithalingam, E., *Mycol. Pap.*, No. **142**: 51 (1979).

Description

Conidiomata pycnidial, 140–190 µm diam. Ostiole papillate. Pycnidial wall made up of thick-walled, brown cells at the outside and thin-walled, hyaline cells at the inner side giving rise to conidiogenous cells. Conidiogenous cells lining the pycnidial cavity, phialidic, hyaline, smooth, cylindrical to doliiform, 10–12 × 1–4 µm. Conidia 1-septate, cylindrical, pale brown to hyaline, smooth, 16–20 × 6–7 µm.

Specimen collected

On dead twigs of *Cynodon dactylon*, L., Courtallam, Tamil Nadu, Jul. 1989, collected by J. Muthumary. IC. NO: 174

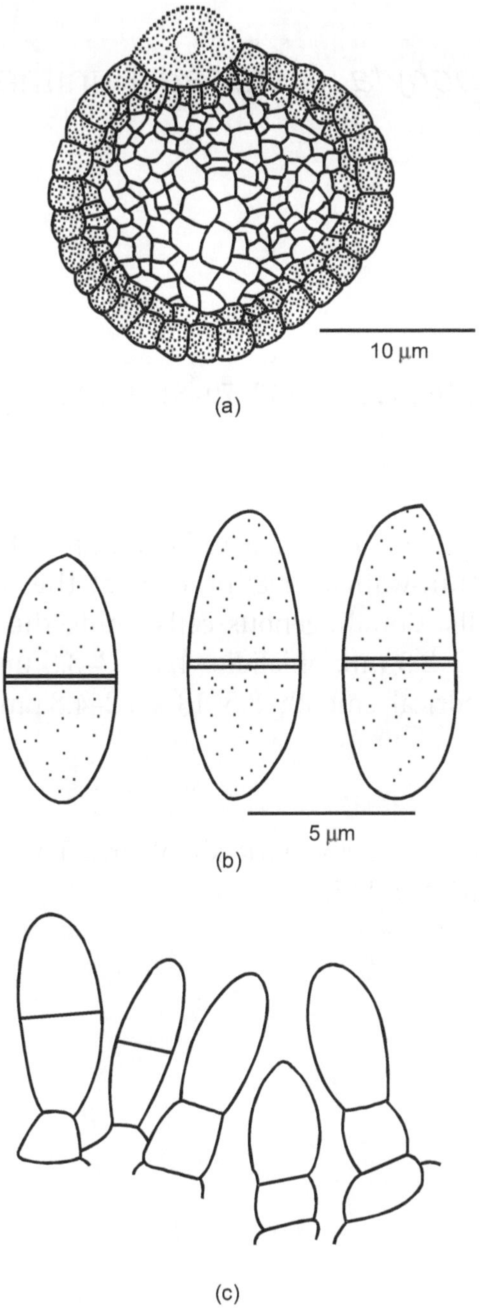

Figure 4.6 *Ascochyta avenae*. (a) V.S. of Conidioma, (b) Mature conidia, (c) Developing conidia

5. *Ascochyta festucae* Punithalingam

Source

Punithalingam, E., *Mycol. Pap.*, **142**: 79–81 (1979).

Description

Conidiomata 150–160 μm. Pycnidial wall made up of thick-walled, brown cells at the outside and thin-walled, hyaline cells at the inner side giving rise to conidiogenous cells. Conidiogenous cells lining the cavity, hyaline, smooth, doliiform, 5–10 × 2–2.5 μm, phialidic, 4.5 × 3–3.5 μm. Conidia pale brown to hyaline, short cylindrical, smooth, 8.5–13 × 3.2–4.8 μm.

Specimen collected

On dead leaflets of *Phoenix sylvestris* Roxb., Madras, Tamil Nadu, Apr. 1988, collected by J. Muthumary. IC. NO: 145.

Figure 4.7 *Ascochyta festucae*. (a–d) Mature conidia

6. *Ascochyta phaseolorum* Saccardo

Source

Saccardo, *Sylloge Fungorum,* III: 398 (1884); Sohi, H.S. & Nayar, S.K., *Indian Phytopath.,* **22**: 462–465.

Description

Conidiomata pycnidial, separate, globose, unilocular, 112–128 µm diam. Pycnidial wall made up of thick-walled, brown cells at the outside and thin-walled, hyaline cells at the inner side giving rise to conidiogenous cells. Conidiogenous cells lining the pycnidial cavity, phialidic, hyaline, discrete, formed from the inner cells of the pycnidial wall 5–8 × 3–5 µm. Conidia cylindrical, hyaline, medianly 1-septate, thin-walled, smooth, 12.8–16 × 2.3 µm.

Specimen collected

On dead fruits of *Pongamia glabra* Vent. Madras, Tamil Nadu, May 1988; on dead petiole of *Carica papaya* L., Madras, Tamil Nadu, Apr. 1988; On dead twigs of *Bambusa arundinacea* Willd., Madras, Tamil Nadu, May 1988, collected by J. Muthumary. IC. NO: 146.

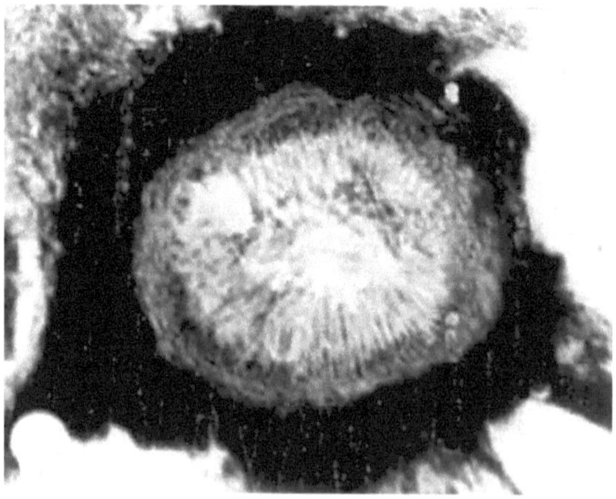

Figure 4.8 V.S. of conidioma (×200) of *Ascochyta phaseolorum*

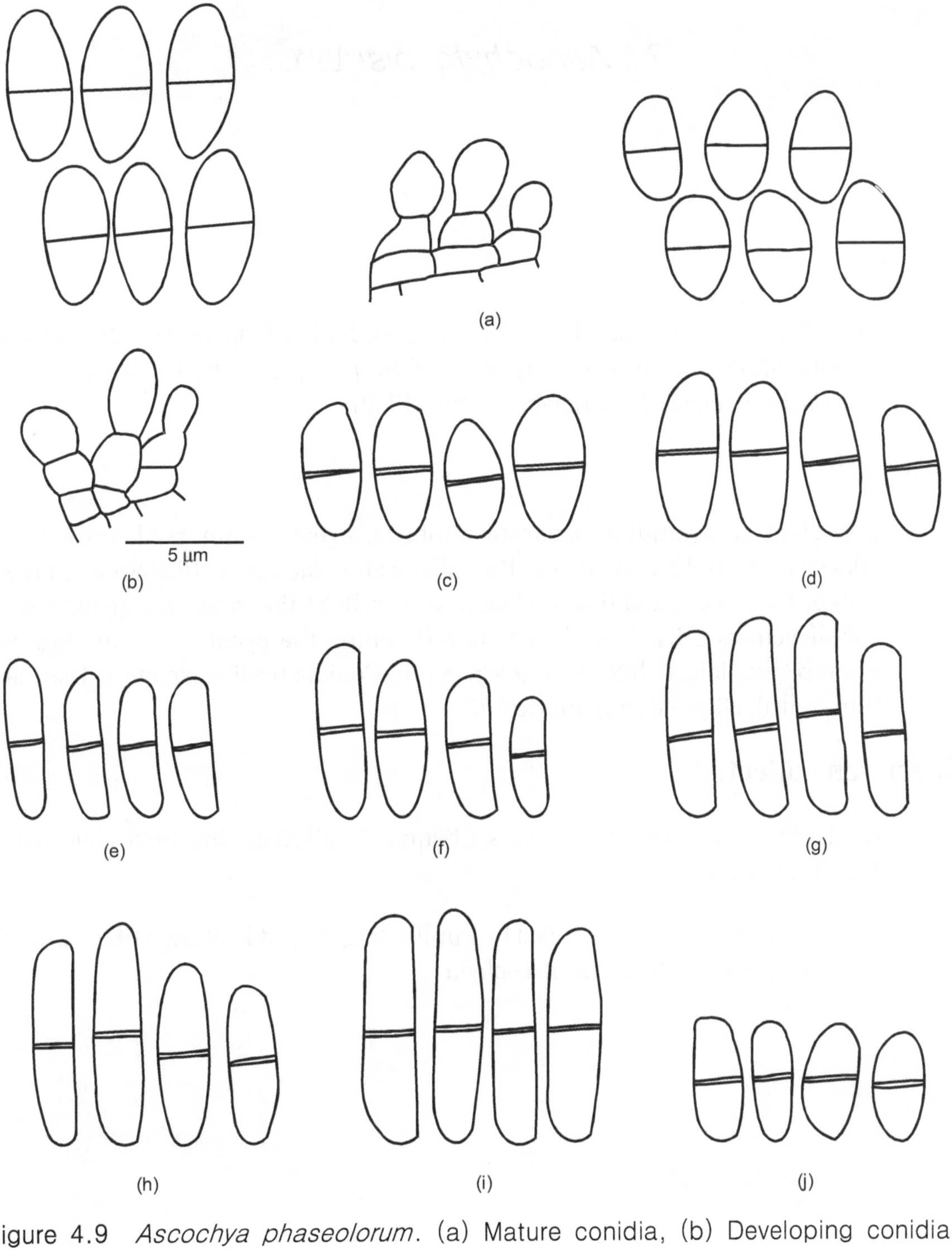

Figure 4.9 *Ascochya phaseolorum*. (a) Mature conidia, (b) Developing conidia, (c–j) Mature conidia

7. *Ascochyta pisi* Lib.

Source

Lib., *Pl. Crypt. Ard. Fasc.,* **1**: 59 (1830); Anon, *Indian J. Agric. Sci.,* **20**: 107–142 (1960); Mathur, *The Coelomycetes of India,* p. 21 (1979); Sutton, B.C., *The Coelomycetes,* CMI, England, p. 409 (1980).

Description

Conidiomata pycnidial, separate, globose, brown, immersed, unilocular, 100–200 µm in diam. ostiolate. Pycnidial wall made up of thick-walled, brown cells at the outside and thin-walled, hyaline cells at the inner side giving rise to conidiogenous cells. Conidiogenous cells lining the pycnidial cavity, hyaline, smooth, phialidic, doliiform, 7–9 × 5–8 µm. Conidia hyaline, smooth, 1-septate, thin-walled, ellipsoid to cylindrical 13 × 3 µm.

Specimen collected

On dead twigs of *Phaseolus vulgaris,* Chennai, Tamil Nadu, Jul. 1988, collected by J. Muthumary. IC. NO: 147.

This species is fully described by Punithalingam and Holliday (1972) and it is the type species of the genus *Ascochyta.*

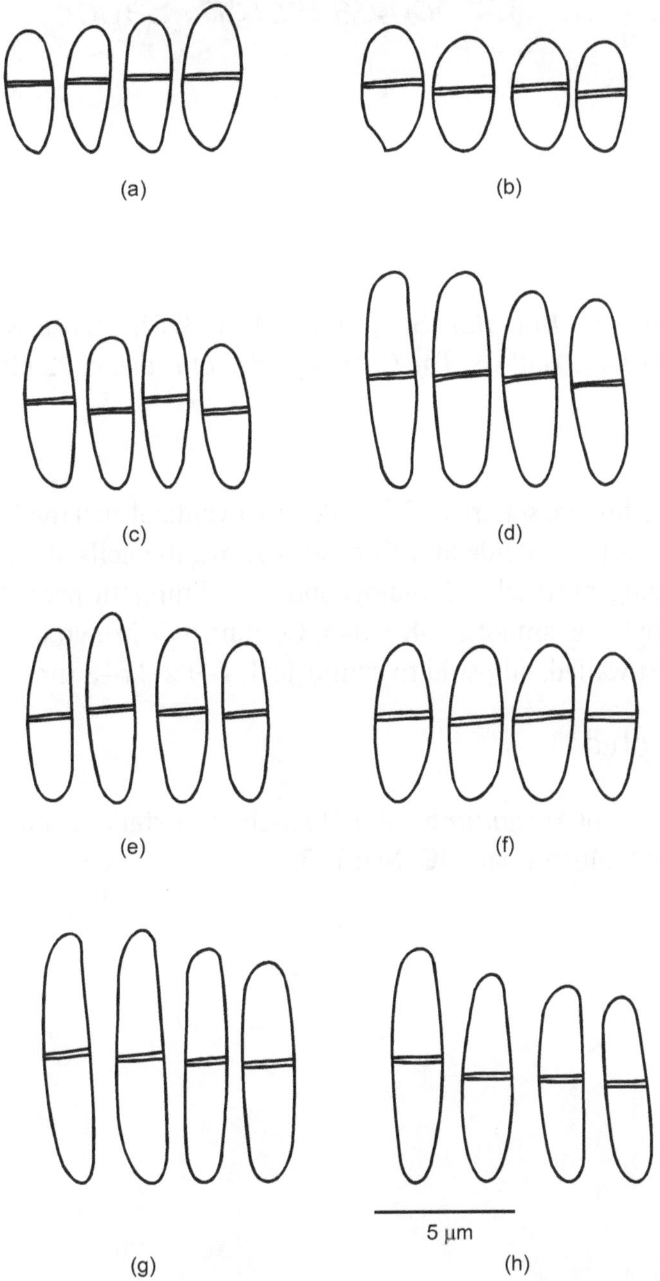

Figure 4.10 *Ascochyta pisi*, (a–h) Mature conidia

8. *Ascochyta sorghi* Sacc.

Source

Sacc., *Nuovo. G. Bot. ital. Ser.*, **II** 7: 302 (1875); Mehta & Bose, *Curr. Sci.*, **15**: 49–50 (1946); Mathur, *The Coelomycetes of India*, p. 22 (1979).

Description

Conidiomata brown, spherical, 160–200 μm. Pycnidial wall made up of thick-walled, brown cells at the outside and thin-walled, hyaline cells at the inner side giving rise to conidiogenous cells. Conidiogenous cells lining the pycnidial cavity, phialidic, doliiform, hyaline, smooth 5–8 × 3–4. Conidia pale brown to hyaline, smooth, 1-septate, thin-walled, ellipsoid to cylindrical, 7–9 × 2.5–3 μm.

Specimen collected

On dead leaves of *Sorghum bicolor* Moench., Courtallam, Tamil Nadu, Jul. 1989, collected by J. Muthumary. IC. NO: 175.

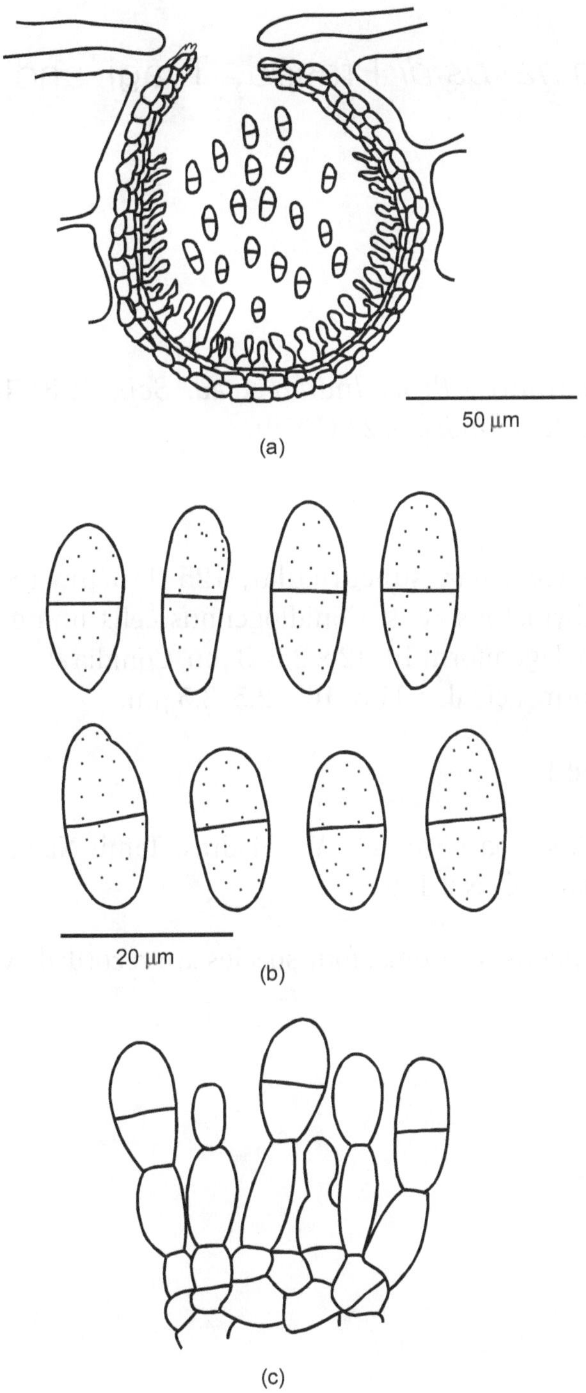

Figure 4.11 *Ascochyta sorghi.* (a) V.S. of Conidioma, (b) Mature conidia, (c) Developing conidia

9. *Asteroma psidi* Prasad, Tyagi and Agnihotri

Source

Tyagi and Agnihotri, *Proc. Indian Acad. Sci.*, **55B**: 118 (1962); Mathur, *The Coelomycetes of India*, p. 23 (1979).

Description

Conidiomata acervular, subcuticular, 125–180 µm in diam. composed of pseudoparenchymatous cells. Conidiogenous cells lining the cavity, phialidic, ampulliform to lageniform 10–12 × 2.5–3 µm. Conidia straight or slightly curved, 1-celled, fusiform, acicular, 11.5–16 × 2.5–3.5 µm.

Specimen collected

On litter of *Dracaena terminalis* W., Madras, Tamil Nadu, Jan. 1989, collected by J. Muthumary, IC. No: 176.

Under the genus *Asteroma*, four species are recorded by Mathur (1979).

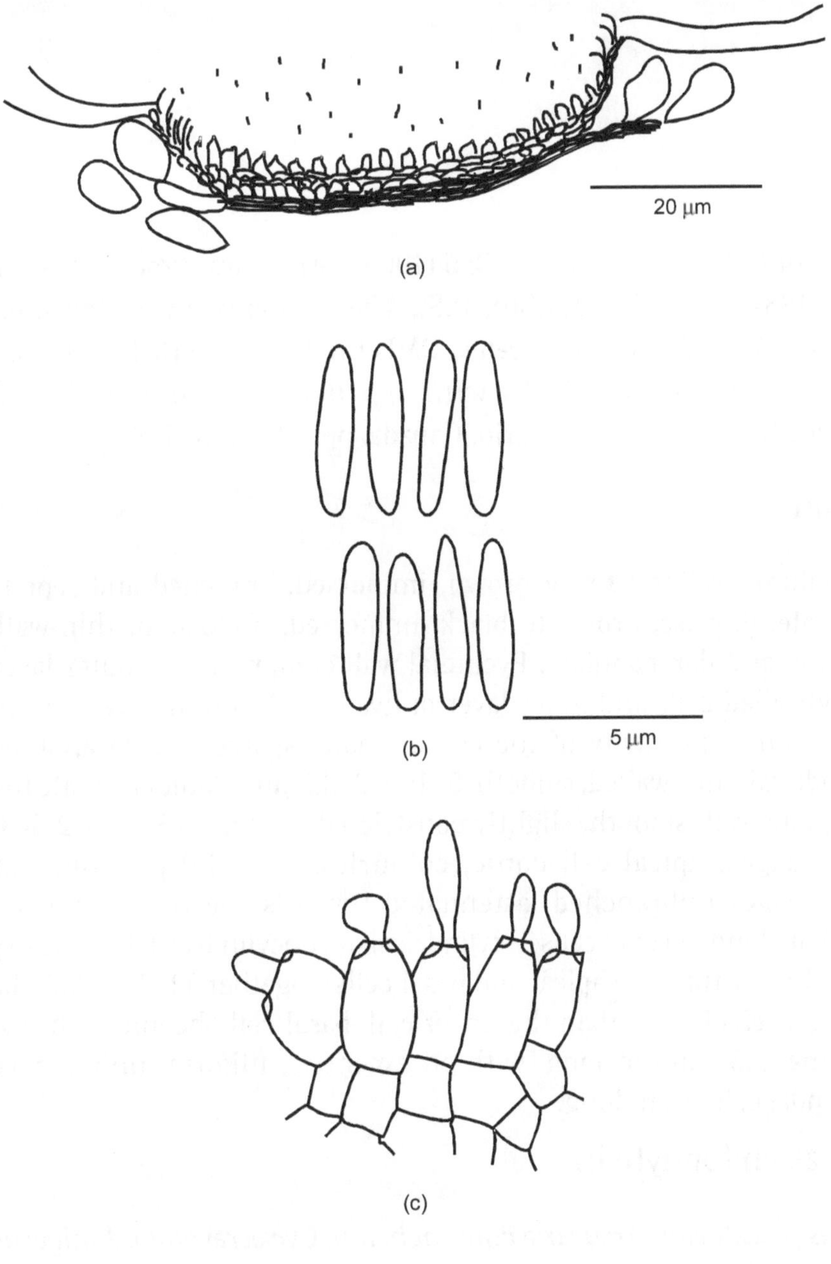

Figure 4.12 *Asteroma psidi*. (a) V.S. of conidioma, (b) Mature conidia, (c) Developing conidia

10. *Bartalinia robillardoides* Tassi

Source

Tassi, *Bull. R. Orto Bot. Siena,* **3**: 5 (1900); Srivastava, *Proc. Nat. Acad. Sci.* India, **34B**: 188–196 (1964); Mathur, R.S., *The Coelomycetes of India,* p. 25 (1979); Sutton, B.C., *The Coelomycetes,* CMI, England, p. 116 (1980); Nag Raj, T.R., *Coelomycetous anamorphs with appendage-bearing conidia,* Mycologue Publications, Waterloo, Ontario, Canada, pp. 141–143 (1993).

Description

Mycelium hyaline to pale brown, immersed, branched and septate. Pycnidia separate, globose, brown to black, immersed, unilocular, thin-walled. Ostiole central, circular, papillate. Pycnidial wall comprising an outer layer of brown, thick-walled cells and inner layer of hyaline, thin-walled cells. Conidiogenous cells lining the cavity of the conidiomata, sparsely septate, ampulliform to cylindrical, thin-walled, smooth, 5–10 × 2–3.5 µm. Conidia clavate to cylindrical, 4-septate, wall smooth, slightly constricted at septa, 15–20 × 2.5–3.5 µm with appendages, apical cell conic, colourless, 2.0–3.0 µm long, three apical appendages unbranched, attenuated towards the tip, flexuous, divergent, 8–18 µm long; median cells 3, cylindrical or subcylindrical, hyaline to pale brown, wall thicker than the apical and basal cells, together 11–15 × 2.5–30 µm; lower median cells longer than the upper cell, basal cell obconic with truncate base, hyaline, 2.5–3.0 µm long, with an excentric, filiform, unbranched, flexuous appendage, 3–5 µm long.

Recorded as endophyte in

Abrus precatorius, Araucaria cunninghamii, Cycas revoluta, Juniperus chinensis, Pinus caribea, K.P. Kannan, 2002, Ph.D. Thesis, "Biodiversity of Endophytic Mycobiota from selected Medicinal plants and some Gymnosperms", MUBL–528.

Aegle marmelos, Cardiospermum, Mimusops elengi, Psidium guajava, V. Gangadevi, 2005, Ph.D. Thesis, "Studies on Fungal Endophytes from some medicinal plants and screening selected coelomycetes for taxol production". MUBL No. 662

Mathur (1979) has recorded more than fifteen different host plants for this fungus.

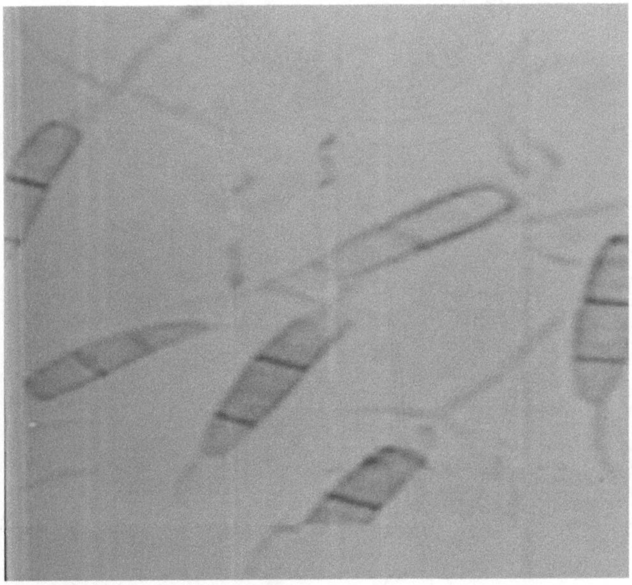

Figure 4.13 Mature conidia (×900) *Bartalinia robillardoides*

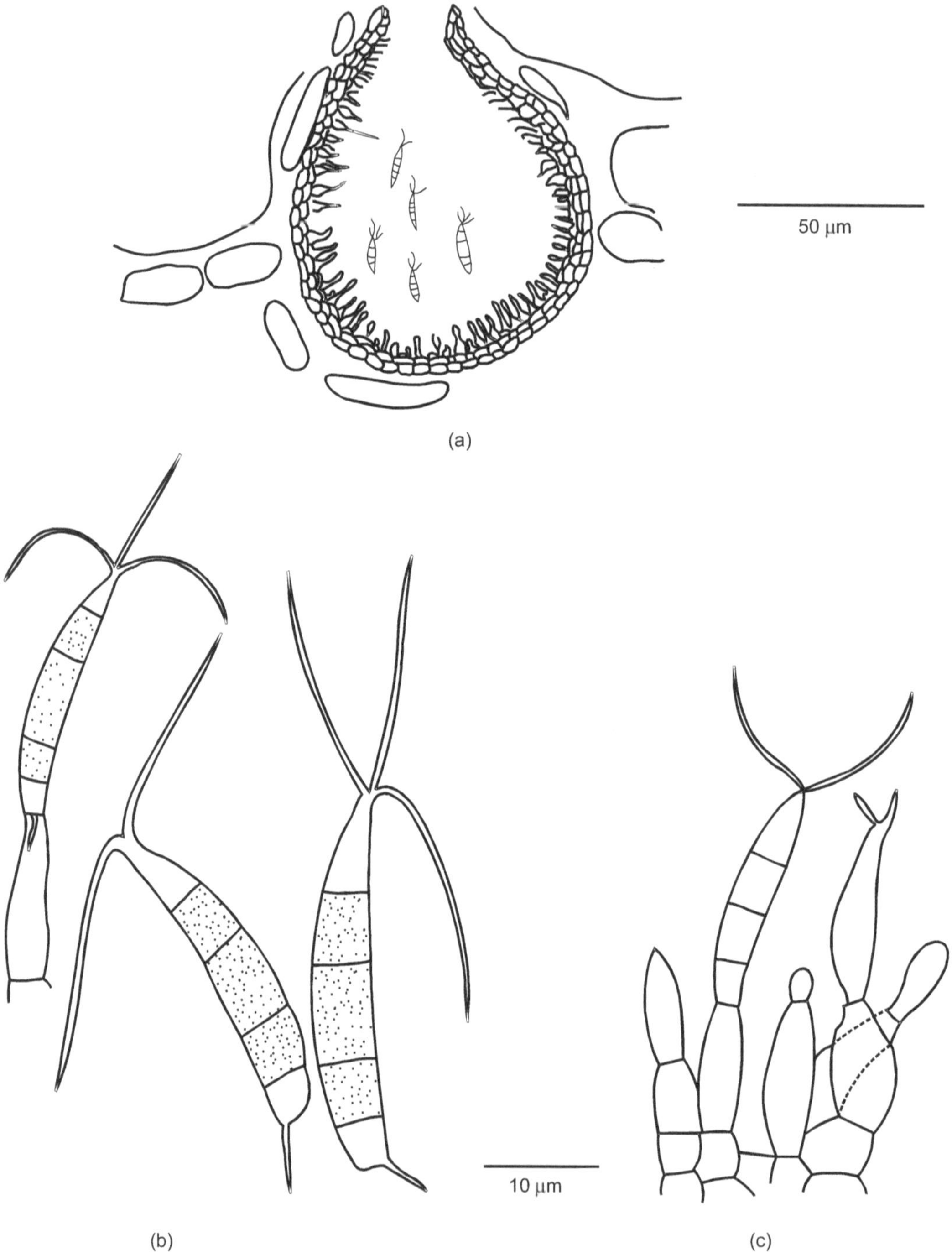

Figure 4.14 *Bartalinia robillardoides*. (a) V.S. of conidioma, (b) Mature conidia, (c) Developing conidia

11. *Botryodiplodia theobromae* Pat.

Source

Pat., *Bull. trimset. Soc. Mycol. Fr.,* **8**: 136 (1892); Petrak, K., *Annls. Mycol.,* **21**: 332 (1923); Tandon, B. and Verma, N., *Curr. Sci.,* **33**: 625–627 (1964); Sohi, H.S. & Mahalay, M.N., *Indian J. Mycol. Pl. Path.,* **4**: 26 (1974); Mathur, *The Coelomycetes of India,* pp.29–33 (1979); Sutton, B.C., *The Coelomycetes,* CMI, England, p. 191 (1980).

Description

Conidiomata eustromatic, immersed, globose, unilocular, 250–500 µm diam. Pycnidial wall 20–25 µm thick, outer layers dark brown, thick-walled, inner layer hyaline and thin-walled giving rise to the conidiogenous cells. Ostiole central, circular. Conidiogenous cells lining the pycnidial cavity holoblastic, cylindrical, hyaline, determinate, 10–15 × 2.5–3 µm intermixed with paraphysis. Paraphysis 20–30 × 1.5–2 µm. Conidia hyaline when young later becoming medianly 1-septate, dark brown, thick-walled, ellipsoid, with longitudinal striations from apex to base 16–20 × 11–12.8 µm.

Specimen collected

On dead leaves of *Polygonum* sp., *Cupressus* sp. and *Rhododendron arboreum* Sm., Kodaikanal, Tamil Nadu, Nov. 1987; on dead fruits of *Theobroma cacao* L., Madras, Tamil Nadu, Jan. 1988; on petiole of *Carica papaya* L., Madras, Tamil Nadu, May 1988; on leaves of *Dracaena terniflora* Roxb., Madras, Tamil Nadu, July 1988; on phyllode of *Cactus indicus* Roxb., Madras, Tamil Nadu, Sep. 1988; on dead twigs of *Pedilanthus tithymaloides* Poit., Madras, Nov. 1988; on dead leaves of *Ficus bengalensis* L., Madras, Tamil Nadu, Nov. 1988; on dead fruit of *Cocos nucifera,* L., Madras, Feb. 1989, collected by J. Muthumary. MUBL No: 663.

Isolated as endophyte from

Abrus precatorius, Canthium dicoccum, Curcuma longa, Cycas revoluta, Glycosmis pentaphylla, Juniperus chinensis, Libocedrus macrolepis, Madhuca indica, Pinus kesiya, Phyllanthus amarus, Phyllanthus maderaspatensis, Solanum surattense, S. torvum, Terminalia arjuna, Withania somnifera, Wrightia tinctoria, K.P. Kannan, 2002, Ph.D. Thesis, "Biodiversity of Endophytic Mycobiota from selected Medicinal plants and some Gymnosperms", MUBL-529. *Aegle marmelos, Andrographis paniculata, Cardiospermum helicacabum, Justicia gendarussa, Mimusops elengi, Ocimum basilicum, Psidium guajava, Rauvolfia tetraphylla, Terminalia arjuna,* V. Gangadevi, 2005, Ph.D. Thesis, "Studies on Fungal Endophytes from some medicinal plants and screening selected coelomycetes for taxol production."

Mathur (1979) has listed more than a hundred host plants from which the fungus had been isolated.

Sutton, had given the name as *Lasiodiplodia theobromae* Pat.

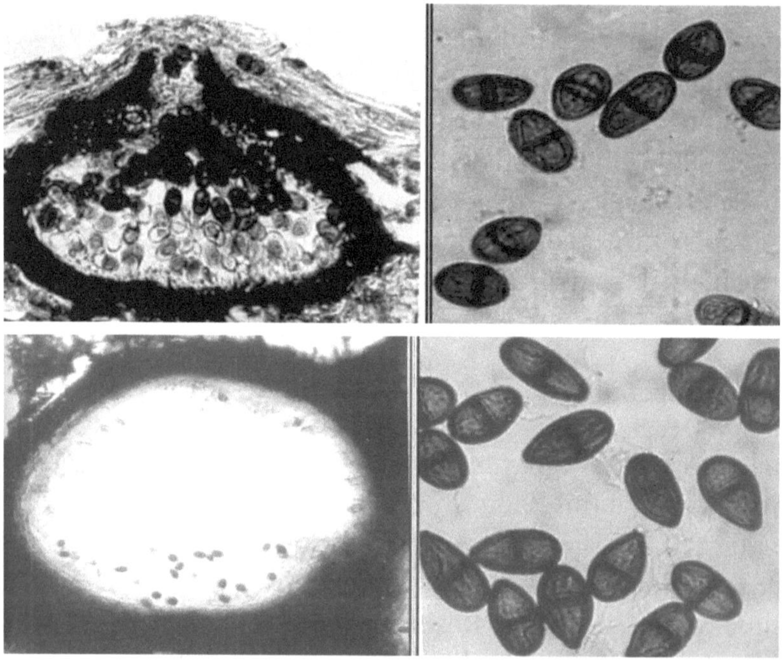

Figure 4.15 *Botryodiplodia theobromae.* (a–b) V.S. of conidiomata (×260), (c–d) Mature conidia (×900)

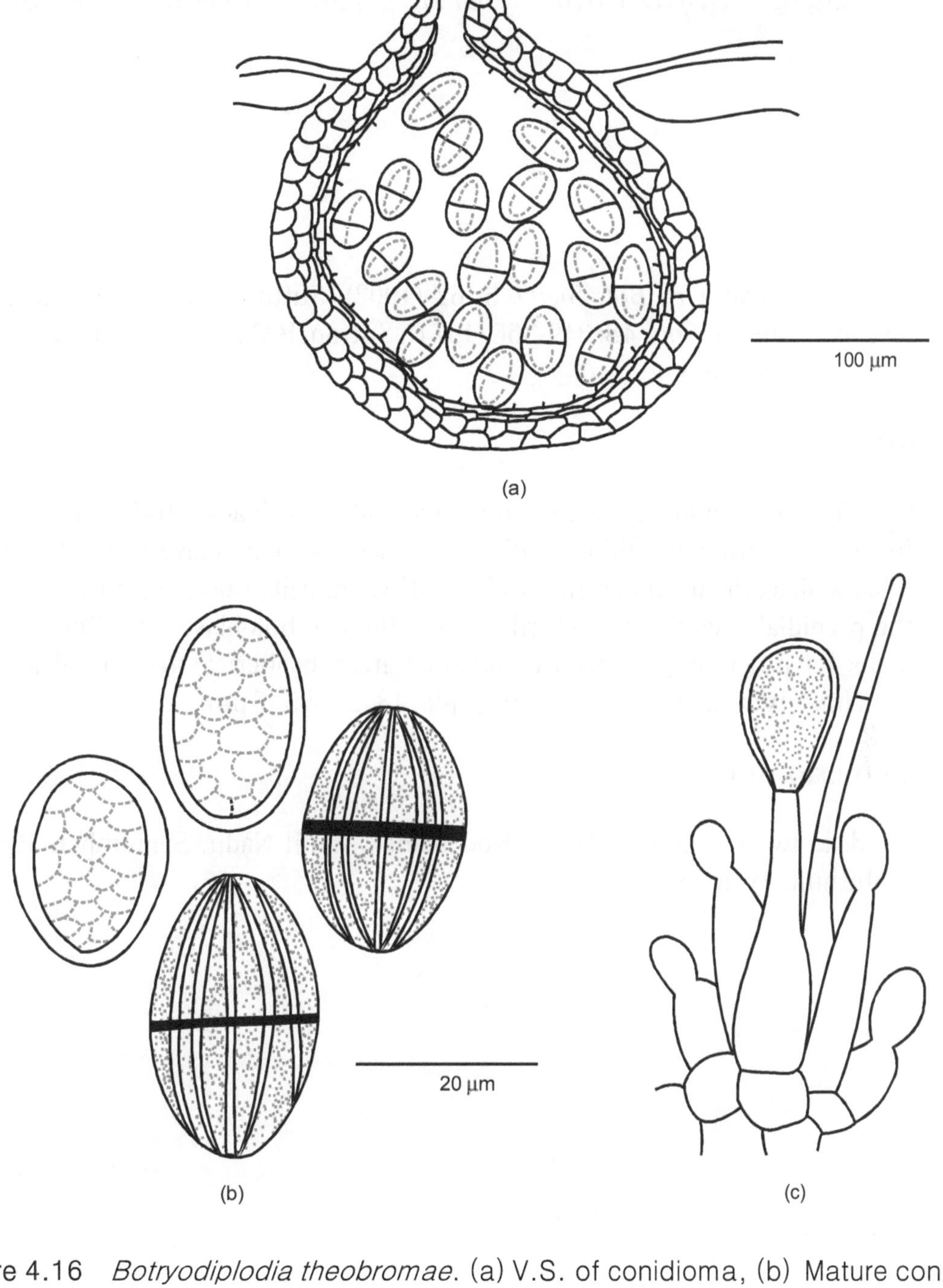

Figure 4.16 *Botryodiplodia theobromae*. (a) V.S. of conidioma, (b) Mature conidia, (c) Developing conidia

12. *Camarosporellum nervisequum* (Tassi) Tassi

Source

Tassi, *Bull. Lab. Ort. Bot. Siena*, **5**: 62 (1902); Sutton, B.C. and Pollack, F.G., *Mycopath. Mycol. Soc.*, **48**: 349-366 (1974); Sutton, B.C., *The Coelomycetes*, CMI, England, p. 80 (1980)

Description

Conidiomata pycnidial, separate, immersed, globose, black. 90–110 μm in diam. In vertical section, pycnidium unilocular, made up of 2–3 layers of outer, brown thick-walled cells and inner thin-walled and hyaline cells. Conidiogenous cells lining the pycnidial cavity, subspherical to ampulliform, holoblastic, hyaline, smooth, 3–5 × 4–6 μm. Conidia muriform, distoseptate, subspherical to ellipsoid, smooth, brown, with transverse and oblique septa, 18–20 × 5–7 μm.

Specimen examined

On dead twigs of *Magnolia* sp., Kodaikanal, Tamil Nadu, September, 1980 by J. Muthumary. IC. No: 100.

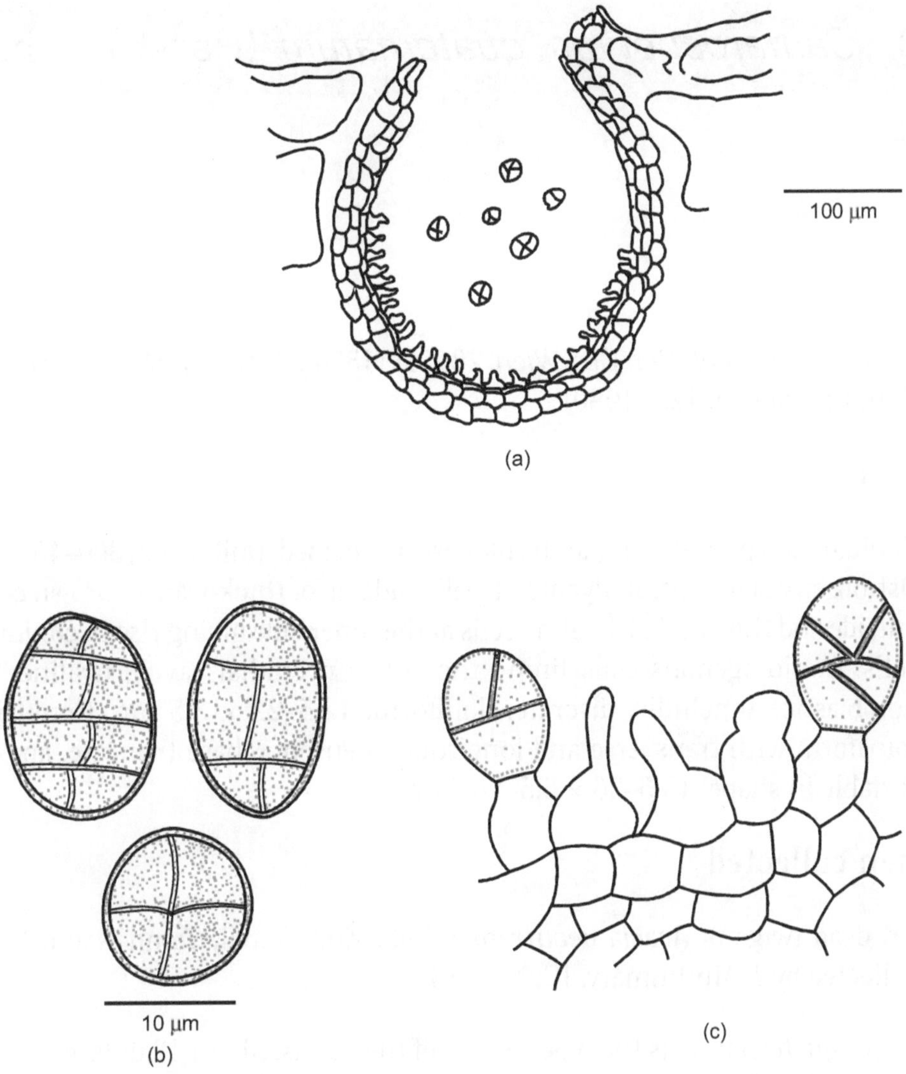

Figure 4.17 *Camarosporellum nervisequum*. (a) V.S. of conidioma, (b) Mature conidia, (c) Developing conidia

13. *Camarosporium quaternatum* (Hazsl.) Schulz

Source

Schulz, *Verh. Zool–Bot. Ges. Wien,* **20**: 641 (1870); Sutton, B.C., *The Coelomycetes,* CMI, England, p. 125 (1980).

Description

Conidiomata pycnidial, separate, globose, immersed, unilocular, 300–450 μm diam. Ostiole circular, central. Pycnidial wall made up of thick-walled, brown cells at the outside and thin-walled, hyaline cells at the inner side giving rise to conidiogenous cells. Conidiogenous cells lining the entire pycnidial cavity, hyaline, smooth, holoblastic, annellidic, discrete, doliiform, 12.5–22 × 3.5 μm. Conidia brown, muriform with transverse and longitudinal eusepta, base truncate, apex obtuse, variable in shape, 15.5–26 × 6.5–10.5 μm.

Specimen collected

On dead twigs of *Acacia decurrens* Willd., Kodaikanal, Tamil Nadu, Nov. 1987, collected by J. Muthumary. IC. No: 113.

C. quaternatum is the type species of the genus. About 12 different species are recorded in this genus from India.

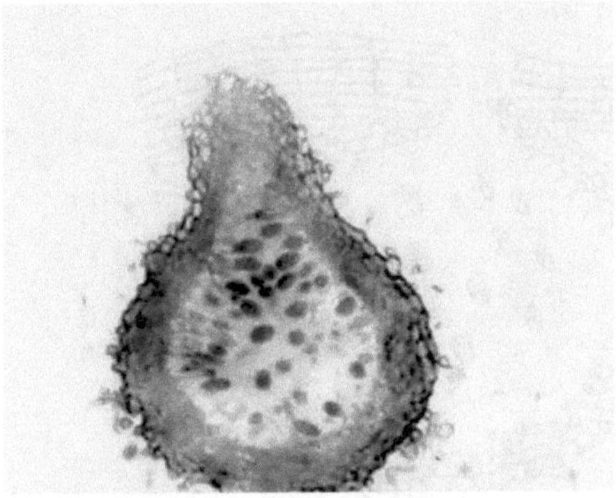

(a)

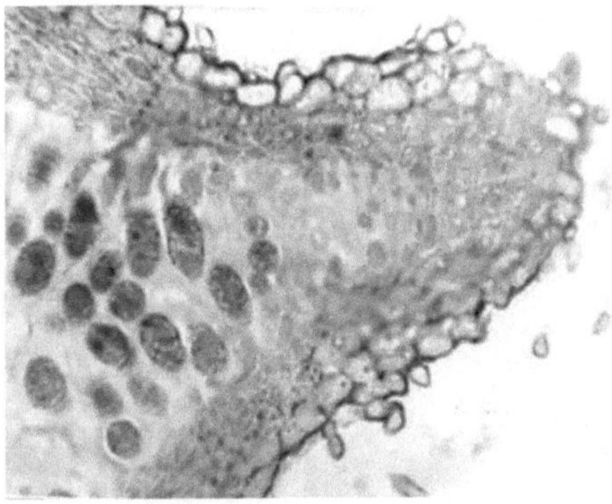

(b)

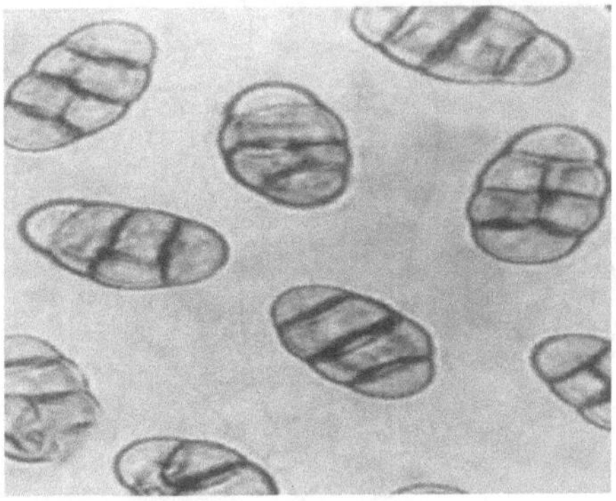

(c)

Figure 4.18 *Camarosporium quaternatum*. (a–b) V.S. of conidiomata (×200), (c) mature conidia (×900)

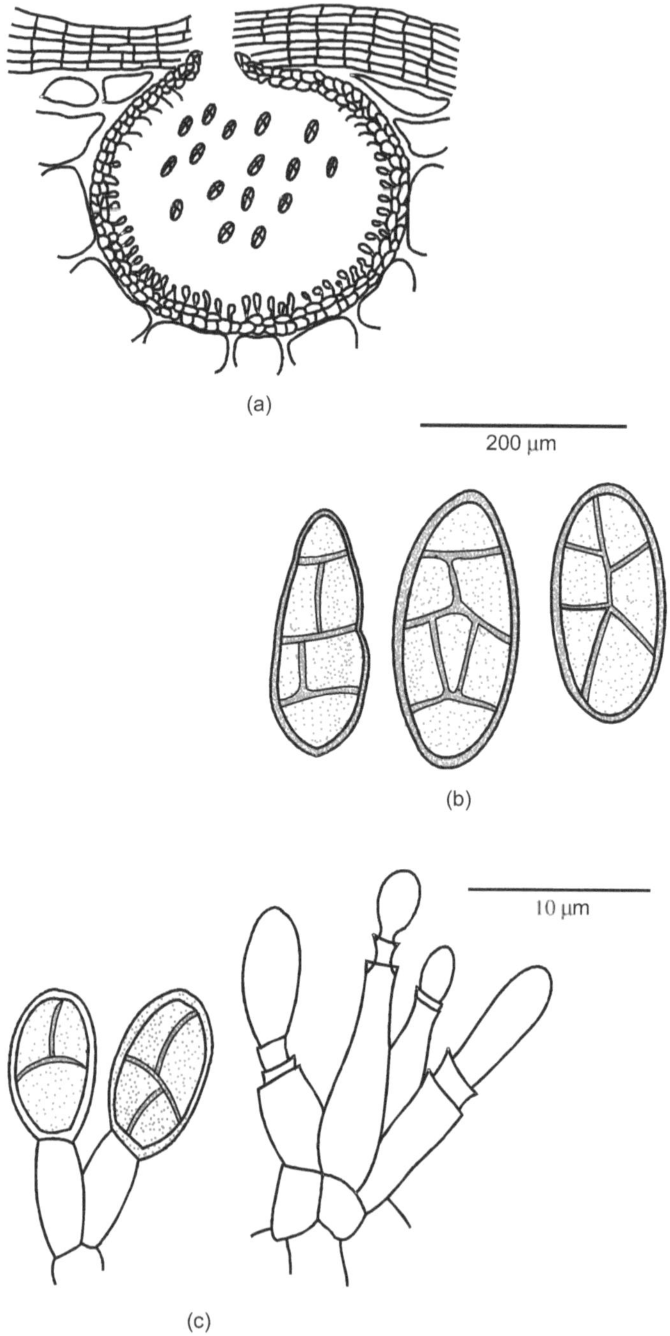

Figure 4.19 *Camarosporium quaternatum*. (a) V.S. of conidioma, (b) Mature conidia, (c) Developing conidia

14. Ceuthospora lauri (Grev.) Grev.

Source

Grev., *Scot. Crypt. Flora,* **5**: 254 (1827); Sutton, B.C., *The Coelomycetes,* CMI, England, p. 473 (1980); Nag Raj, T.R., *Coelomycetous anamorphs with appendage bearing conidia,* Mycologue Publications, Canada. pp. 172–173 (1993); Masilamani, S. and J. Muthumary, *J. Indian Bot. Soc.,* **72**: 311–312 (1993).

Description

Conidiomata pseudostromatic, immersed, black, spherical, 500–800 µm diam., wall composed of brown, pseudoparenchymatous cells, 5–10 µm thick. Conidiogenous cells lining the base and sides of the stroma, phialidic, determinate, integrated, cylindrical, hyaline, smooth, 20–3.5 µm. Conidia hyaline, cylindrical, aseptate, smooth, eguttulate, with an apical and basal fan-shaped, gelatinous appendage, disappearing at maturity, 11.5–15 × 2.5 µm.

Specimen collected

On dead leaves of *Prunus avium* L., Kodaikanal, Tamil Nadu, Nov. 1987, collected by J. Muthumary, IC. No: 114

According to Dicosmo *et al.* (1984) *C. lauri* had presumptive connection by association with *Phacidium multivalve* (DC: St. Amans) Fries.

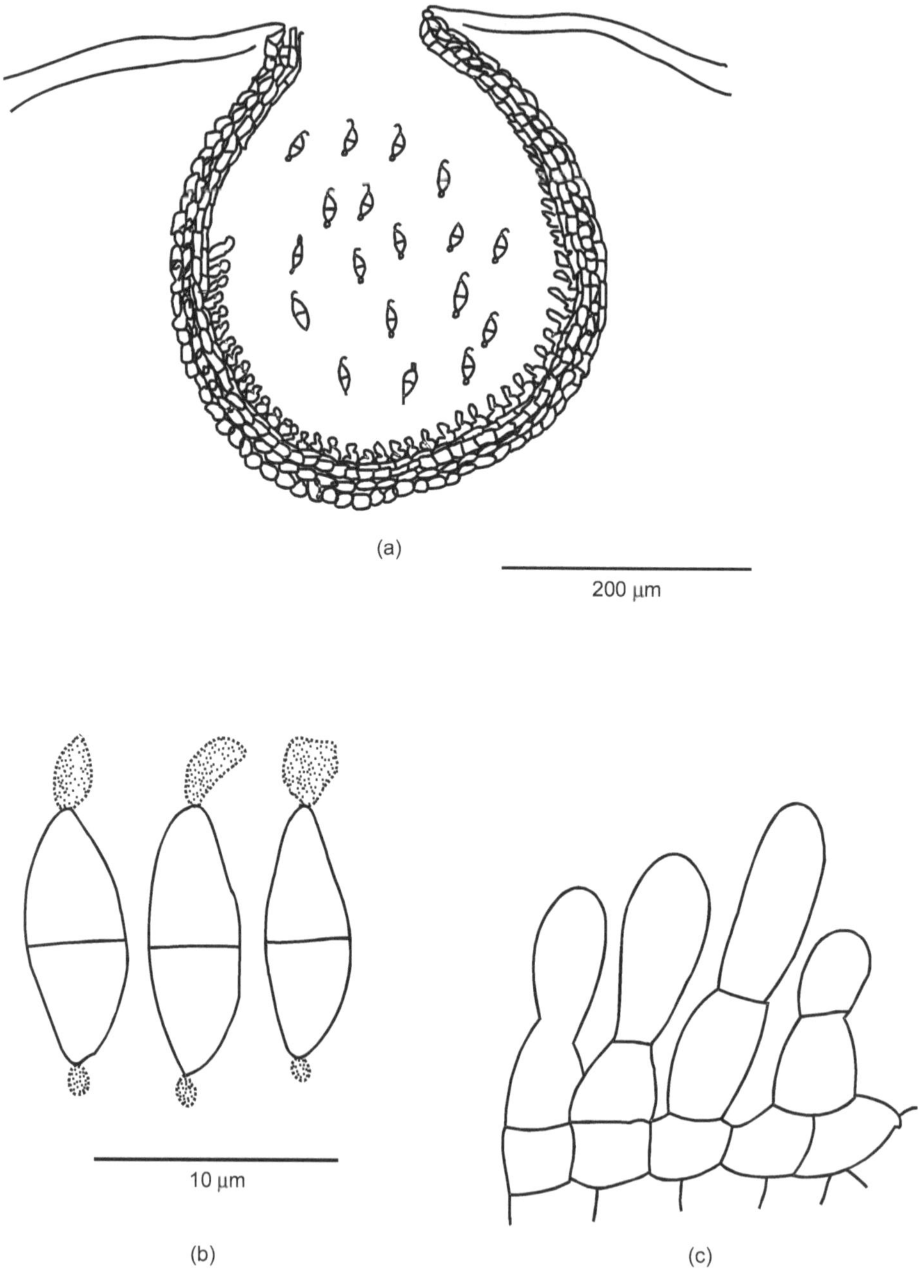

Figure 4.20 *Ceuthospora lauri*. (a) V.S. of conidioma, (b) Mature conidia, (c) Developing conidia

15. *Chaetoconis polygoni* (Ell. & Ev.) Clements

Source

Clements, *Genera of fungi,* 125 (1909); Sutton, B.C., *The Coelomycetes,* CMI, England, p. 412 (1980); Nag Raj, T.R., *Coelomycetous anamorphs with appendage bearing conidia,* Mycologue Publications, Canada, pp.189–191(1993).

Description

Conidiomata stromatic, immersed, dark brown, thick-walled, 250 µm diam. Ostiole circular. Conidiomatal wall made up of thick-walled, brown cells at the outside and thin-walled, hyaline cells at the inner side giving rise to conidiogenous cells. Conidiogenous cells lining the entire cavity, phialidic, cylindrical, hyaline, smooth, 9–11 × 3–3.5 µm. Conidia hyaline, 1-euseptate, base obtuse, 20–25 × 3.5–4 µm, apex prolonged into a filiform, cellular, unbranched appendage, 10–15 µm.

Specimen collected

On dead twigs of *Flacourtia indica,* Kodaikanal, Tamil Nadu, Nov. 1987, collected by J. Muthumary. IC. No: 116.

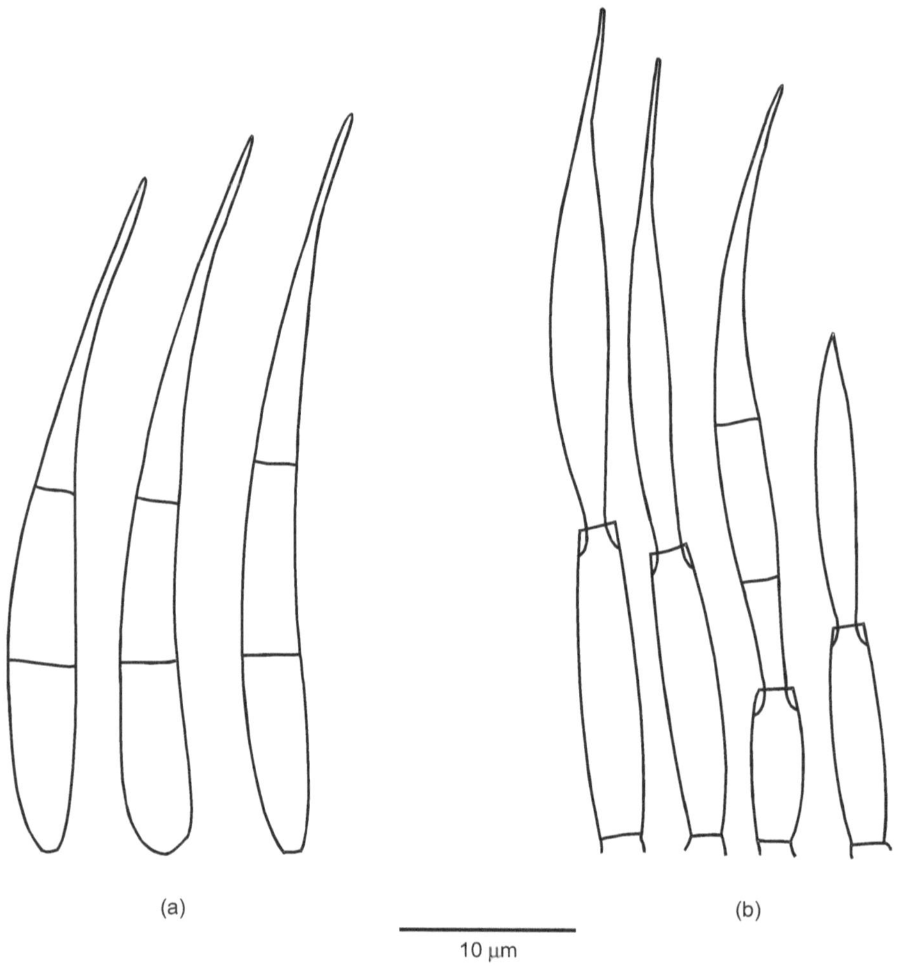

Figure 4.21 *Chaetoconis polygoni*. (a) Mature conidia, (b) Developing conidia

16. *Chaetodiplodia* sp.

Source

Mathur, *The Coelomycetes of India*, p. 36 (1979); Sutton, B.C., *The Coelomycetes*, CMI, England, p. 429 (1980).

Description

Conidiomata pycnidial, superficial, dark brown to black, 100–200 µm in diam. globose with numerous setae, pycnidial wall composed of thin-walled, smooth, pale brown pseudoparenchymatous cells. Setae pale brown, straight, 30–60 × 5–6 µm diam. Conidiogenous cells lining the pycnidial cavity, phialidic, ampulliform to doliiform, 5–6 × 3–4 µm. Conidia 1-septate, pale brown, thin-walled with prominent septum, smooth, apex obtuse, base rounded to truncate, 10–13 × 4.5–5.0 µm.

Specimen collected

On dead twigs of unidentified plants, Kambakkam, Chennai, Tamil Nadu, Nov. 1987, collected by J. Muthumary. IC. No: 119.

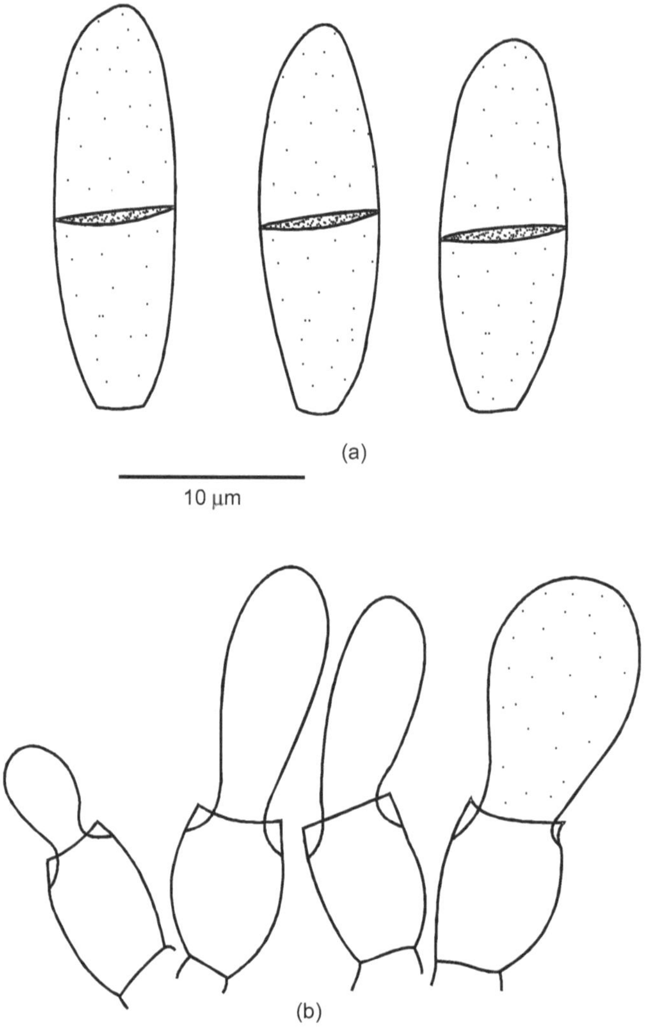

Figure 4.22 *Chaetodiplodia* sp. (a) Mature conidia, (b) Developing conidia

17. *Chaetomella acutiseta* Sutton & Sarbhoy

Source

Sutton and Sarbhoy, *Trans. Br. Mycol. Soc.,* **66**: 299 (1976); Sutton, B.C., *The Coelomycetes,* CMI, England, p. 548 (1980).

Description

Conidiomata pycnidial, superficial, dark brown, 500–700 × 400 µm. Setae evenly formed over the pycnidia, long, straight, rigid, 4–10 septate, 500 × 10–15 µm, with acuminate to subacute apex. Pycnidial wall made up of thick-walled, brown cells at the outside and thin-walled, hyaline cells at the inner side giving rise to conidiogenous cells. Conidiogenous cells lining the pycnidial cavity, phialidic, lateral or terminal, usually the central branch remaining sterile, 20–32 × 2–3 µm. Conidia hyaline, smooth, aseptate, fusiform, straight, 7–10 × 1.5–2 µm.

Specimen collected

On dead leaves of *Pyrus communis* L., Nilgiris, Tamil Nadu, Feb. 1988, collected by J. Muthumary, IC. No: 148.

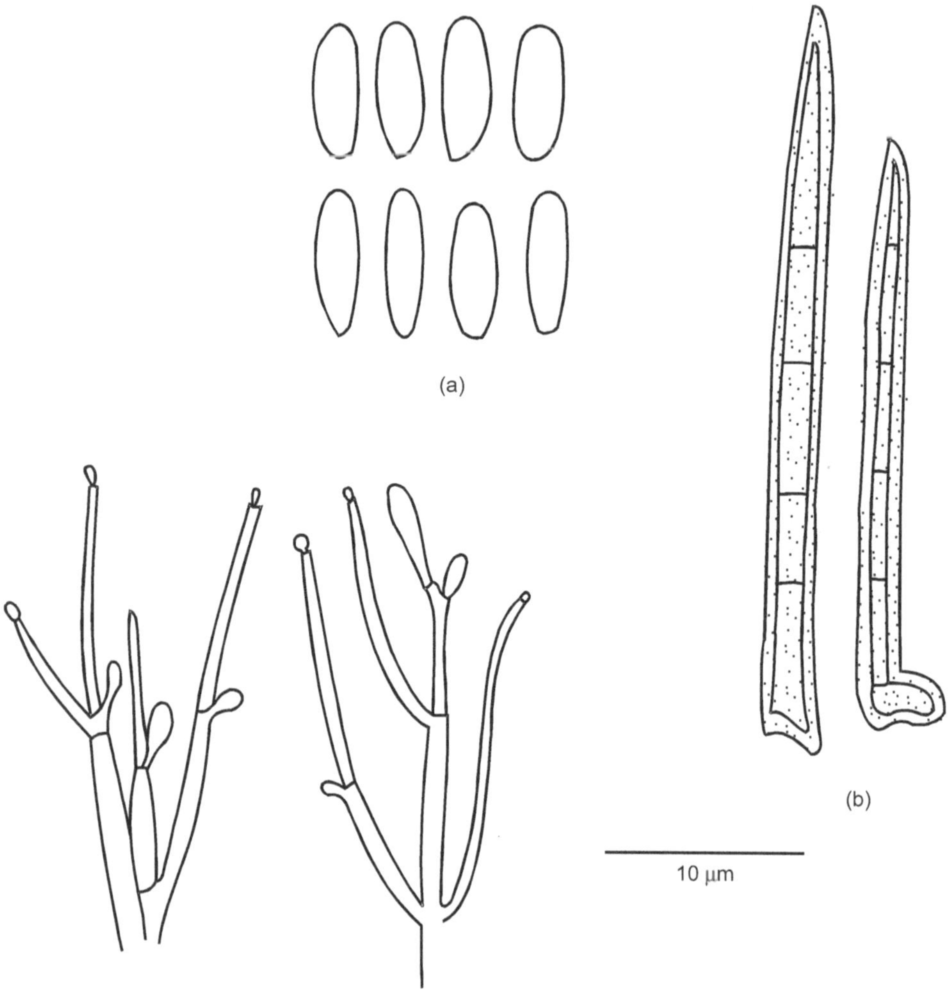

Figure 4.23 *Chaetomella acutiseta*. (a) Mature conidia, (b) Setae, (c) Developing conidia

18. *Chaetomella raphigera* Swift

Source

Swift, *Mycologia*, **22**: 165 (1930); Stolk, A.C., *Trans. Br. Mycol. Soc.*, **46**: 409–425 (1963); Sutton, B.C. & Sarbhoy, A.K., *Trans. Br. Mycol. Soc.*, **66**: 297–303 (1976); Sutton, B.C., *The Coelomycetes*, CMI, England, p. 548 (1980).

Description

Conidiomata pale white to black, superficial, separate, initially globose, 150–450 × 100–200 µm, opening widely at maturity, sessile. In vertical section, unilocular, thick-walled, brown, outer wall layers arranged radially, pseudoparenchymatous, inner cells thin-walled and hyaline. Raphe longitudinal, single and dehiscing at maturity, 90–100 µm long. Setae pale brown, often hooked at the apices, 50–150 µm. Conidiophores hyaline, smooth, branched irregularly, filiform, formed from the inner cells of the pycnidia. Conidia formed apically from the conidiogenous cell, hyaline, smooth, one–celled, allantoid, 4–5 × 0.5–2.0 µm.

Specimen collected

On dead leaves of several host plants and soil, Tamilnadu, Jul. 1988, collected by J. Muthumary. IC. No: 148.

Recorded as endophyte in

Aegle marmelos, Rauvolfia tetraphylla and *Terminalia arjuna*, V. Gangadevi, 2005, Ph.D. Thesis, "Studies on Fungal Endophytes from some medicinal plants and screening selected coelomycetes for taxol production". MUBL N0: 665.

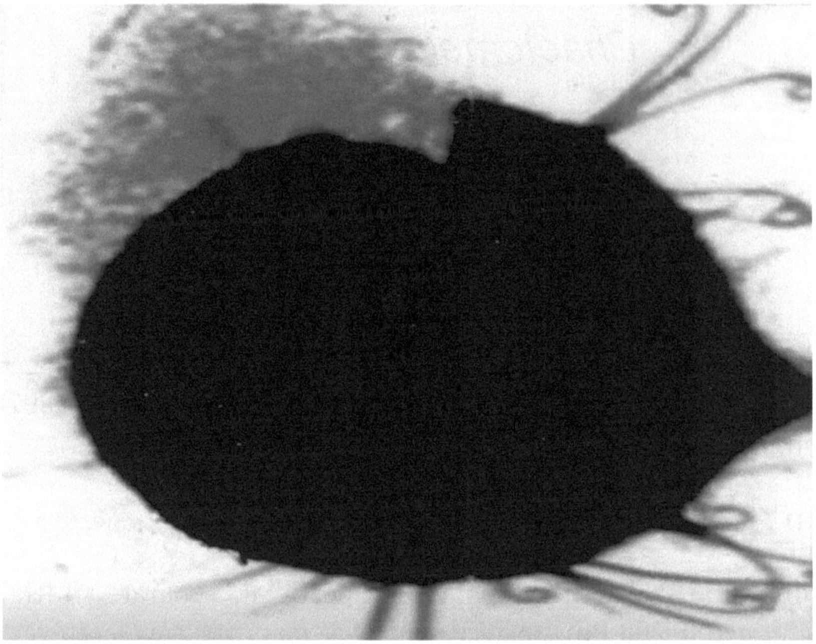

Figure 4.24 *Chaetomella raphigera*, mature conidioma (×200)

19. *Chaetopatella indica* Vasant Rao and Verghese

Source

Vasant Rao & I.M. Verghese, *Norw. J. Bot.*, **24**(1): 45–47 (1977); R.S. Mathur, *The Coelomycetes of India*, p. 38 (1979); Nag Raj, T.R., *Coelomycetous anamorphs with appendage bearing conidia*, Mycologue Publications, Canada. p.725 (1993).

Description

Conidiomata scattered to gregarious, 250–450 μm diam, superficial, cupulate when wide open, wall made up of brown, pseudoparenchymatous cells. Conidiogenous cells subcylindrical, lageniform, hyaline, smooth, 7–10 × 1.5–2 μm. Conidia long cylindrical, straight to slightly curved, 3-septate, smooth, hyaline, apex obtuse, base truncate, 20–30 × 2–2.5 μm. Appendage one at each end, unbranched, filiform, excentric, 3–4 μm long.

Specimen collected

On *cultures* of *Bambusa* sp., Kambakkam, Tamilnadu, Jul. 1988, collected by J. Muthumary. IC. No: 149.

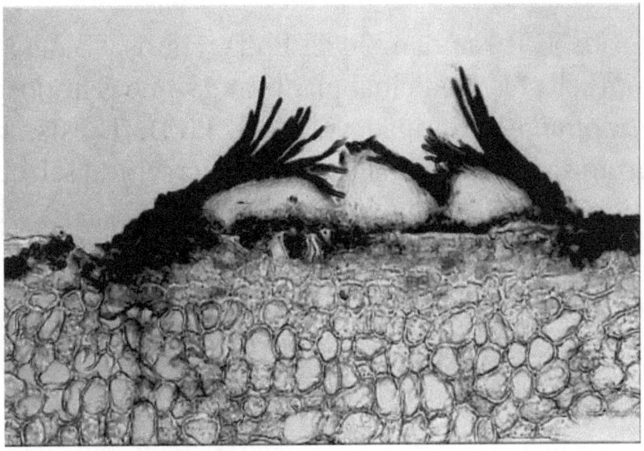

Figure 4.25 V.S. of Conidioma (×200) *Chaetopatella indica*

20. *Chaetospermum carneum* Tassi

Source

Tassi, *Bull. Lab. Ort. Bot., Siena,* **3**: 130 (1900); Sutton, B.C., *The Coelomycetes,* CMI, England, p. 152 (1980); Nag Raj, T.R., *Coelomycetous anamorphs with appendage bearing conidia,* Mycologue Publications, Waterloo, Ontario, Canada, p. 198 (1993).

Description

Conidiomata eustromatic, aggregated and confluent, initially closed, later rupturing irregularly to expose an orange yellow conidial mass, unilocular, 200–300 µm diam, wall at the base of conidioma hyaline, dehiscence irregular. Conidiophores up to 10–12 × 2–3 µm, hyaline, branched, smooth, formed from the inner cells of the conidiomatal wall. Conidiogenous cells holoblastic, discrete, determinate, cylindrical, smooth, hyaline. Conidia 26–45 × 5–8 µm, hyaline, smooth, aseptate, thin-walled, eguttulate, with 5–9 cellular, unbranched, filiform, appendages at the ends.

Recorded as endophyte in

Abrus precatorius, K.P. Kannan, 2002, Ph.D. Thesis, "Biodiversity of Endophytic Mycobiota from selected Medicinal plants and some Gymnosperms", MUBL–530; *Rauvolfia tetraphylla* V. Gangadevi, 2005, Ph.D. Thesis, "Studies on Fungal Endophytes from some medicinal plants and screening selected coelomycetes for taxol production", MUBL No: 666.

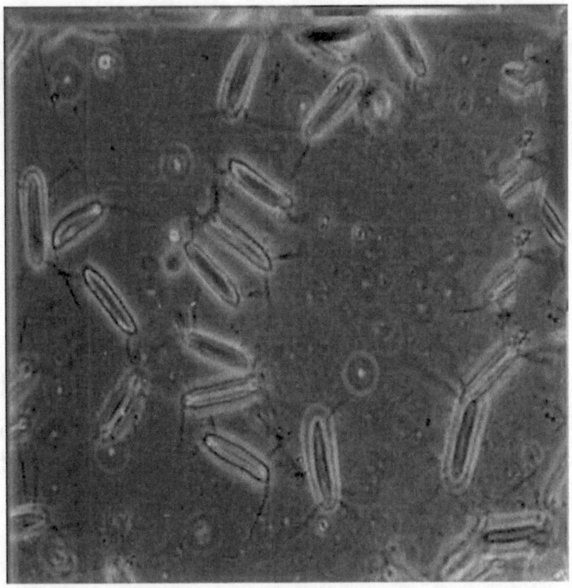

Figure 4.26 Mature Conidia (×900) *Chaetospermum carneum*

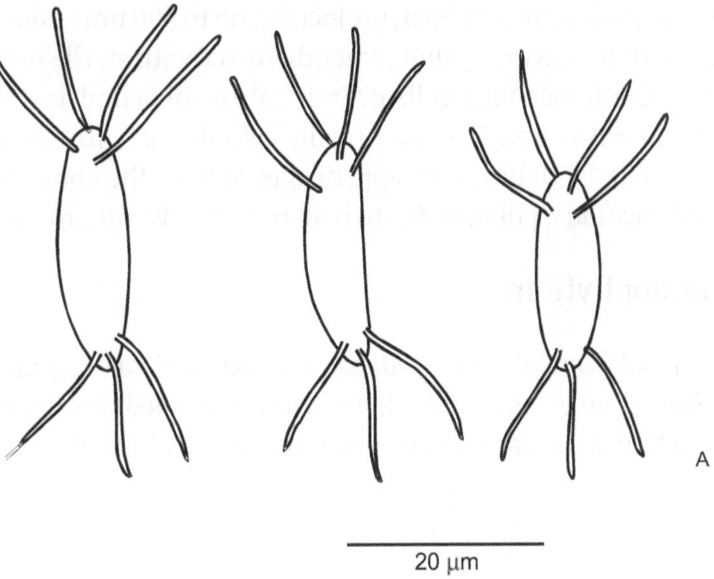

Figure 4.27 Mature conidia

21. *Chaetospermum chaetosporum* (Patuillard) Smith and Ramsbottom

Source

Smith & Ramsbottom, *Trans. Br. Mycol. Soc.*, **4**: 328 (1914); Sutton, B.C., *The Coelomycetes*, CMI, England, p.152 (1980); Nag Raj, T.R., *Coelomycetous anamorphs with appendage bearing conidia*, Mycologue Publications, Canada, pp. 196–198 (1993).

Description

Culture isolated as endophytes from leaves of *Rauvolfia tetraphylla* on PDA, mycelium white to grey. Conidiomata pycnidial, colourless to pale brown, scattered to gregarious, globose to irregular, unilocular, up to 500 µm in diam. Conidiophores lining the pycnidial cavity, hyaline, smooth, mixed with sterile mycelium, branched and septate. Conidiogenous cells cylindrical to subcylindrical, hyaline, smooth, blastic, 10–20 × 3–4 µm. Conidia broadly cylindrical, with obtuse ends, hyaline, smooth, 20–30 × 8–10 µm, with appendages at both the ends. Appendages 10–12 at each end, flexuous, polar, unbranched, filiform, 30–40 µm long.

Recorded as endophyte in

culture isolated from leaves of *Andrographis paniculata* V. Gangadevi, 2005, Ph.D. Thesis, "Studies on Fungal Endophytes from some medicinal plants and screening selected coelomycetes for taxol production", MUBL NO: 667.

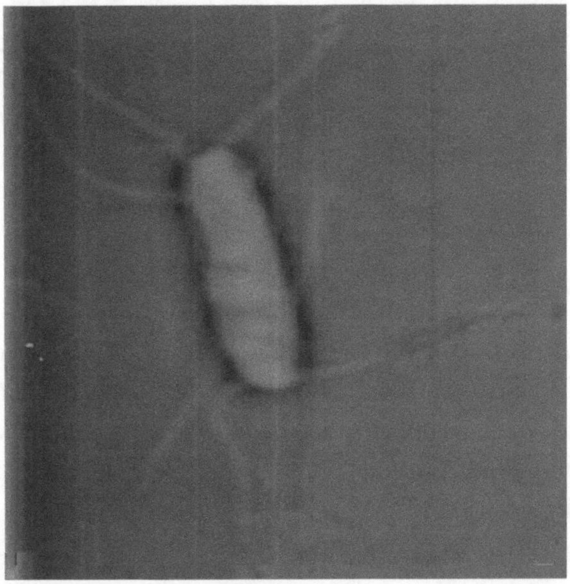

Figure 4.28 Mature Conidia (×900) *Chaetospermum chaetosporum*

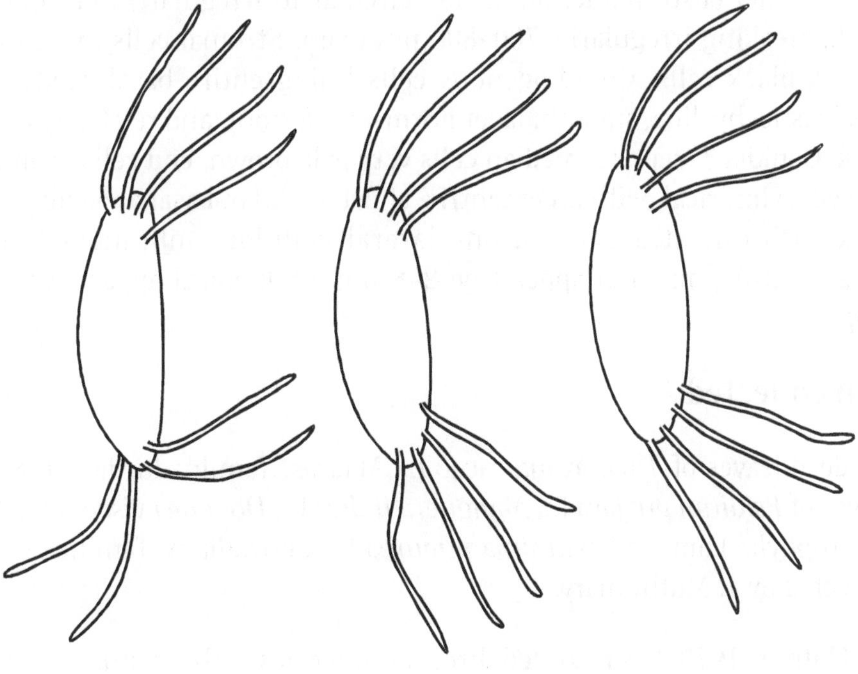

Figure 4.29 Mature conidia *C. chaetosporum*

22. *Ciliochorella mangiferae* Sydow

Source

Plagionema, Subram. & Ramakr. *J. Indian. Bot. Soc.* **32**: 135 (1953).Sydow and Mitter, *Annls mycol.,* **33**: 63 (1935); R.S. Mathur, *The Coelomycetes of India*, pp. 40–41 (1979); Sutton, B.C., *The Coelomycetes*, CMI, England, p. 480 (1980); Nag Raj, T.R., *Coelomycetous anamorphs with appendage bearing conidia*, Mycologue Publications, Canada, pp. 210–213 (1993); Sharma, N.D. and N. Sushma, *Indian Phytopath.*, **43**(3): 477–481(1900); Masilamani S. and Muthumary, J., *Mycol. Res.*, **98**(8): 857–861 (1994).

Description

Conidiomata eustromatic, dimidiate, circular to irregular in outline, glabrous, black, breaking irregularly. 700–800 μm diam. Stromal cells made up of thick-walled, black cells. Conidiogenous cells lining entire basal cavity of stroma, holoblastic, hyaline, smooth, lageniform with a stout, and relatively long, narrow neck. Conidia 3- septate, median cells very pale brown, end cells hyaline, slightly curved, cylindrical, with an excentric, short, cellular, basal appendage and at the apex with one straight and one lateral cellular, unbranched appendage, 20–22 × 2–4 μm; basal appendage 2–8 μm long; apical appendages 15–30 μm long.

Specimen collected

On dead leaves of *Theobroma cacao* L., Madras, Tamilnadu, Jan. 1988; On dead leaves of *Psidium guajava* L., *Mangifera indica* L., *Dodonea viscosa* L., *Artocarpus heterophylla* Lam. and *Bauhinia pupurea* L., Courtallam, Tamilnadu, Jul. 1989, collected by J. Muthumary.

Mathur (1979) has recorded three more species in this genus from India. More than seventy host plants are known for this species in India.

Recorded as endophyte in

Mimusops elengi, Psidium guajava, V. Gangadevi, 2005, Ph.D. Thesis, "Studies on Fungal Endophytes from some medicinal plants and screening selected coelomycetes for taxol production", MUBL No: 667. IC. No:121.

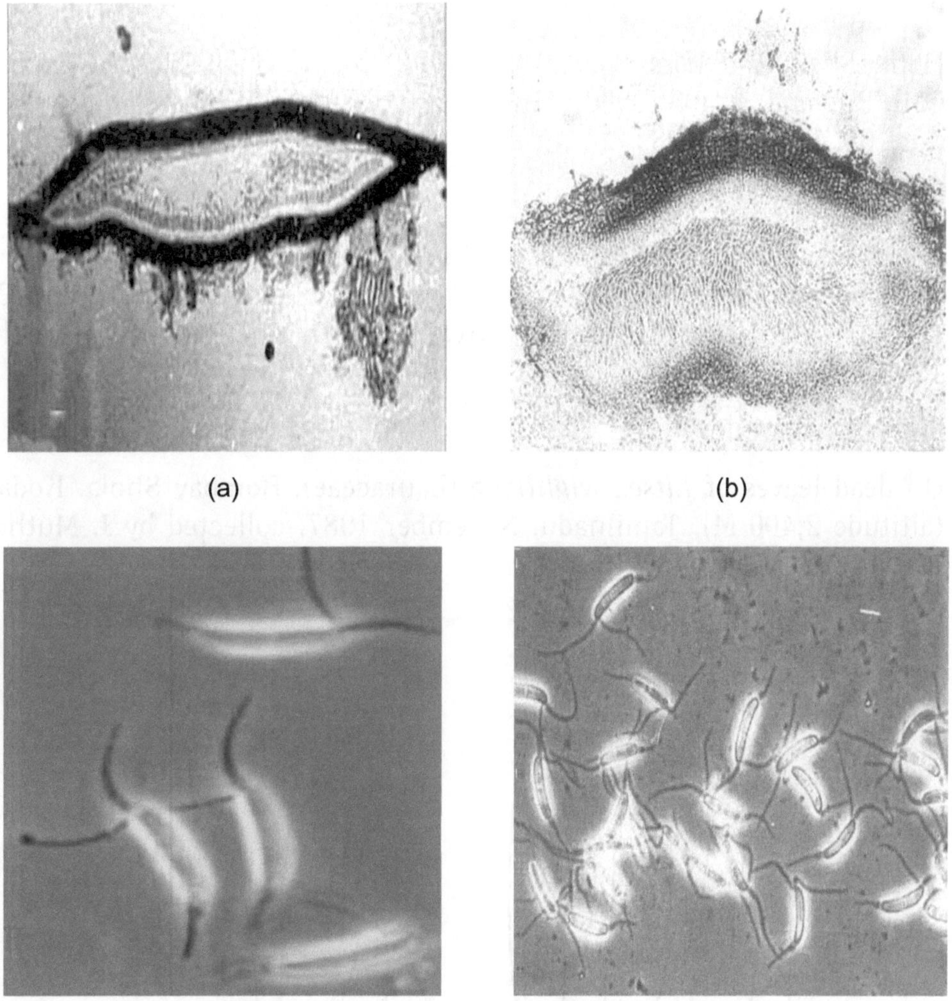

(a) (b) (c) (d)

Figure 4.30 *Ciliochorella mangiferae,* (a–b) V.S. of Conidiomata (×200); (c) Mature conidia (×900), (d) Mature conidia (×700)

23. *Coleophoma crateriformis* (Dur. & Mont.) Höhn.

Source

Hohn, *Sher. Akad. Wiss. Wien.*, **116**: 637 (1907); Sutton, B.C., *The Coelomycetes*, CMI, England, p. 403 (1980); Muthumary, J. and S. Masilamani, *J. Indian Bot. Soc.*, **68**: 131–133 (1989).

Description

Conidiomata pycnidial, 100–150 μm in diameter: ostiole 15–20 μm in diameter, conidiogenous cells 3–5 × 2–3 μm. Paraphysis like structures absent. Conidia cylindrical, aseptate, straight, hyaline with one guttule at each end, 10–13 × 2.0–2.5 μm.

Specimen examined

On dead leaves of *Litsea wightiana* (Lauraceae), Bombay Shola, Kodaikanal (altitude 2,400 M), Tamilnadu, November 1987, collected by J. Muthumary. IC. No: 132.

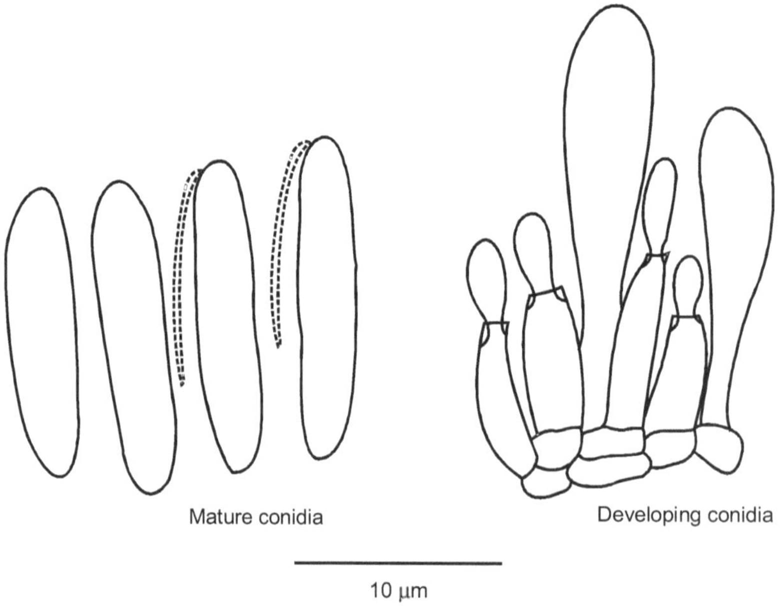

Figure 4.31 *Coleophoma crateriformis*

24. *Coleophoma cylindrospora* (Desm.) Höhn.

Source

Hohn, *Ber. Dtsch. Bot. Ges.*, **37**: 14 (1919); Sutton, B.C., *The Coelomycetes*, CMI, England, p. 403 (1980); Muthumary, J. and S. Masilamani, *J. Indian bot. Soc.*, **68**: 131–133 (1989).

Description

Conidiomata 200–250 µm. Pycnidial wall made up of thick-walled, brown cells at the outside and thin-walled, hyaline cells at the innerside giving rise to conidiogenous cells. Conidiogenous cells lining the pycnidial cavity 5–7 × 3–4 µm. Conidia 20–26 × 2.5 µm.

Specimen collected

On the litter of *Ilex wightiana*, Kodaikanal, Tamilnadu, Nov. 1987, collected by J. Muthumary. IC. No: 124.

(a)

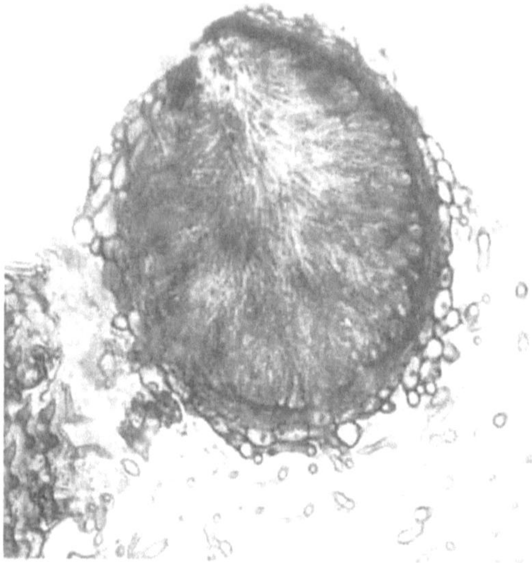

(a)

Figure 4.31 Coleophoma cylindrospora, (a–b) V.S of condioma (×200)

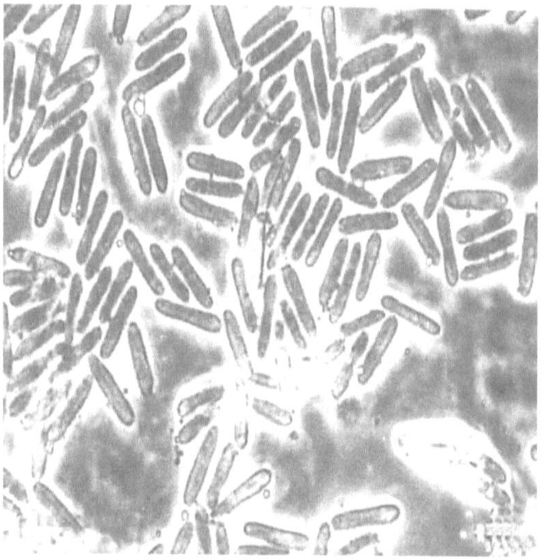

(c)

Figure 4.32 (c) Mature conidia (×900)

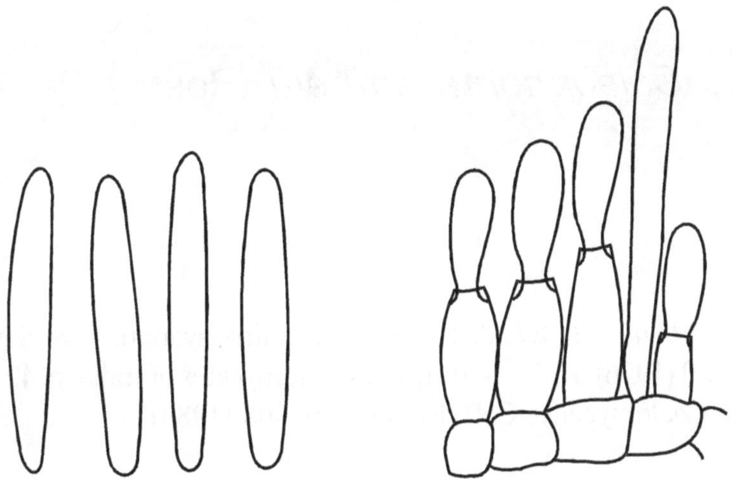

Figure 4.33 Developing conidia (b) *Coleophoma cylindrospora*

25. *Coleophoma empetri* (Rostr.) Petrak

Source

Petrak, *Annls mycol.*, **27**: 331 (1929); Laxminarayanan, P. & S.M. Reddy, *Curr. Sci.*, **45**: 530 (1976); R. S. Mathur, The Coelomycetes of India, p. 42 (1979); Sutton, B.C., *The Coelomycetes*, CMI, England, p. 403 (1980).

Description

Conidiomata aceruular, 200–250 μm in diam. Aceruular wall made up of thick-walled, brown cells at the outside and thin-walled, hyaline cells at the innerside, cells at the basal cavity giving rise to conidiogenous cells. Conidiogenous cells lining the aceruular cavity, hyaline, smooth, cylindrical, phialidic, 5–8 × 3–5 μm. Conidia cylindrical, aseptate, guttulate, 12–20 × 2–3 μm.

Specimen collected

On dead leaves of *Yucca gloriosa* L., Kodaikanal, Tamilnadu, Nov. 1987, collected by J. Muthumary. IC. No: 120.

Figure 4.34 *Coleophoma empetri*, V.S. of Conidioma (×200)

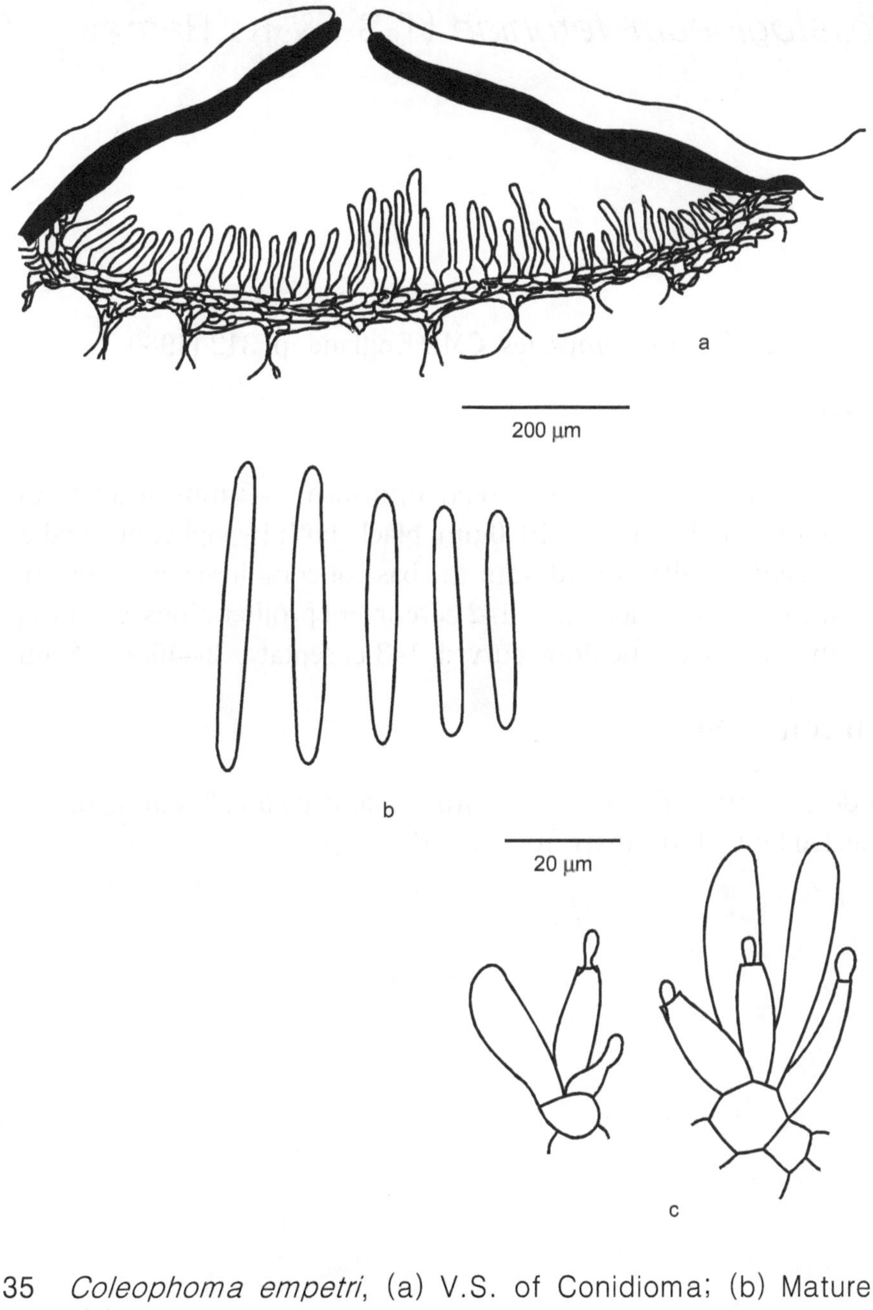

Figure 4.35 *Coleophoma empetri*, (a) V.S. of Conidioma; (b) Mature conidia; (c) Developing conidia

26. *Colletogloeum feroniae* (T.S. & K. Ramakr.) Sutton

Source

Sutton, B.C., *The Coelomycetes*, CMI, England, p. 312 (1980).

Description

Lesions on leaves circular to irregular, brown, 1–2 mm in diam. Conidiomata irregular to confluent, 700–1000 μm, black, both hypophyllous and epiphyllous. Conidiogenous cells formed from the base of conidioma, pale brown to hyaline, cylindrical 25 × 4 μm long with 1–2 percurrent proliferations. Conidia pale brown, smooth to finely verruculose, curved, 1–3 euseptate, 30–40 × 3–4 μm.

Specimen collected

On dead leaves of *Feronia elephantum*, Kambakkam, Tamilnadu, October 1990, collected by J. Muthumary. IC. No: 204.

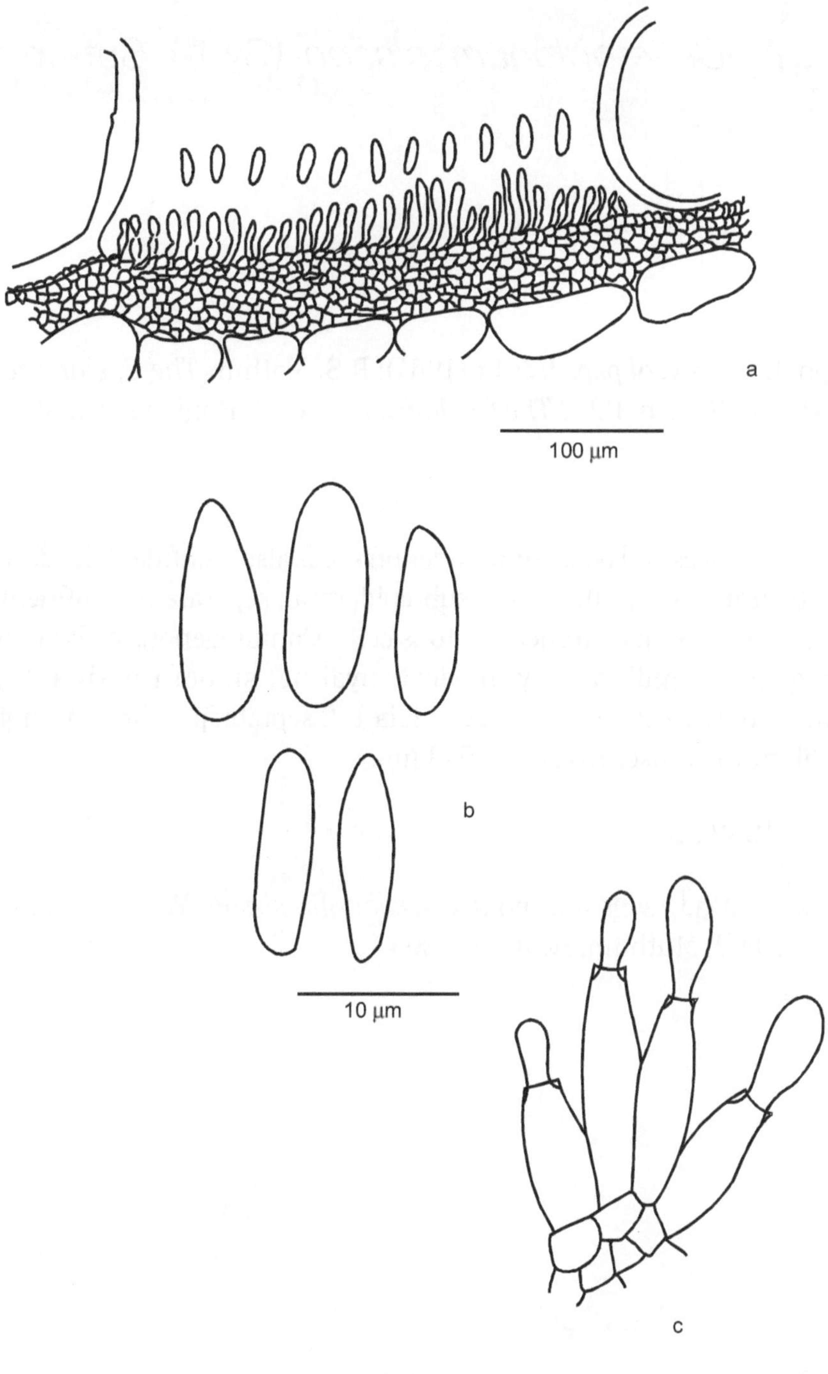

Figure 4.36 *Colletogloeum feroniae*, (a) V.S. of Conidioma; (b) Mature conidia; (c) Developing conidia.

27. *Colletogloeum sissoo* (Syd.) Sutton

Source

Sutton, B. C., *Mycol pap.*, **97**: 14 (1964); R.S. Mathur, *The Coelomycetes of India*, p.42 (1979); Sutton, B.C., *The Coelomycetes*, CMI, England, p. 318 (1980).

Description

Lesions on leaves and pods, sometimes on seeds also, confluent, black, conidiomata 100–200 µm diam. epidermal to sub-epidermal, separate to confluent, composed of pale, brown pseudoparenchymatous cells. Conidiogenous cells arising from the floor of the conidioma, cylindrical, hyaline, smooth with 1–2 percurrent proliferations, 10–20 × 2–3 µm. Conidia 1–2 septate, pale brown, slightly curved, apex blunt to obtuse, 15–20 × 2.5–3 µm.

Specimen collected

On leaves, dead twigs and pods of *Dalbergia sissoo*, Vandalur Zoo, Nov. 1990, collected by J. Muthumary. IC. No: 205.

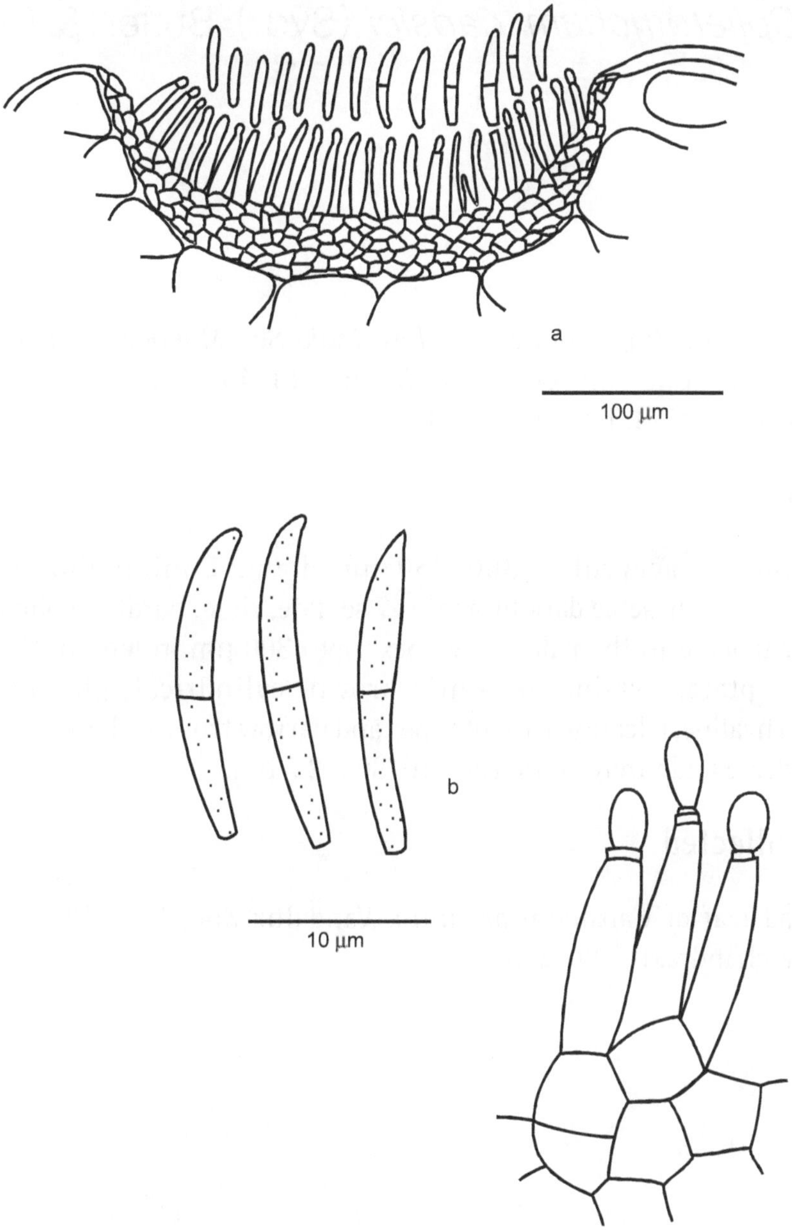

Figure 4.37 *Colletogloeum sissoo*, (a) V.S. of Conidioma; (b) Mature conidia; (c) Developing conidia.

28. *Colletotrichum capsici* (Syd.) Butler & Bisby

Source

Butler & Bisby, *Imp. Counc. Agr. Res. India Sci. Monogr.*, **1**: 152 (1931); R.S. Mathur, *The Coelomycetes of India*, pp. 43–45 (1979); Sutton, B.C., *The Coelomycetes*, CMI, England, p. 537 (1980).

Description

Conidiomata acervular, 300–350 µm diam. Conidiophores branched, 25–40 × 3–5 µm, setae dark brown, 1–7 septate, rigid, hardly swollen at the base, slightly tapered to the pale acute apex, upto 300 µm in length. Conidiogenous cells aseptate, hyaline to faintly brown, cylindrical, phialidic, smooth. Conidia hyaline, falcate with acute apex and narrow truncate base, aseptate, mostly uninucleate and rarely binucleate, 16–30 × 2.5 µm.

Specimen collected

On dead leaf of *Capsicum annuum*, Vandalur Zoo, Dec. 1991, collected by J.A. Jayachandra. IC. No: 223.

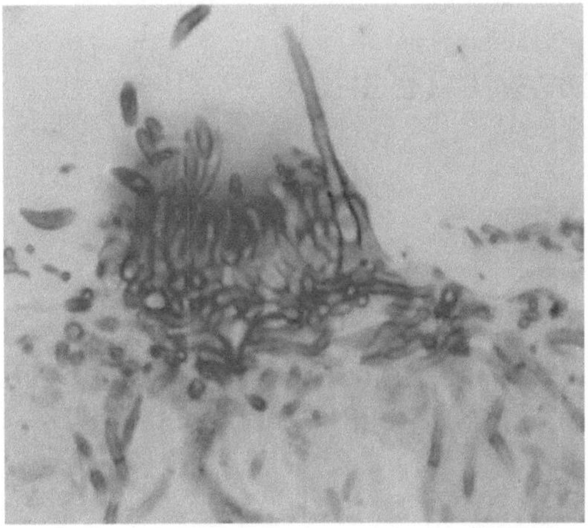

(a)

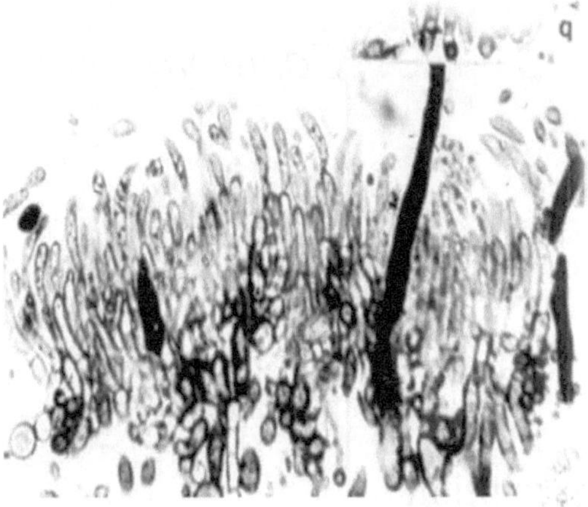

(b)

(c)

Figure 4.38 *Colletotrichum capsici,* (a–b) V.S. of Conidiomata (×200); (c) Mature conidioma (×900)

29. *Colletotrichum circinans* (Berk.) Vogl.

Source

Vogl, *Ann. R. Acad. Agr. Torino*, **49**: 175 (1907); Sutton, B.C., *The Coelomycetes*, CMI, England, p. 537 (1980); De, A.B., *Curr. Sci.*, **56**(23):1242–1243 (1987).

Description

Conidiomata acervular, globose, black. Setae abundant. Conidiogenous cells phialidic, hyaline. Conidia falcate fusiform, apices acute, guttulate, 19–21 × 3.5 µm. Appressoria abundant, clavate 10–14.5 × 6–6.5 µm.

Recorded as endophyte in

Abrus precatorius–Leaf lamina, Leaf midrib and Rachis, K.P. Kannan, 2002, Ph.D. Thesis, "Biodiversity of Endophytic Mycobiota from selected Medicinal plants and some Gymnosperms", MUBL–532; *Ocimum basilicum*- Young, mature and senescent leaves, V. Gangadevi, 2005, Ph.D. Thesis, "Studies on Fungal Endophytes from some medicinal plants and screening selected coelomycetes for taxol production". MUBL No: 668.

Figure 4.39 Mature conidioma (×200), *Colletotrichum circinans*

30. *Colletotrichum dematium* (Pers. Ex Fr.) Grove

Source

Grove, *J. Bot., Lond,* **56**: 341 (1918); Narendra, D.V. & V.G., Vasant Rao, *Sydowia,* **26**: 282–287 (1972); R.S. Mathur, *The Coelomycetes of India,* pp. 47–49 (1979); Sutton, B.C., *The Coelomycetes,* CMI, England, p. 535 (1980); Menta, M., *Indian Phytopath.,* **49**(2): 191(1996).

Description

Conidiomata acervular, dark brown, made up of brown pseudoparenchyma, 100–250 µm diam. conidiogenous cells enteroblastic, phialidic, cylindrical, hyaline, smooth, 7–10 × 2–2.5 µm. Conidia falcate to fusiform, apices acute, hyaline, smooth, 10–15 × 2–3 µm.

Specimen collected

On dead twig of *Dalbergia paniculata* Roxb., Madras, Tamilnadu, Jul. 1988, collected by J. Muthumary.

Recorded as endophyte in

Gymnema sylvestre, Solanum nigrum, Thuja orientalis, K.P. Kannan, 2002, Ph.D. Thesis, "Biodiversity of Endophytic Mycobiota from selected Medicinal plants and some Gymnosperms", MUBL–534; *Cardiospermum helicacabum, Justicia gendarussa, Mimusops elengi, Terminalia arjuna, Vitex negundo,* V. Gangadevi, 2005, Ph.D. Thesis, "Studies on Fungal Endophytes from some medicinal plants and screening selected coelomycetes for taxol production". MUBL No: 669

This species has been recorded on at least 90 different hosts from India (Mathur 1979).

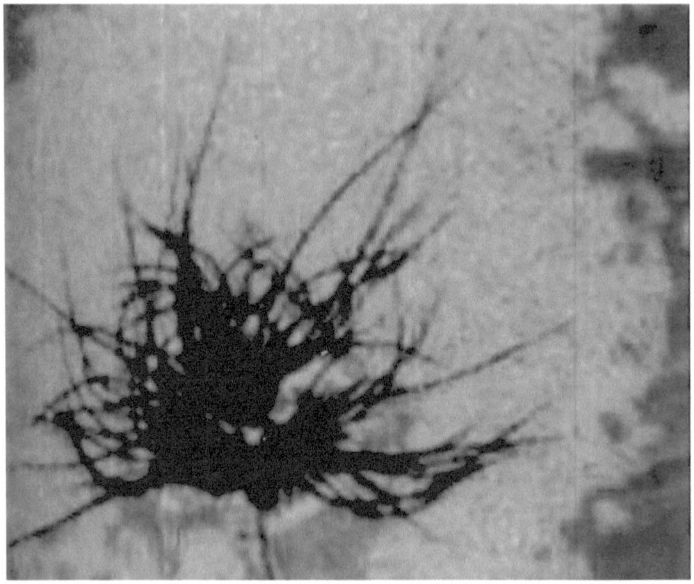

Figure 4.40 Mature conidioma (×200), *Colletotrichum dematium*

31. *Colletotrichum falcatum* Went.

Source

Went., *Arch. Java Suikerindustrie,* **1**: 265 (1893); Ramakrishnan, T.S., *Proc. Indian Acad. Sci.*, Sec. B., **29**: 289–294(1949); Sutton, B.C., *The Coleomycetes,* CMI, England, p. 531(1980).

Description

Conidiomata acervular, subcuticular, subepidermal, separate or confluent, made up of pale brown to dark brown, thick-walled pseudoparenchymatous cells. Setae in conidiomata brown, smooth, septate, tapered at the apex. Conidiogenous cells formed from the base of the conidiomata, hyaline, smooth, phialidic. Conidia falcate, single celled, 20–25 × 3.5–4 µm.

Recorded as endophyte in

Solanum surattense–Leaf midrib, Petiole and Stem, K.P. Kannan, 2002, Ph.D. Thesis, "Biodiversity of Endophytic Mycobiota from selected Medicinal plants and some Gymnosperms", MUBL–530; *Aegle marmelos, Cardiospermum helicacabum, Mimusops elengi, Ocimum basilicum, Rauvolfia tetraphylla, Terminalia arjuna and Vitex negundo*, V. Gangadevi, 2005, Ph.D. Thesis, "Studies on Fungal Endophytes from some medicinal plants and screening selected coelomycetes for taxol production". MUBL No: 670.

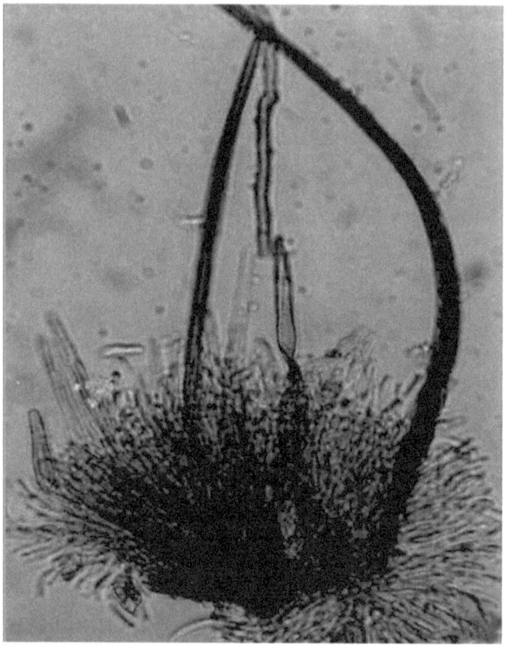

(a)

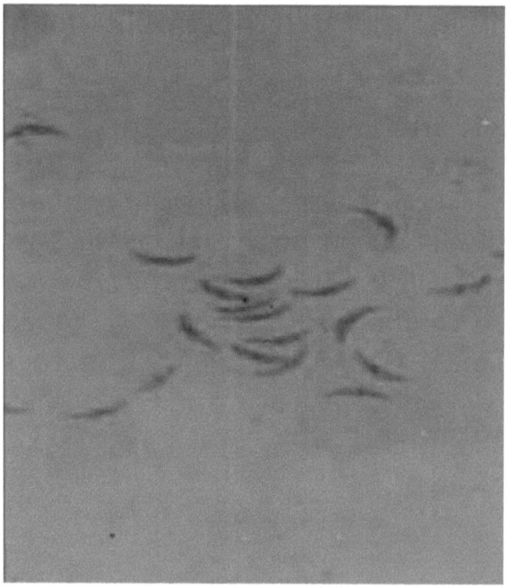

(b)

Figure 4.41 *Colletotrichum falcatum*, (a) Mature Conidioma (×200); (b) Mature conidia (×900)

32. Colletotrichum gloeosporioides (Penz.) Sacc.

Source

Sacc., *Fung. Agrum.*, **2**: 6 (1882); Tandon, B. and N. Verma, *Proc. Natl. Acad. Sci. India.*, **33**B: 411–417 (1963); R.S. Mathur, *The Coelomycetes of India*, pp. 51–59 (1979); Sutton, B.C., *The Coelomycetes*, CMI, Kew, pp. 530–531(1980); Sinha, J.N. and A.P. Singh, *Indian Phytopath.*, **47**(2): 214 (1994).

Description

Conidiomata acervular, subcuticular, subepidermal, separate or confluent, made up of pale brown to dark brown thick-walled pseudoparenchymatous cells. Setae in conidiomata brown, smooth, septate, tapered at the apex. Conidiogenous cells formed from the base of the conidiomata, hyaline, smooth, phialidic. Conidia subglobose, 10–20 × 3–3.5 µm.

Recorded as endophyte in

Abrus precatorius, Araucaria columanaris, Canthium dicoccum, Glycosmis pentaphylla, Gymnema sylvestre, Libocedros macrolepis, Melia azedarach, Phyllanthus amarus, P. maderaspatensis, Santalum album, Solanum surattense, S. torvum, S. trilobatum, Terminalia arjuna and *Wrightia tinctoria*, K.P. Kannan, 2002, Ph.D. Thesis, "Biodiversity of Endophytic Mycobiota from selected Medicinal plants and some Gymnosperms", MUBL–536; *Aegle marmelos, Andrographis paniculata, Cardiospermum helicacabum, Justicia gendarussa, Mimusops elengi, Ocimum basilicum, Psidium guajava, Rauvolfia tetraphylla, Terminalia arjuna* and *Vitex negundo*, V. Gangadevi, 2005, Ph.D. Thesis, "Studies on Fungal Endophytes from some medicinal plants and screening selected coelomycetes for taxol production". MUBL No: 671.

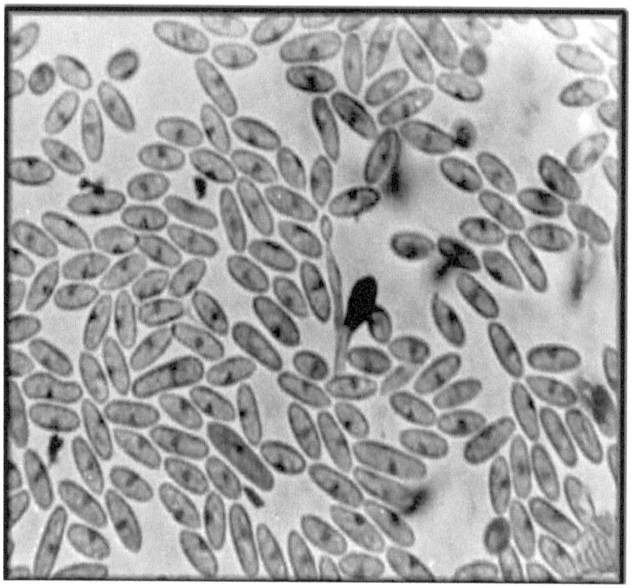

(a)

Figure 4.42 Mature conidia (×900), *Colletotrichum gloeosporioides*

33. *Colletotrichum lindemuthianum* (Sacc. & Magn.) Br. & Cav.

Source

Br. & Cav., *Fungi Parass,* p. 50 (1889); Sahni, V.P. *Mycopath. Mycol. Appl.,* 27: 342–356 (1965); R.S. Mathur, *The Coelomycetes of India,* pp. 61-62 (1979); Sutton, B.C., *The Coelomycetes,* CMI, England, p. 528 (1980).

Description

Conidiomata acervular, subcuticular, separate, composed of dark brown, thick-walled psudoparenchyma, 100–200 μm in diam. Setae brown, smooth, tapered to the apices, 25–30 × 3–4 μm. Conidiogenous cells enteroblastic, phialidic, hyaline, smooth, cylindrical, discrete 10–15 × 2–3 μm. Conidia cylindrical, hyaline, aseptate, straight, smooth, thin-walled, 10–12 × 3.5–4.9 μm.

Specimen collected

On dead leaves of *Pongamia glabra* Vent., Madras, Tamilnadu, Nov. 1988; On dead twigs of *Bambusa arundinacea* Willd. collected from Madras, Tamilnadu, Jan. 1988, collected by J. Muthumary.

Recorded as endophyte in

Solanum torvum, Araucaria cunninghamii, K.P. Kannan, 2002, Ph.D. Thesis, "Biodiversity of Endophytic Mycobiota from selected Medicinal plants and some Gymnosperms". MUBL No: 537.

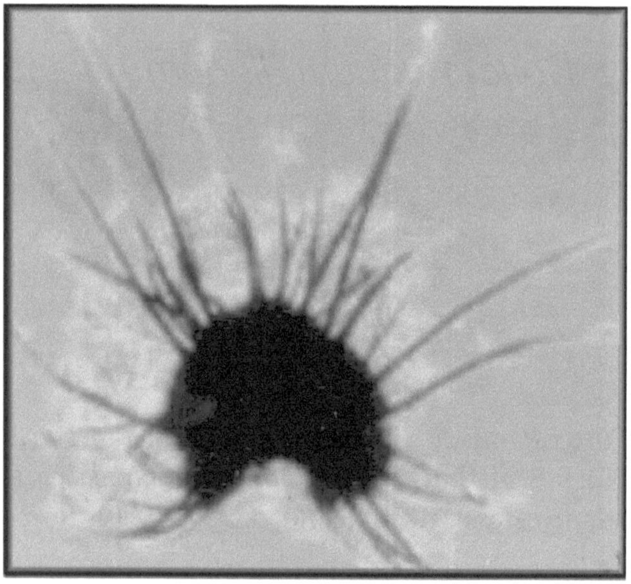

Figure 4.43 Mature conidioma (×200), *Colletotrichum lindemuthianum*

34. *Coniella castaneicola* (Ell. & Ev.) Sutton

Source

Sutton, B.C., *The Coelomycetes,* CMI, England, p. 420 (1980); Muthumary, J. and S. Masilamani, *J. Indian bot. Soc.* **68**: 131–133 (1989).

Description

Conidiomata pycnidial, spherical, scattered, unilocular, 300–350 µm diam. Ostiole central, 20–25 µm diam. Pycnidial wall made up of thick-walled, brown cells at the outerside and thin-walled, hyaline cells at the innerside giving rise to conidiogenous cells. Conidiogenous cells lining the pycnidial cavity, phialidic, ampulliform, 20–22 × 2.5–3 µm. Conidia cylindrical, aseptate, with evanescent mucoid appendage, 25–30 × 2–3.5 µm.

Specimen collected

On litter of *Elaeocarpus* sp., Bombay Shola, Kodaikanal (altitude 2400 m), Tamilnadu, November 1987, collected by J. Muthumary. IC. No: 134.

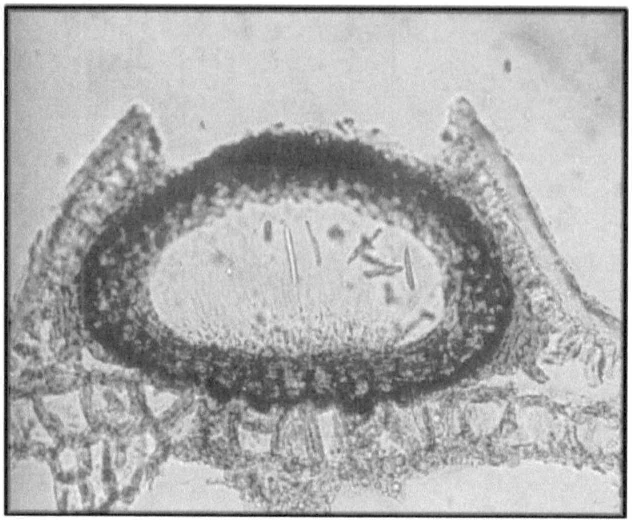

(a)

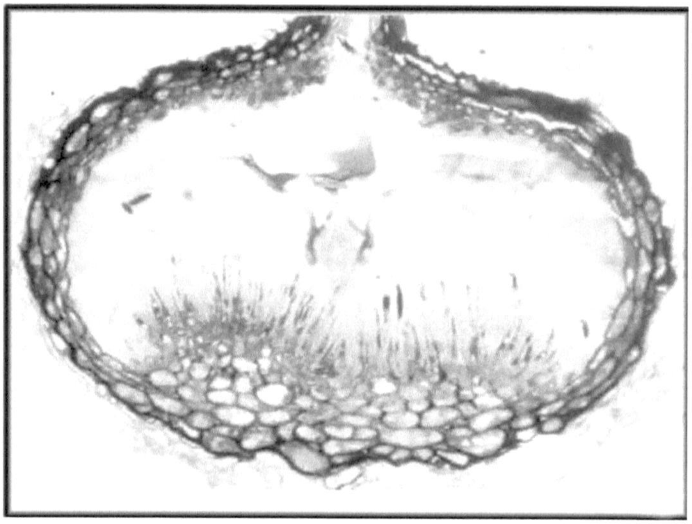

(b)

Figure 4.44 *Coniella castaneicola,* (a–b) V.S. of conidiomata (×200).

35. *Coniella fragariae* (Oud.) Sutton

Source

Sutton, *Mycol. Pap.,* **141**: p. 47 (1977); R.S. Mathur, *The Coelomycetes of India,* pp-65 (1979); Sutton, B.C. *The Coelomycetes,* CMI, England, p. 422 (1980); Muthumary, J. and R. Vanaja, *Trans. Br. Mycol. Soc.,* **87**(1): 109–113 (1986).

Description

Conidiomata, measuring 400-600 ìm diam. on oatmeal agar. In vertical section, the conidiomata spherical to subspherical, unilocular and ostiolate. Stroma consisting of 6-8 layers of thin-walled, pale brown textura-angularis surrounds the conidioma. The conidiomatal wall consists of 3-4 layers of brown, thick-walled pseudoparenchyma that produces the conidiophores is situated at the base of the conidioma. The conidiophores consist of one or two cylindrical hyaline cells, each of which gives rise to one or more conidiogenous loci. The conidiogenous cells are simple, cylindrical to filiform, hyaline, smooth, tapered at the apices, 14–25 (X = 20) × 2.5–3.0 ìm. The conidia are brown, one-celled, ellipsoidal, 7.0–12.5 (X = 10.5) × 5.0–7.5 (X = 5.5) ìm.

Specimen collected

On dead leaves of *Eucalyptus globulus* Labill., Bryant Park, Kodaikanal, Tamilnadu, Jul. 1988, collected by J. Muthumary. IC. No: 150.

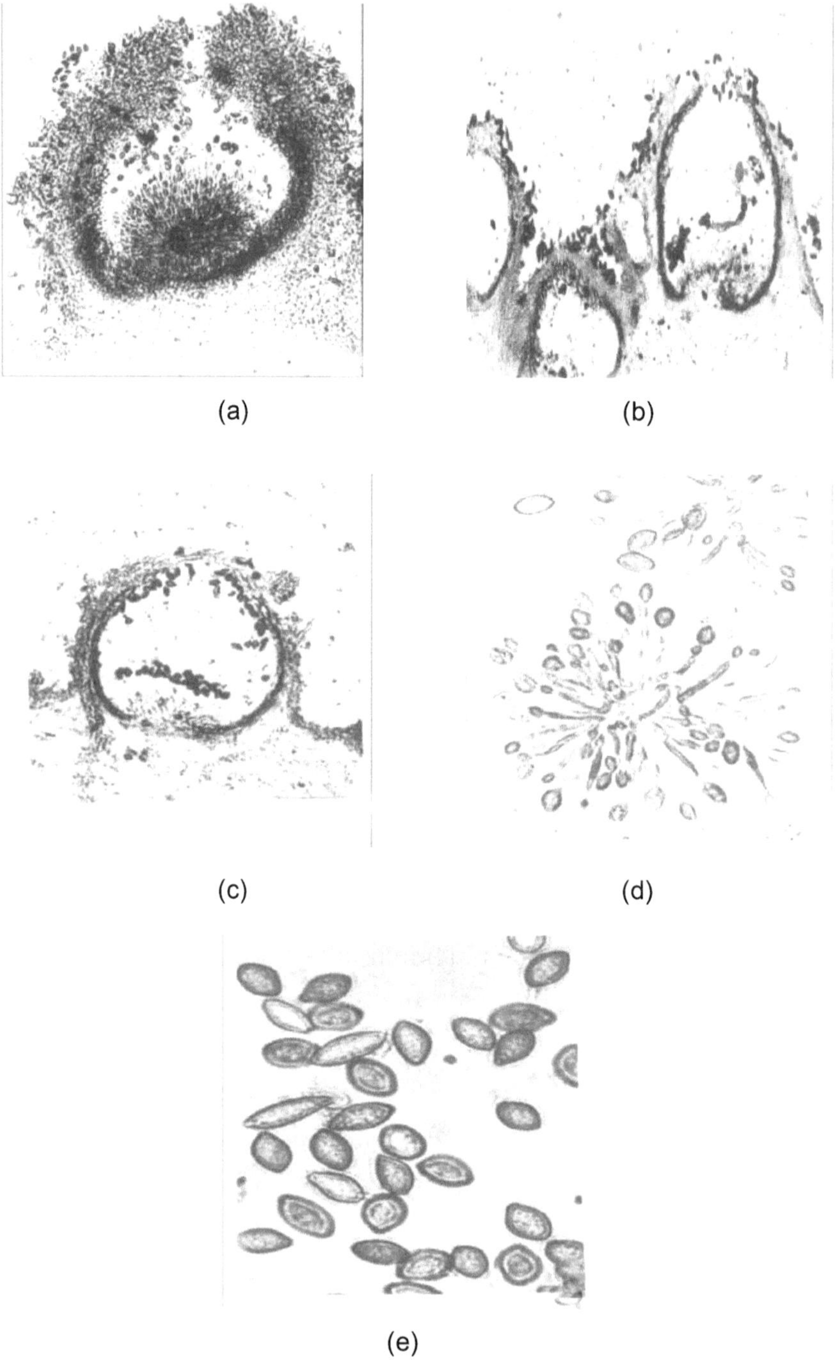

Figure 4.45 *Coniella fragariae,* (a–c) V.S. of conidiomata (×200); (d) Developing conidia (×200); (e) Mature conidia (×900).

36. *Coniothyrium fuckelii* Sacc.

Source

Sacc. In *Ven. novi vel critici Ser.*, **5**: 200; Sahni, V.P., *Mycopath. Mycol. appl.*, **29**: 226–229 (1966). Mathur, *The Coelomycetes of India*, p. 68 (1979).

Description

Conidiomata globose, pale brown, immersed, unilocular, 60–150 µm diam. Ostiole circular, central. Pycnidial wall made up of thick-walled, brown cells at the outside and thin-walled, hyaline cells at the innerside giving rise to conidiogenous cells. Conidiogenous cells lining the pycnidial cavity, hyaline, smooth, holoblastic, annellidic, discrete, doliiform, 5–8 × 3–4 µm. Conidia brown, thick-walled, aseptate, occasionally 1-septate, spherical to subspherical, 2–5 × 1.5–3 µm.

Specimen collected

On dead leaves of *Mangifera indica* L., Theosphical society, Madras, Tamilnadu, Nov. 1987, collected by J. Muthumary. IC. No. 121

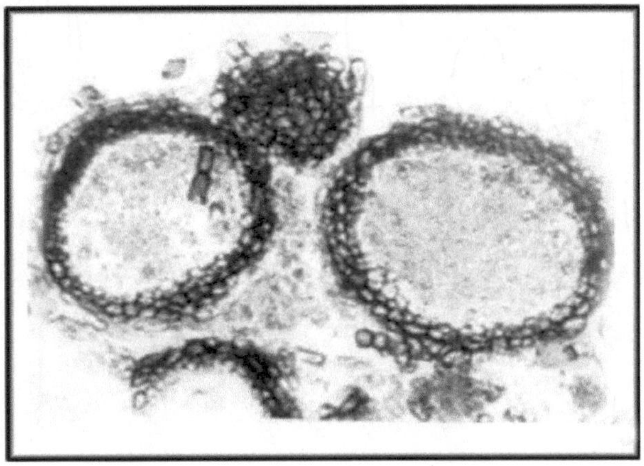

Figure 4.46 *Coniothyrium fuckelii,* (a) V.S. of conidioma (×200).

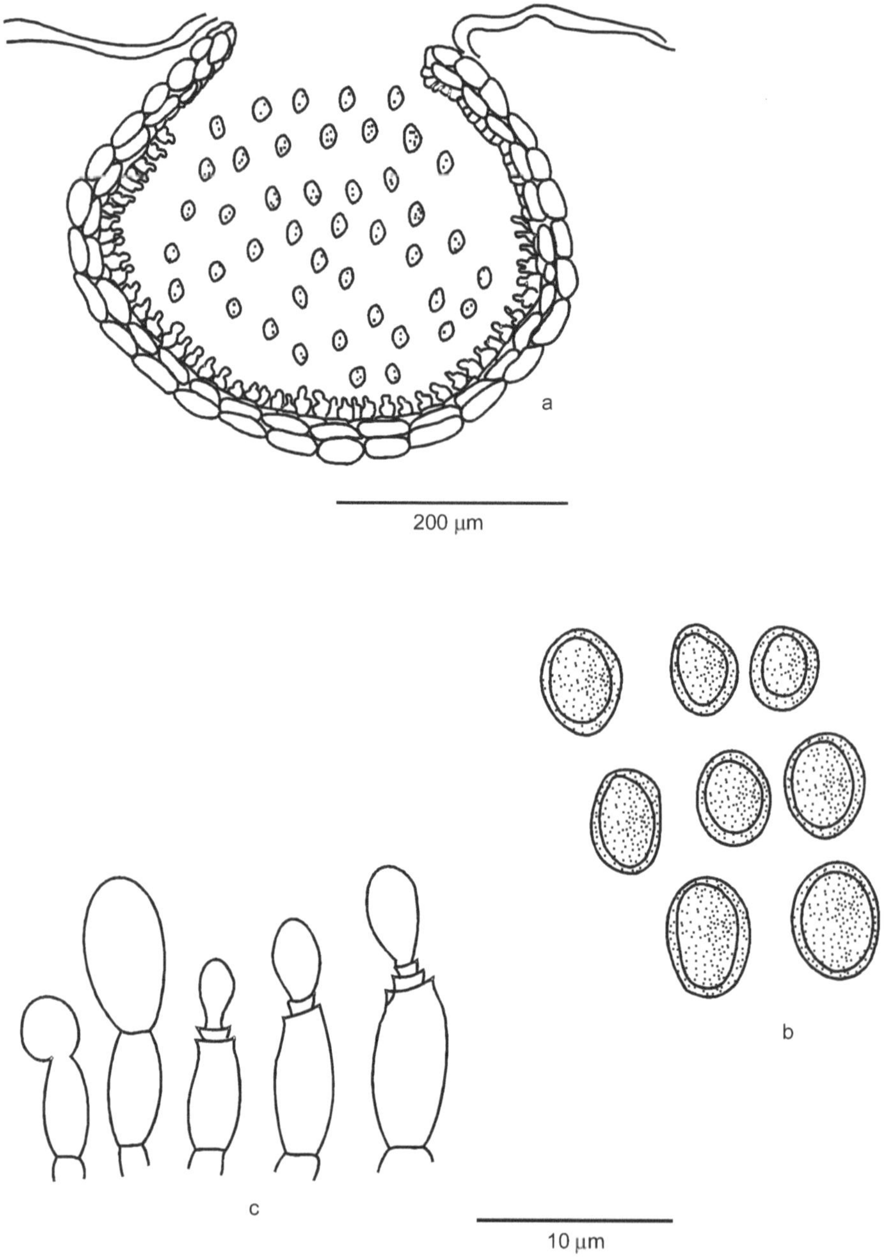

Figure 4.47 *Coniothyrium fuckelii*, (a) V.S. of Conidioma; (b) Mature conidia; (c) Developing conidia.

37. *Coniothyrium palmarum* Cda.

Source

Cda., *Icones fungorum,* **4**: 38 (1840); Sutton, B.C., *The Coelomycetes,* CMI, England, p. 123 (1980).

Description

Conidiomata foliicolous, pycnidial, separate, globose, dark brown, immersed, 150–200 µm in diam. In vertical section, pycnidium unilocular, pycnidial wall made up of thick-walled, pseudoparenchyma, and inner layer hyaline thin-walled, giving rise to conidiogenous cells. Conidiogenous cells ampulliform, smooth, hyaline, annellidic with 5–4 annellations, 4–10 × 3–4 µm. Conidia brown, thick-walled, 0–1 euseptate, apex obtuse, base truncate to blunt, globose to subcylindrical, 8–10 × 6–7 µm.

Specimen collected

On dead leaves of unidentified palm., Kambakkam, Tamilnadu, India, Jul. 1988, collected by J. Muthumary. IC. No: 151.

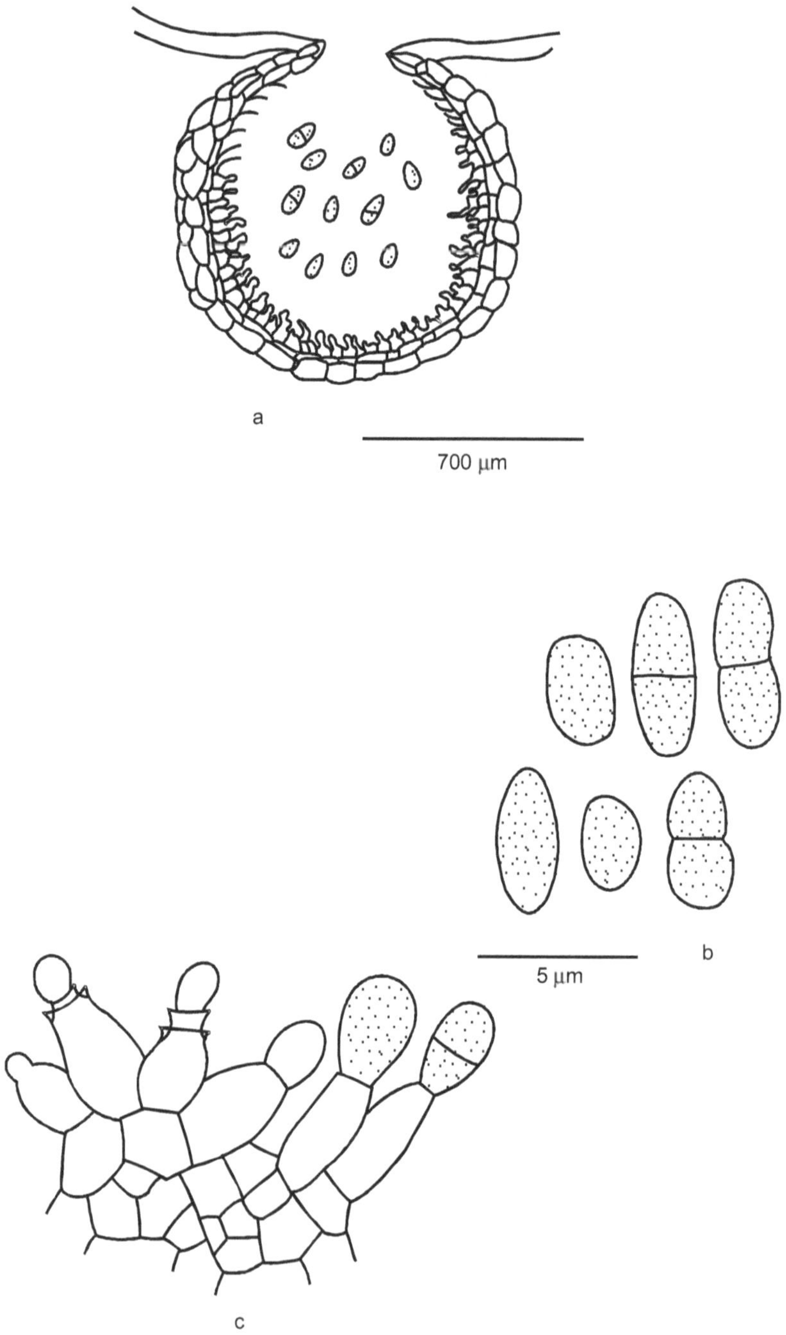

Figure 4.48 *Coniothyrium palmarum*, (a) V.S. of Conidioma; (b) Mature conidia; (c) Developing conidia.

38. *Conostroma quercicola* Muthumary

Source

Muthumary, J., *Current Science,* **55**(21): pp 1081–1082, (1986)

Description

Conidiomata stromatic, dark brown, sub-peridermal, 800–1000 µm in diam., multilocular, situated around a central column of sterile tissue composed of thin-walled, pale brown, textura-angularis with somewhat larger cells at the centre; locular wall tissue and basal stroma of thin-walled pseudoparenchyma. Dehiscence by irregular rupturing of the central column of tissue to release the conidia. Conidiogenous cells line the entire locular cavity, simple, branched at the base, hyaline, smooth, cylindrical, tapering towards the tip, blastic, sympodial, occasionally with 2–3 annellations, 15–22 (x = 20) × 2.5–3.5 (x = 3.0) µm. Conidia fusiform, medianly 1-euseptate, hyaline, orange coloured in mass, smooth, thin-walled, guttulate, attenuated towards the truncate base and pointed apex, 17.5–20.5 (x = 18) × 2.5–3.0 (x = 2.8) µm.

Specimen collected

On fallen twigs of *Quercus alba,* Bryant Park, Kodaikanal, Tamilnadu, December 1983, collected by J. Muthumary. MUBL No: 2914.

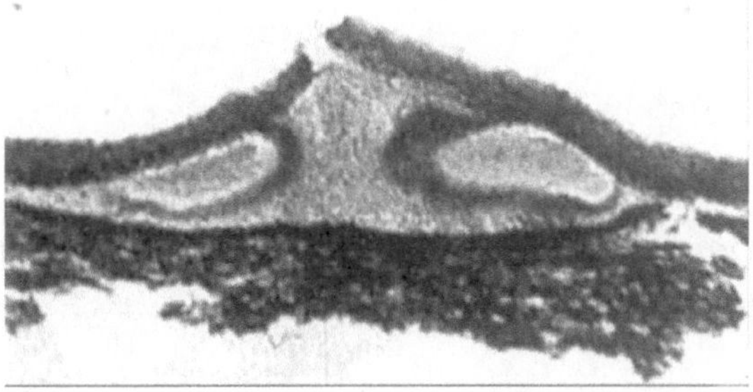

Figure 4.49 *Conostroma quercicola,* (a) V.S. of conidioma (×200).

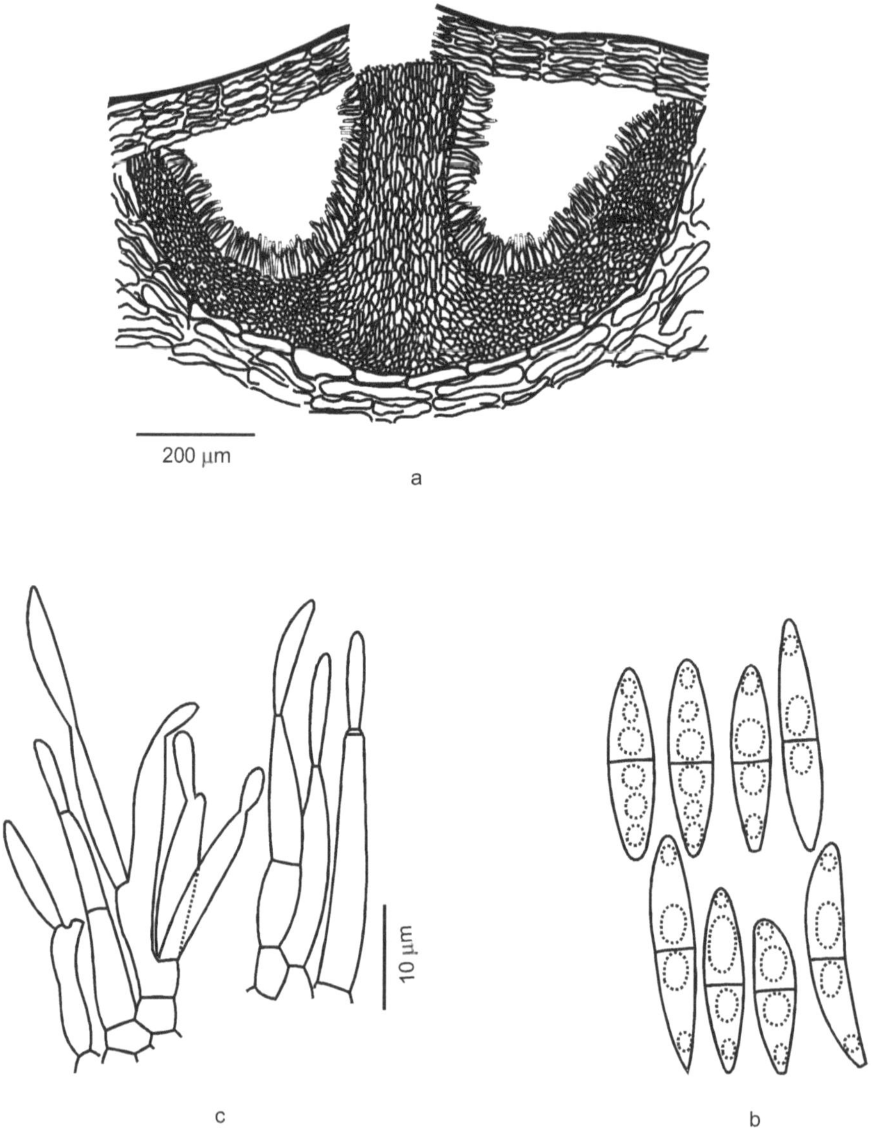

Figure 4.50 *Conostroma quercicola*, (a) V.S. of Conidioma; (b) Mature conidia; (c) Developing conidia.

39. *Coryneum indicum* Sutton & Rizwi

Source

Sutton, B. C. & Rizwi, M.A., *Nova Hedwigia*, **32**: 341–346 (1980).

Description

Conidiomata acervular, up to 800 μm diam. Conidiogenous cells cylindrical, 50–80 × 3–4 μm. Conidia fusiform, straight, apex obtuse, base truncate, 6–7 distoseptate, median cells brown, 117–135 × 17.5–19 μm; Paraphyses like structures, filiform, septate, pale brown, 70–150 × 3–4 μm.

Specimen collected

On dead twigs of *Quercus* sp., Kodaikanal, Tamilnadu, Nov. 1987, collected by J. Muthumary. IC. No: 151

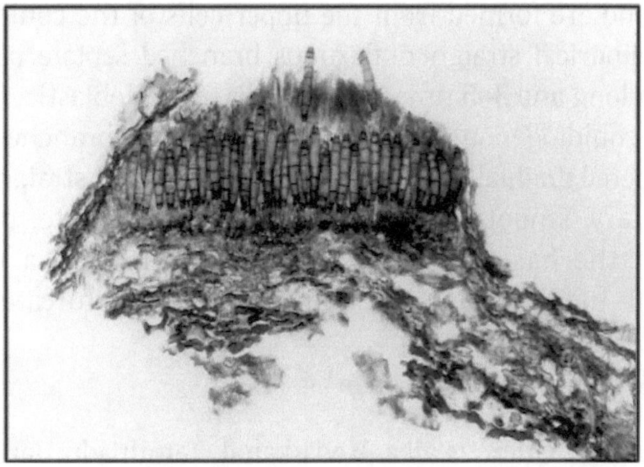

Figure 4.51 V.S. of conidioma (×200), *Coryneum indicum*

40. *Coryneum quercinum* Muthumary

Source

Muthumary, J. and B.C. Sutton, *Trans. Br. Mycol. Soc.*, **86**(3): 512–515 (1986).

Description

It forms abundant, blackish green, acervular conidiomata on dead twigs. The conidiomata are scattered, rarely confluent, at first immersed, later erumpent by rupturing the bark, finally punctiform, 500–1000 µm diam. In vertical section, the conidiomatal stroma is composed of irregular, thick-walled, brown textura-angularis at the base becoming more vertically orientated towards the conidiogenous layer. From the basal tissue, groups of thin-walled, pale brown, compact, septate hyphae arise which become paler towards the apices and eventually give rise to conidiophores. The conidiogenous cells are either apical or intercalary and are formed from the upper cells of the conidiophores. They are filiform to cylindrical, straight or flexuous, branched, septate, pale brown to hyaline, 200–300 µm long and 3–5 µm wide. Conidia are holoblastic, formed either at the apex of the conidiogenous cell or on short lateral branches. They are broadly fusiform, tapered gradually towards both apex and base, straight to slightly curved, brown, solitary, smooth, 6–7 transversely distoseptate, obtuse at the apex, truncate at the base, apical cell dark brown with a much paler apex, 45–60 × 14.0–16.0(X = 15.0) µm. Some conidia are bifurcate at the apices.

Specimen collected

On fallen twigs of *Quercus alba,* Kodaikanal, Tamilnadu, Jan. 1986, collected by J. Muthumary. MUBL No: 2872.

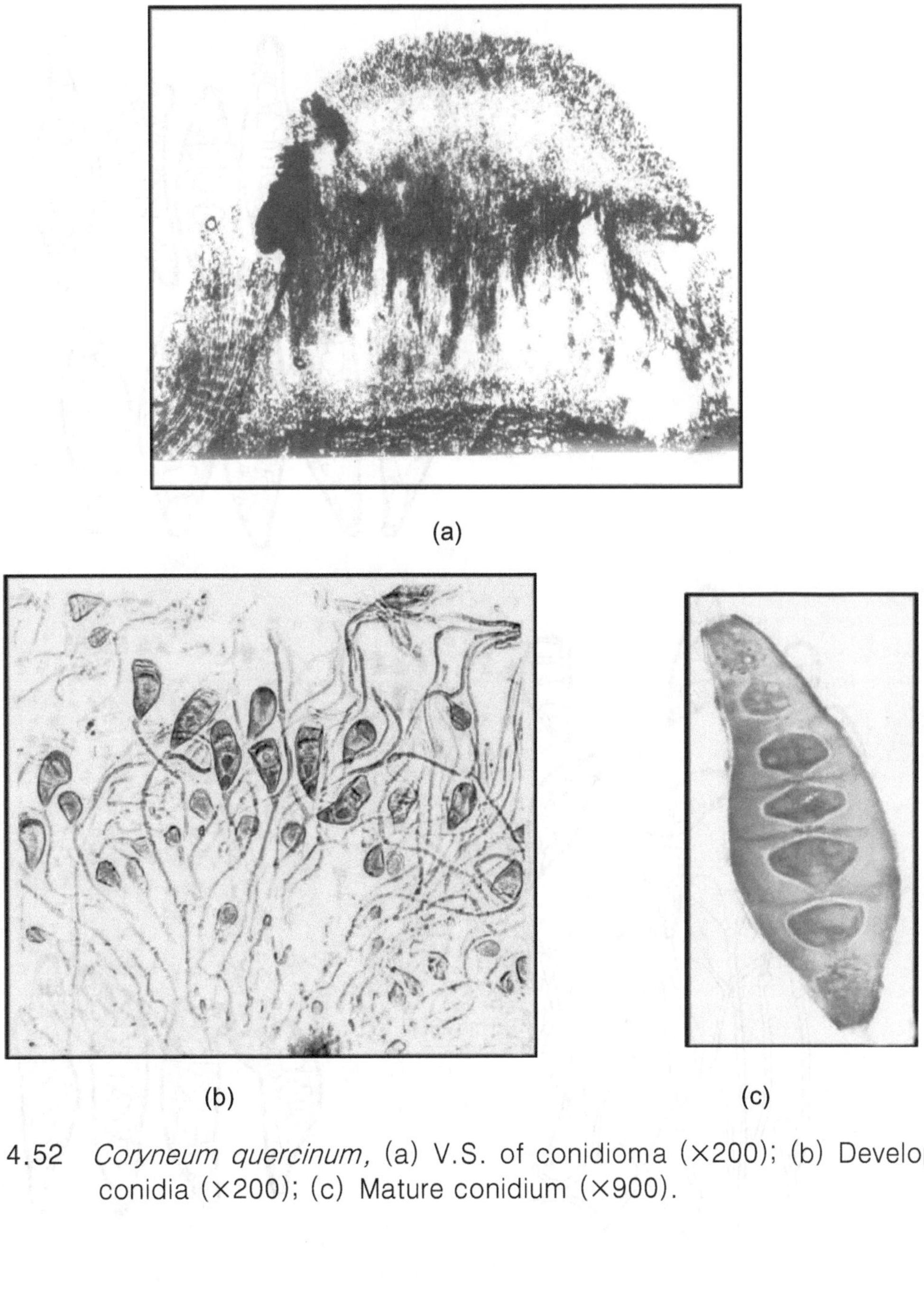

(a)

(b)

(c)

Figure 4.52 *Coryneum quercinum,* (a) V.S. of conidioma (×200); (b) Developing conidia (×200); (c) Mature conidium (×900).

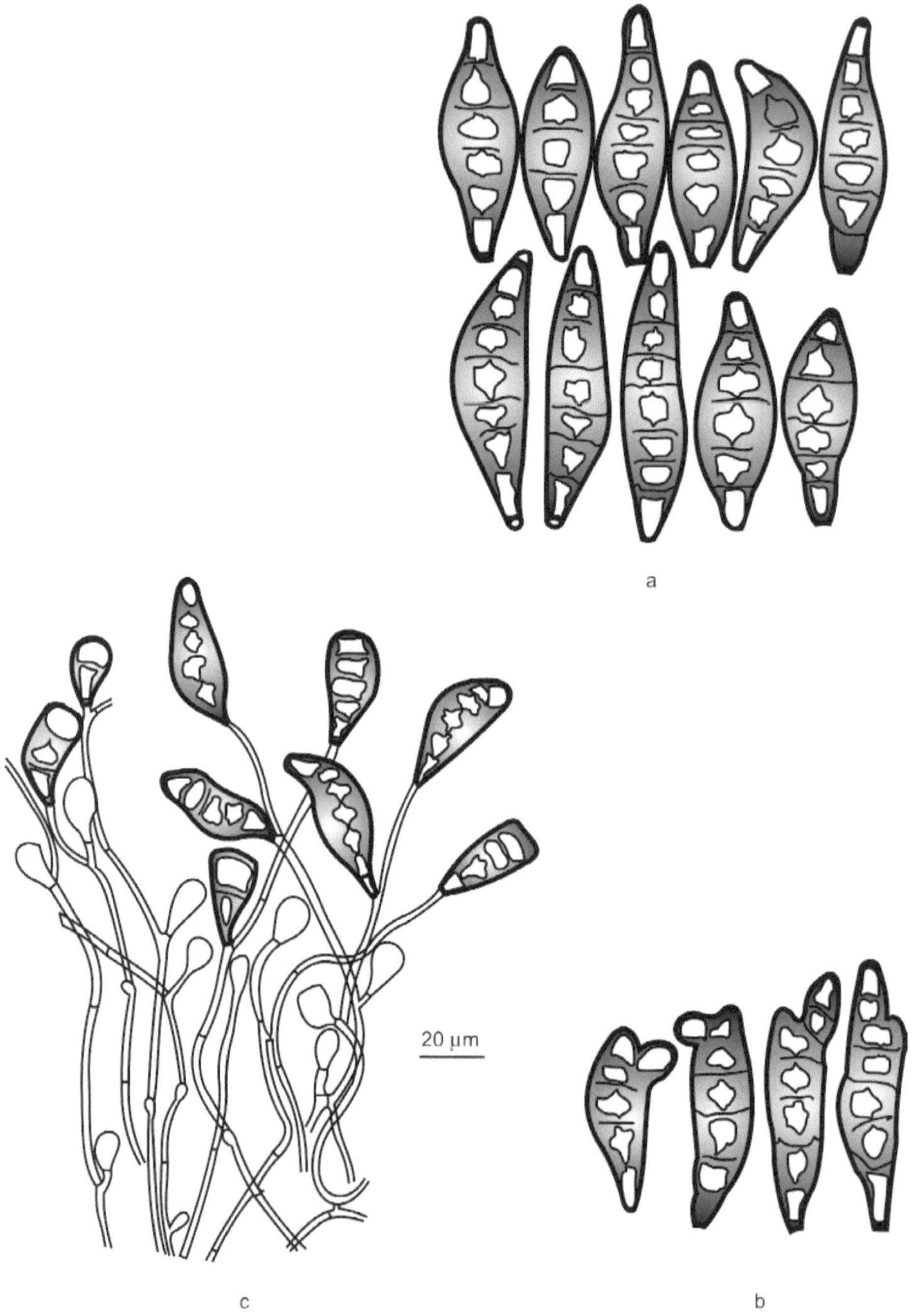

Figure 4.53 *Coryneum quercinum*, (a) Mature conidia; (b) Abnormal conidioma; (c) Developing conidia.

41. *Coryneum modonium* (Sacc.) Griff. & Maubl.

SourceSource

Griff. & Maubl., *Bull. Trimmest. Soc. Mycol. Fr.,* **26:** 379 (1910); Sutton, B. C., *Mycological Papers.,* **138:** 38–40 (1975). Sutton, B.C., *The Coelomycetes,* CMI, England, p. 354, (1980); Muthumary, J., *Current Science,* **56**(23): pp. 1245–1247 (1987).

Description

Conidiomata acervular, abundant, sub-peridermal, scattered, rupturing the periderm, flat and effuse, 800–1000 μm in diameter, dark brown, composed of thick-walled, brown, textura-angularis at the base and light brown, comparatively thin-walled, vertically arranged, elongated cells towards the conidiogenous region. Conidiophores formed from the upper cells of the acervuli, cylindrical, erect, branched at the base, septate, pale brown, becoming hyaline towards the apices, 50–60 × 3–6 μm, often merging imperceptibly into the pseudoparenchyma. Conidiogenous cells holoblastic, annellidic, hyaline to pale brown, cylindrical, smooth, with usually a single annellation. Conidia formed singly from the apices of the conidiogenous cells, variable in shape, clavate, narrowly fusiform, very rarely more broadly fusiform, tapered gradually towards the obtuse apex, straight, slightly curved or with a tendency to be sigmoid, also tapered gradually towards the truncate base which often carries away part of the conidiogenous cell at seccession, brown, smooth, 5–8 distoseptate, apical cell sometimes darker than the rest and invariably with a hyaline tip, 45–67 (65) × 14–20 (19) μm.

Specimen collected

On dead twigs of *Ficus benghalensis,* Theosphical Society, Adayar, Tamilnadu, April 1980, MUBL No. 2916; Srivaikuntam, Tirunelvely District, Tamil nadu, October 980, collected by J. Muthumary.

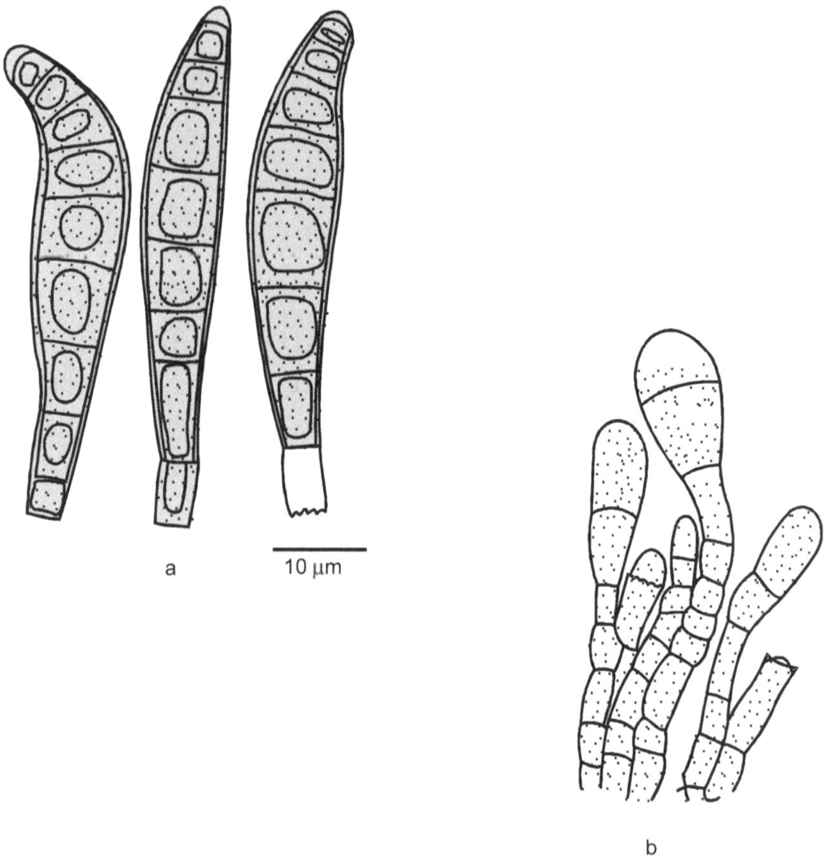

Figure 4.54 *Coryneum modonium*, (a) Mature conidia; (b) Developing conidia.

Figure 4.55 V.S. of conidioma (×200), (b) *coryneum modonium*

42. *Cryptosporiopsis malicorticis* (Cordl.) Nannfeldt

Source

Nannfeldt, *Nova Acta Reg. Soc. Scient. Upsal.,* ser. 4, **8**: 91 (1932); Sutton, B.C., *The Coelomycetes,* CMI, England, p. 513 (1980).

Description

Conidiomata cupulate, subepidermal, 250 µm, unilocular, separate. Conidiophores absent. Conidiogenous cells enteroblastic, phialidic, discrete, cylindrical, hyaline, smooth. Conidia comparatively large, 8-13 x 3-4 µm, hyaline, thin-walled, smooth, eguttulate, apex obtuse.

Specimen collected

On fresh leaves of *Plumeria obtusa* (Apocyanaceae), Theosophical society, Adayar, Chennai, Tamil Nadu, May 2000, collected by Senthil Priya. IC No. 260

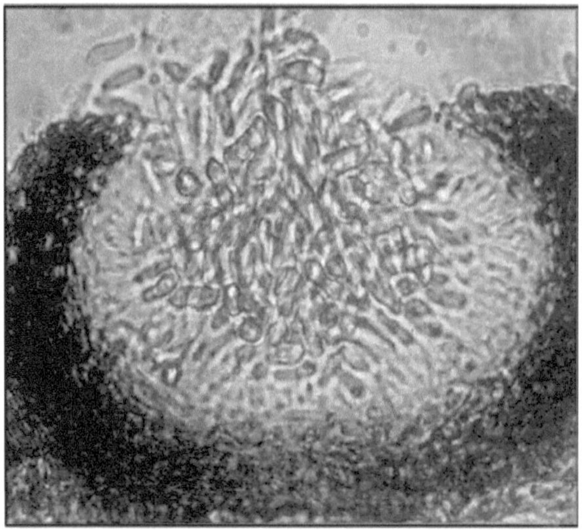

(a)

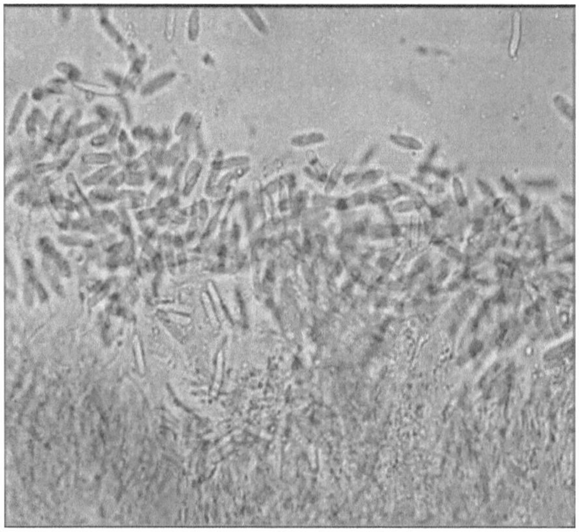

(b)

Figure 4.56 *Cryptosporiopsis malicorticis*, (a) V.S. of conidioma (×200); (b) Mature conidia (×900)

43. *Cylindrogloeum trillii* (Ell. & Ev.) Arx

Source

Arx, *Verh. K. Ned. Akad, Wet. Afd. Natuurk.,* **51**: 143 (1957); Sutton, B.C., *The Coelomycetes,* CMI, England, p. 596 (1980).

Description

Conidiomata acervular, 160–200 µm, separate. Conidiogenous cells enteroblastic, phialidic. 4–8 × 2–3 µm. Conidia hyaline, straight, cylindrical, 0–2 septate, 16–17 × 5.8 µm.

Specimen collected

On dead leaves of *Citrus limonum* L., Madras, TamilNadu, Mar. 1989, collected by J. Muthumary. IC. No: 177.

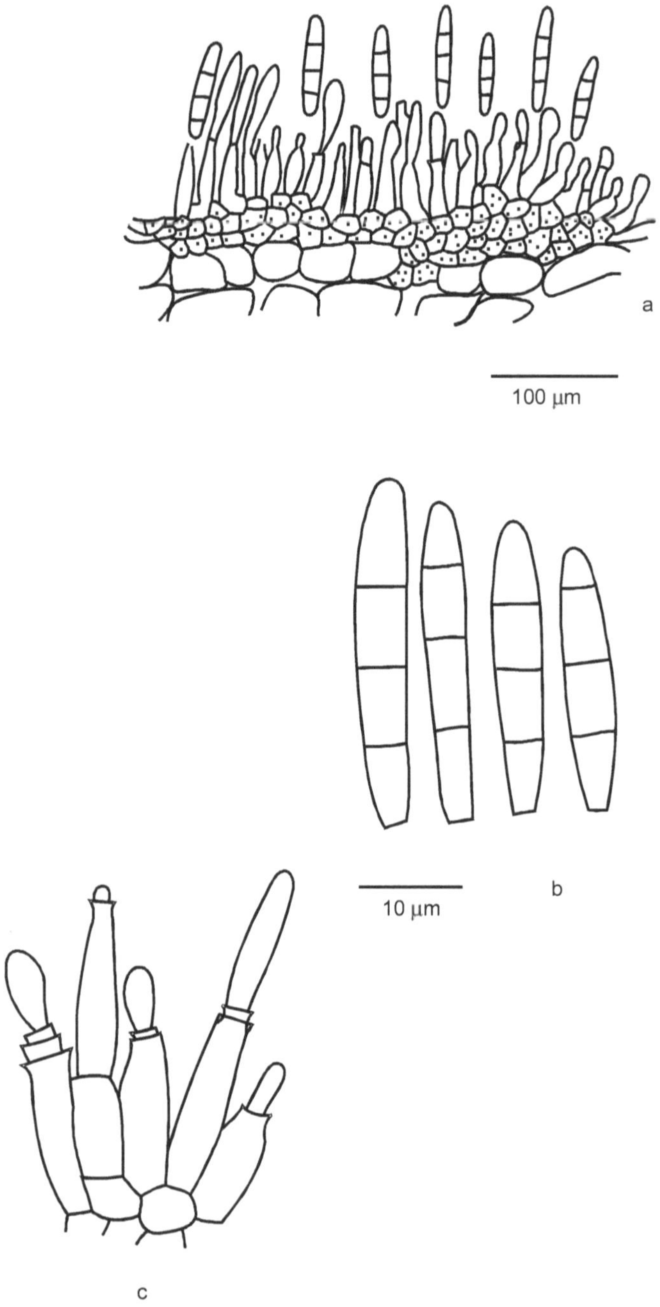

Figure 4.57 *Cylindrogloeum trillii*, (a) V.S. of conidioma; (b) Mature conidia; (c) Developing conidia.

44. *Cytospora chrysosperma* Pers.

Source

Persoon, C.H., *Ex Fr. Syst, Mycol.*, **2**(2): 542 (1823); Singh, J., *Curr. Sci.*, **5**: 589 (1937); Mathur R.S., *The Coelomycetes of India*, p-74 (1979); Sutton, B.C., *The Coelomycetes*, CMI, England, p. 580 (1980).

Description

Conidiomata eustromatic, multilocular, stroma made up of pseudoparenchymatous cells, 800–950 µm diam. Conidiogenous cells lining the entire cavity, phialidic, hyaline, smooth, 18–20 × 1.5–2 µm. Conidia aseptate, hyaline, smooth, allantoid, 4.5–5 µm long.

Specimen collected

On dead twigs of *Rosa leschenaultiana* W. & A., Narkanda, Shimla, Himachala Pradehs, Oct. 1989, collected by J. Muthumary. IC No: 178.

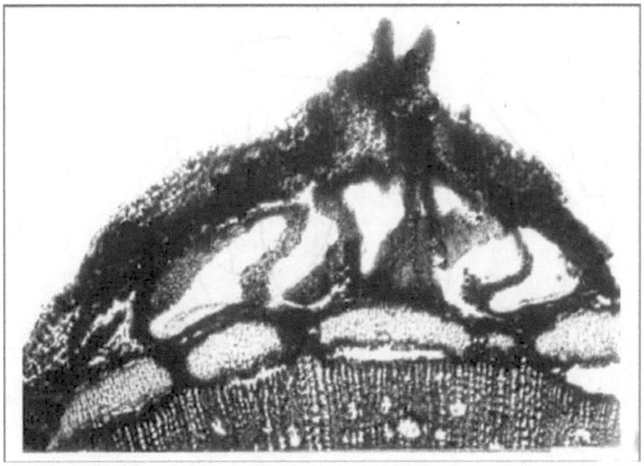

Figure 4.58 *Cytospora chrysosperma;* (a) V.S. of conidioma (×200)

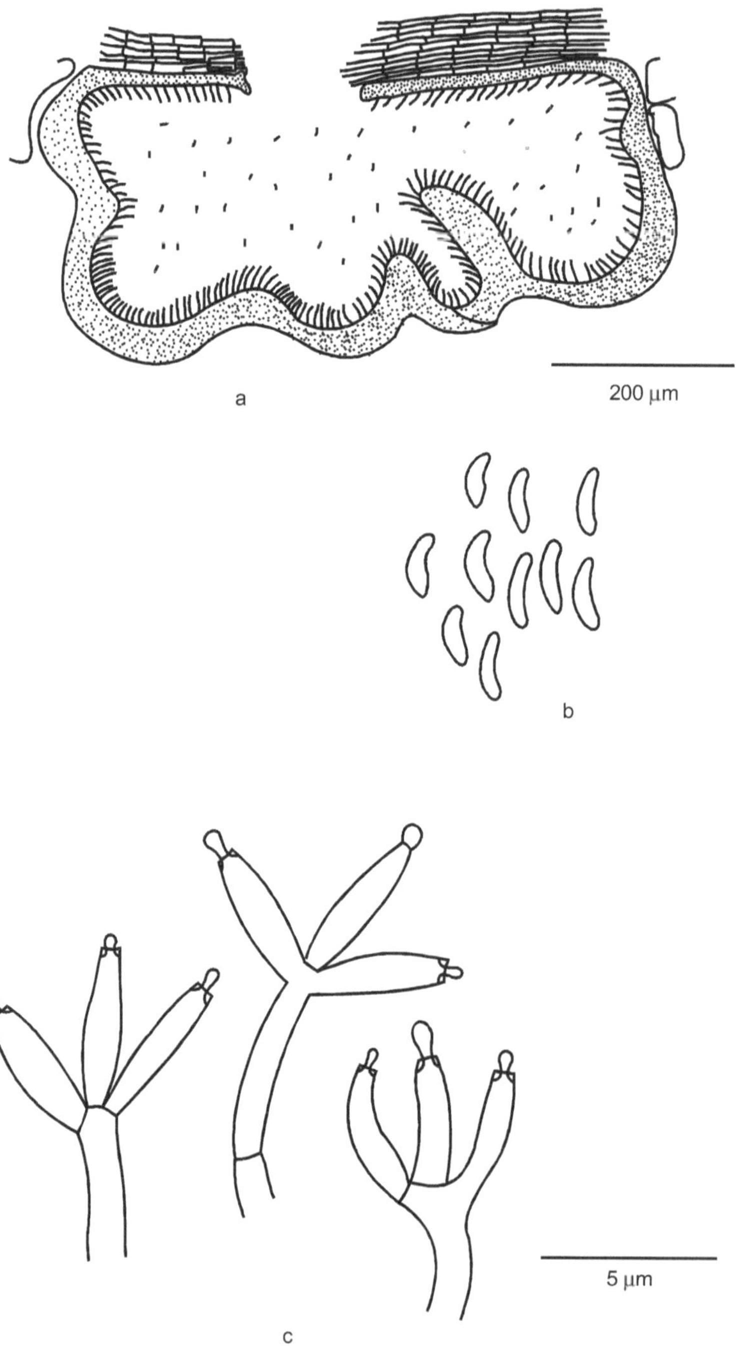

Figure 4.59 *Cytospora chrysosperma*, (a) V.S. of conidioma; (b) Mature conidia; (c) Developing conidia.

45. *Cytosphaera mangiferae* Died.

Source

Died., *Annls mycol.*, **14**: 205 (1916); R.S. Mathur, *The Coelomycetes of India*, pp.73 (1979); Sutton, B.C., *The Coelomycetes*, CMI, England, p. 161 (1980); Muthumary, J., *Current Science*, **56**(9), pp. 426-427 (1987).

Description

Conidiomata eustromatic, separate, pulvinate, dark brown, erumpent at maturity, multilocular, globose, locules peripheral at the upper part of stromata, 1300–1500 µm. Conidiomata producing macroconidia are larger when compared to those producing microconidia, measuring 1500–1800 µm. Conidiogenous cells are holoblastic, 8–10 × 3–5 µm. Macroconidia are spherical to subspherical, aseptate, thick-walled hyaline, smooth, guttulate, base truncate, 18–28 × 10–13. Conidiomata producing microconidia 500–600 µm in diameter with 3–5 locules. Conidiogenous cells phialidic, 6–14 × 2–2.5 µm. Microconidia ellipsoid to oblong, aseptate, hyaline, smooth, thin-walled, 3–4 × 1.0–1.5 µm.

Specimen collected

On dead twigs of *Acrocarpus fraxinifolius*, Magod Falls, Karnataka, November 1979. FFSI No: 2788.

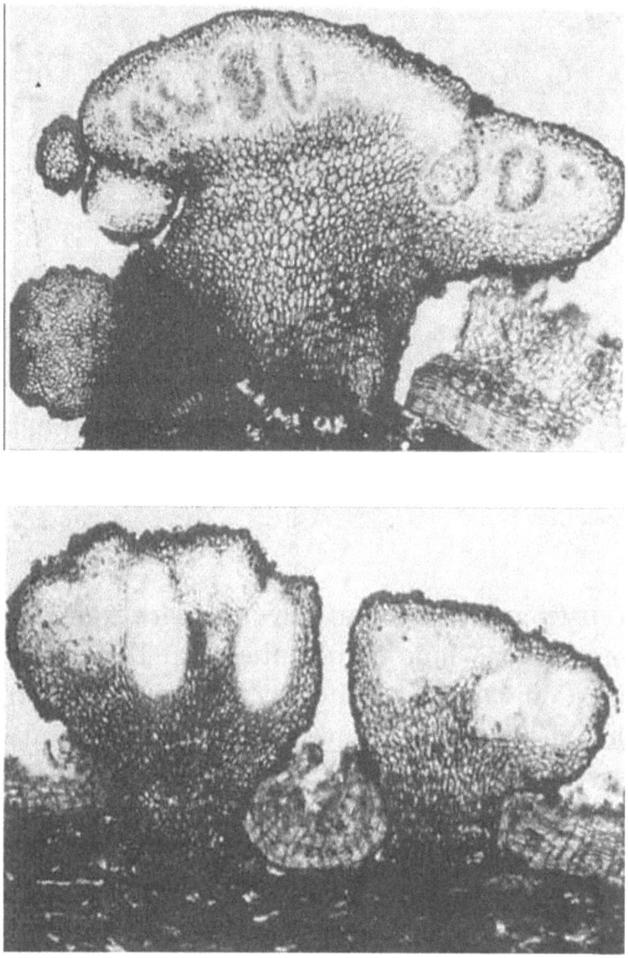

Figure 4.60 *Cytosphaera mangiferae*, (a–b) V.S. of conidiomata (×200)

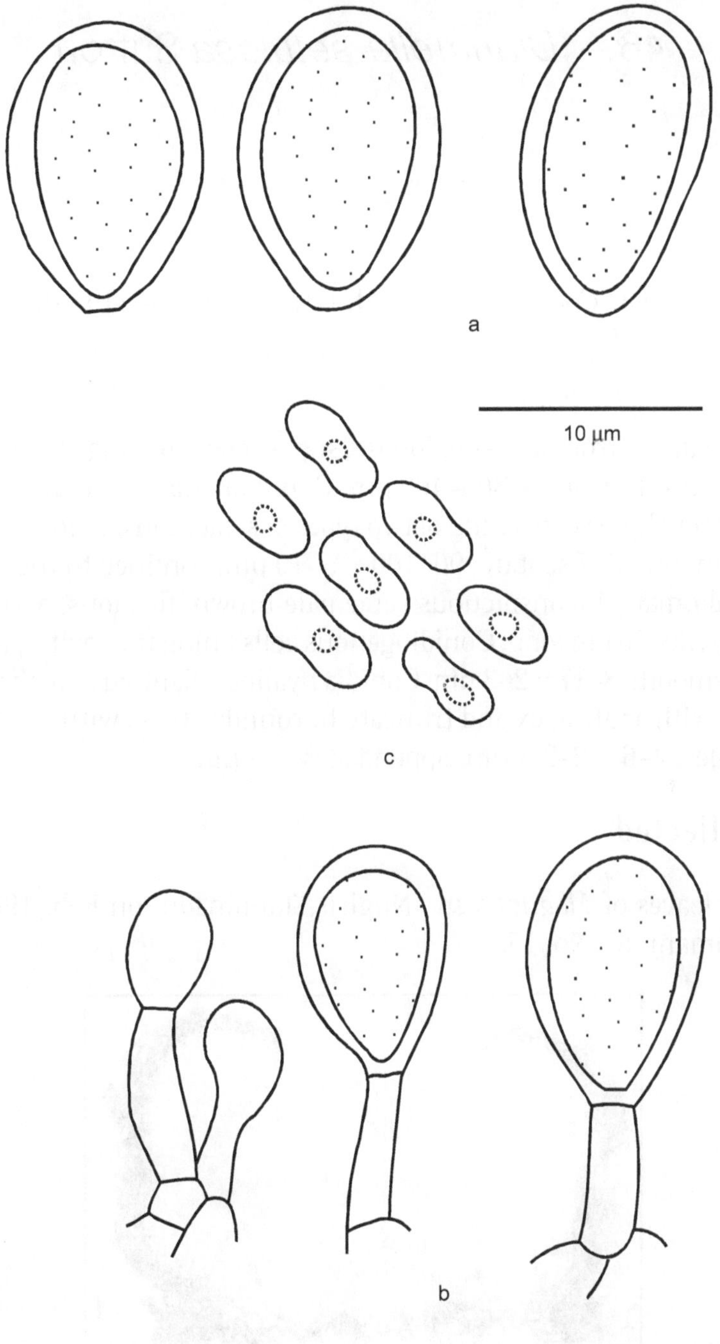

Figure 4.61 *Cytosphaera mangiferae*, (a) Mature conidia; (b) Developing conidia; (c) Microconidia.

46. *Diarimella setulosa* Sutton

Source

Sutton, B.C., *The Coelomycetes*, CMI, England, p. 454 (1980).

Description

Conidiomata eustromatic, subglobose, superficial, opening at maturity into a cup-like structure 100–500 × 300–400 μm. Conidiomatal wall made up of thin-walled pseudoparenchyma consisting of two types of setae; conspicuous setae, rigid, dark-brown, smooth, 3–7 septate, 90–100 × 5–10 μm, confined to the base and sides of the conidiomata; inconspicuous setae pale brown, flexuous, aseptate, tapered to the apices, 30–40 μm long. Conidiogenous cells lining the cavity, philaidic, filiform, hyaline, smooth, 8–17 × 2–3 μm. Conidia hyaline, allantoid to fusiform, biguttulate, aseptate, with acute apex and truncate to rounded base, with 1–3 apical and basal appendages, 4–6 × 2–2.5 μm; appendages 3–8 μm.

Specimen collected

On dead leaves of *Magnolia* sp., Nilgiris, Tamilnadu, on Feb. 1988, collected by S. Masilamani. IC. No: 152.

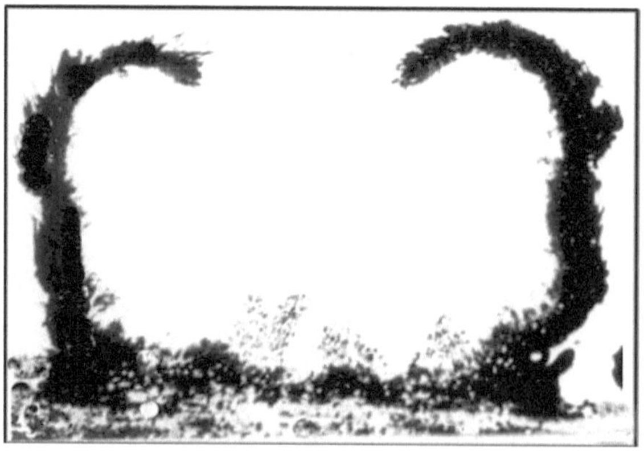

Figure 4.62 *Diarimella setulosa*, (a) V.S. of conidioma (×200)

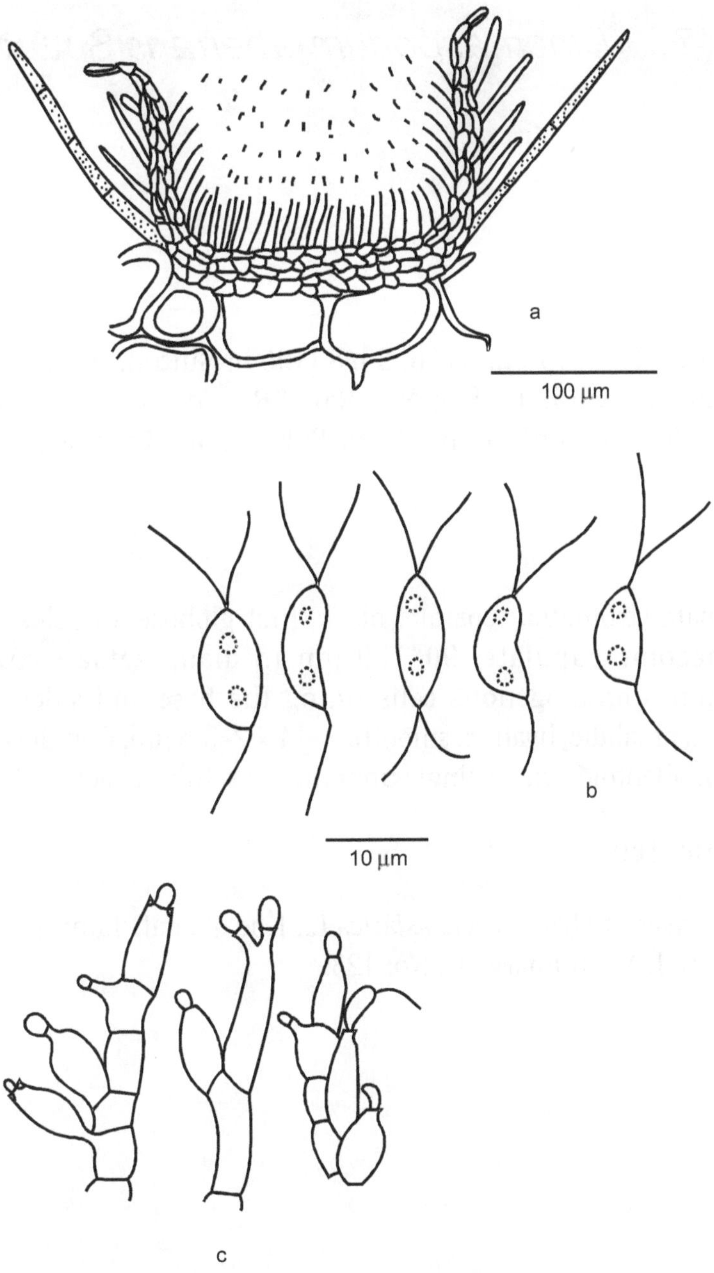

Figure 4.63 *Diarimella setulosa*, (a) V.S. of conidioma; (b) Mature conidia; (c) Developing conidia.

47. *Dinemasporium aberrans* Sutton

Source

Sutton, B.C., *Can. J. Bot.,* **47**: p. 2098 (1969); Sutton, B.C., *The Coelomycetes,* CMI, England, p. 457 (1980); Nag Raj, T.R., *Coelomycetous anamorphs with appendage bearing conidia,* Mycologue Publications, Canada. pp. 261–262 (1993).

Description

Conidiomata stromatic, separate, black, first globose and closed, later opening out to become cupulate, 90–150 µm in diam. setae sparse, acuminate, 75–150 µm. Conidiogenous cells lining the base and sides of the cupulate conidioma, phialidic, hyaline, smooth, 8–14 × 2–3.5 µm. Conidia hyaline, aseptate, straight or allantoid, with a single unbranched setula at each end, 4–6 × 2–2.5 µm.

Specimen collected

On dead leaves of *Hydrocotyle asiatica* L., Kodaikanal, Tamilnadu, on Nov. 1987, collected by J. Muthumary. IC. No: 123.

Dinemasporium aberrans, (a) V.S. of conidioma (×200)

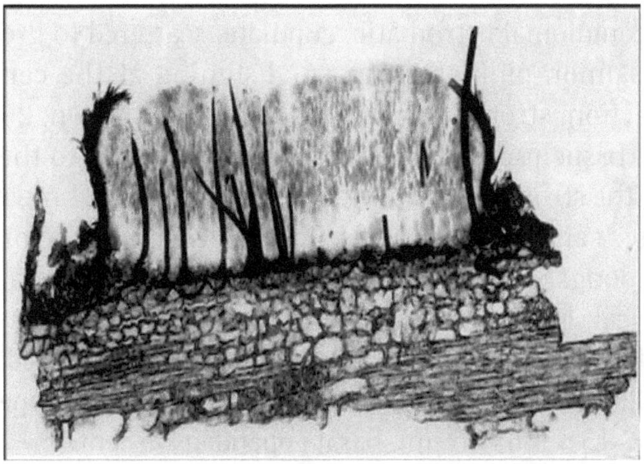

Figure 4.64 *Dinemasproium aberrans*, (b) V.S. of conidioma (×200)

48. *Dinemasporium strigosum* (Persoon:Fries) Saccardo

Source

Saccardo, *Michelia*, **2**: 281 (1881); Sutton, B.C., *The Coelomycetes*, CMI, England, p. 458 (1980); Nag Raj, T.R., *Coelomycetous anamorphs with appendage bearing conidia*, Mycologue Publications, Canada. pp. 270–272 (1993).

Description

Foliicolous conidiomata stromatic, cupulate, scattered to gregarious, erumpent, unilocular, salmon-pink spore mass glistening at the centre of the stroma. In vertical section, stroma cupulate, to black, dark brown, 200–250 µm in diam. Setose, basal tissue pseudoparenchymatous, extending to the sides and upwards into making the stroma with elongated, hyaline to pale brown cells, interspersed with, septate, setae made up of thick-walled brown, elongated cells. Setae 300–800 µm long. 5–10 µm at the base and 2–3 µm at the apex. Conidiogenous cells cylindrical, hyaline, thin-walled, smooth, 13–15 × 2.0–2.5 µm, phialidic. Conidia fusiform to ellipsoid, unicellular, hyaline, thin-walled, smooth, guttulate, 8–10 × 1.5–2.0 µm, with a single, unbranched, filiform appendage at each end; appendages 8–15 × 1.0–1.5 µm. basal appendage excentric.

Specimen collected

On leaves and culms of *Bambusa* sp., *Cynodon dactylon*, Kodaikanal, TamilNadu, Jul. 1987, collected by J. Muthumary. IC No: 153.

Mehrotra (1979) has recorded *D.erysiphoides*, Ell. & Ev., *D. graminum* (Berk) Leu. *D. hispidulum* (Schrad.) Sacc. and *D. microsporum* Sacc. from India.

Figure 4.65 *Dinemasproium strigosum*, (b) V.S. of conidioma (×200)

49. *Diplodia sycina* Mont.

Source

Mont, In cast, *cat. Pl. Mars. Suppl.*, p. 64 (1851) Saccardo, *Sylloge Fungorum*, 3: 350 (1884); Sydow & Butler, *Annls. Mycol.*, **14**:198 (1916); R.S. Mathur, *The Coelomycetes of India*, p. 86 (1979); Masilamani, S., Ph.D. thesis, University of Madras, (1994).

Description

Conidiomata pycnidial, 128-160 x 80 - 96 μm diam., separate, globose, immersed, unilocular, wall composed of thick-walled pseudoparenchyma. Pycnidial wall 20–25 μm, thick, outer layers dark brown, thick-walled, inner layer hyaline and thin-walled giving rise to the conidiogenous cells.Ostiole single, circular, central. Conidiogenous cells lining the pycnidial cavity holoblastic, determinate, cylindrical, hyaline, smooth, 5–6 × 3–4 μm. Conidia brown, smooth, thick-walled, medianly 1- euseptate, apex obtuse, base truncate, 12–16 × 4.8–6.5 μm.

Specimen collected

On dead leaves of *Ficus benghalensis* L., Madras, Tamilnadu, May 1988; On dead leaves of *Lagerstroemia parviflora* Roxb., Madras, Tamilnadu, on Sep. 1988, collected by J. Muthumary. IC No: 154.

More than seventy species were recorded from India in *Diplodia* from various hosts (Mathur, 1979). *Diplodia mutila* Fr. & Mont. is the type species of this genus (Sutton, 1980).

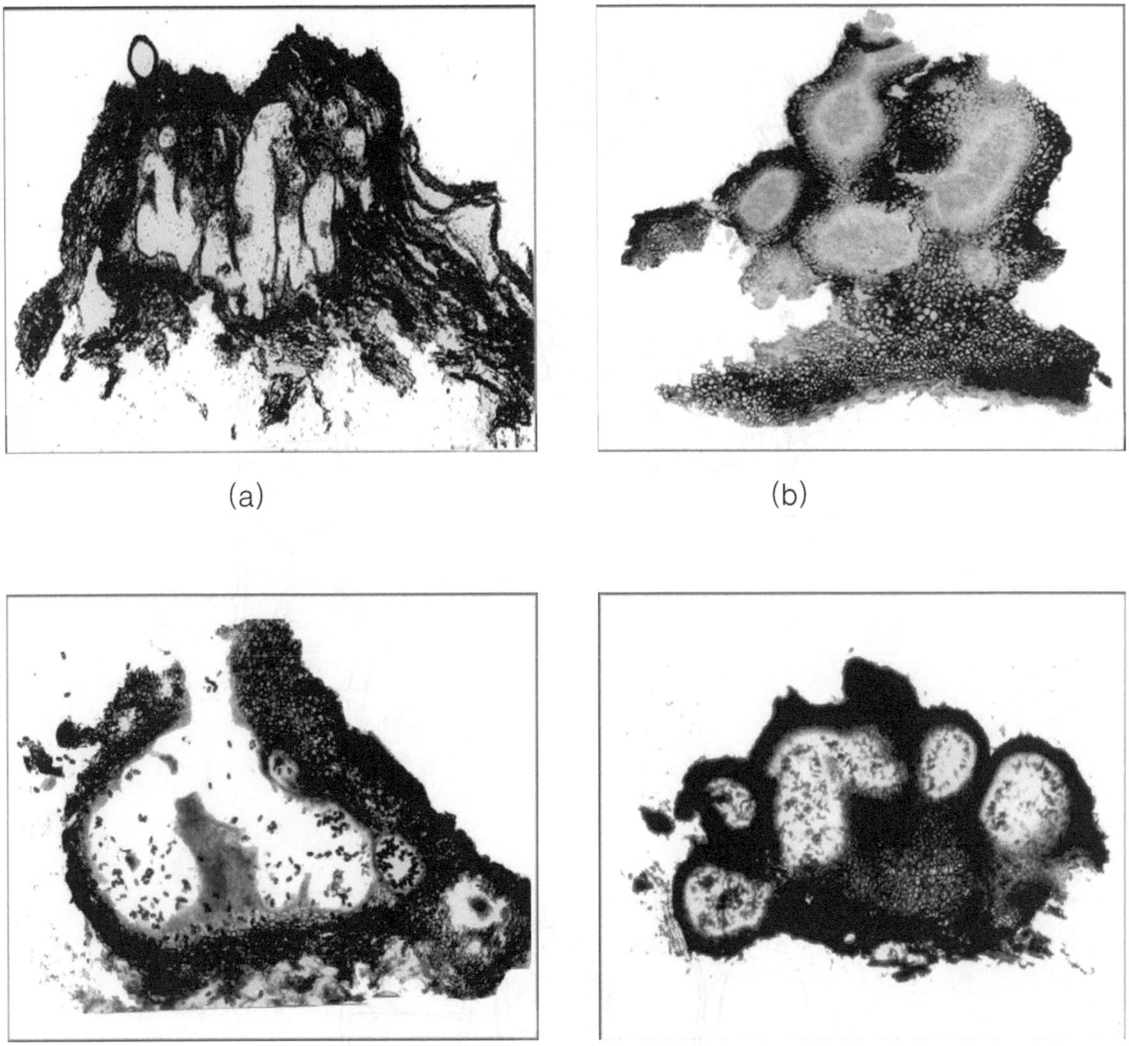

(a) (b)

(c) (d)

Figure 4.66 *Diplodia* sp., (a–d) V.S. of conidiomata (×260)

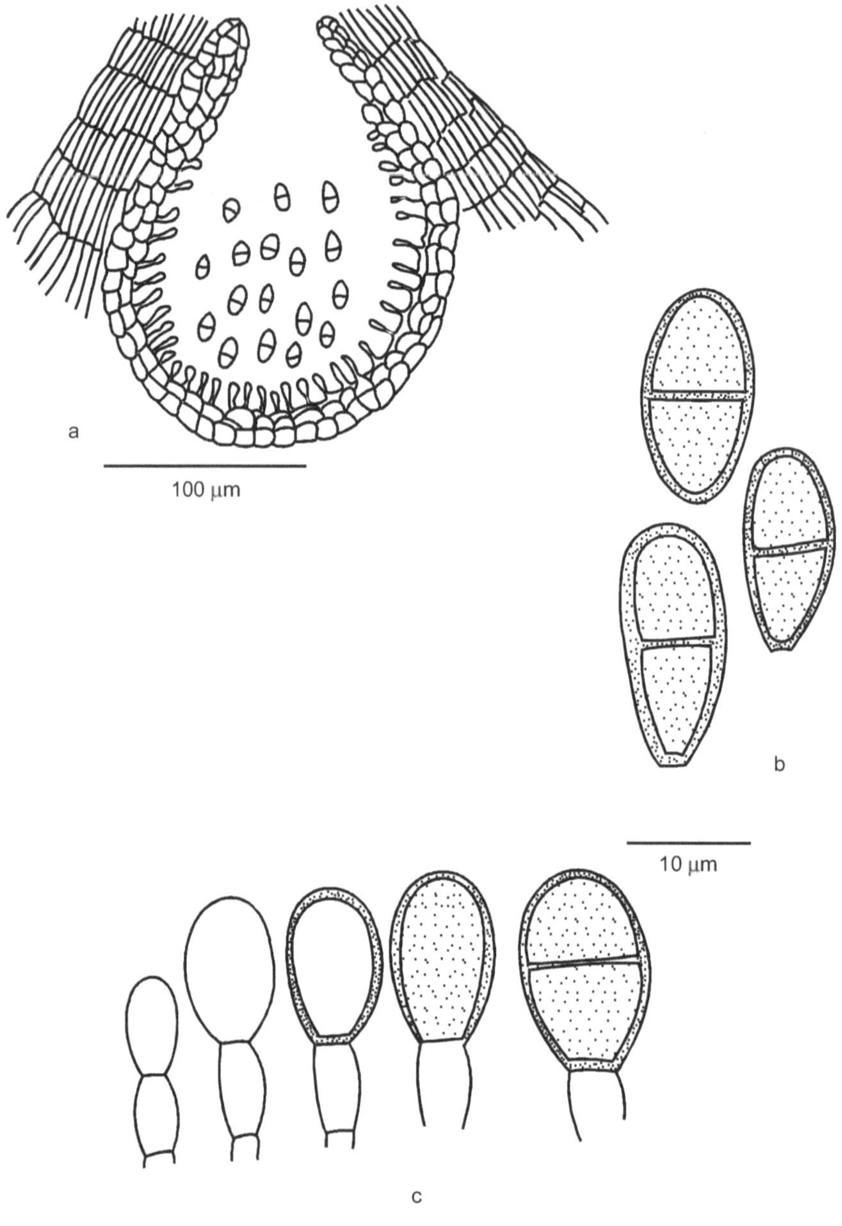

Figure 4.67 Diplodia *sycina*, (a) V.S. of conidioma; (b) Mature conidia; (c) Developing conidia.

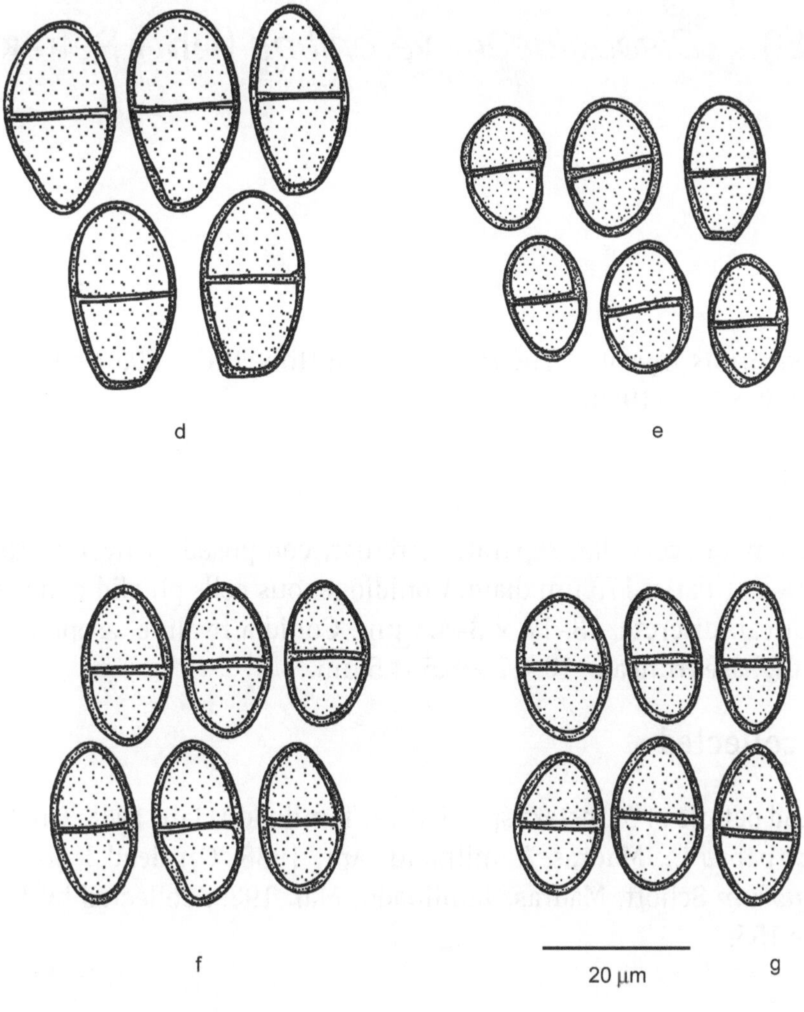

Figure 4.68 Diplodia *sycina*, (d–g) Mature conidia;

50. *Discogloeum veronicae* (Lib.) Petrak

Source

Petrak, *Annls mycol.*, **27**: 370 (1929); Sutton, B.C., *The Coelomycetes*, CMI, England, p. 505 (1980).

Description

Conidiomata acervular, separate, circular, composed of hyaline to pale brown textura-angularis, 175 µm diam. Conidiogenous cells phialidic, hyaline, smooth, cylindrical, discrete, 13–17 × 3–4.5 µm. Conidia hyaline, aseptate, thin-walled, smooth, base truncate, 10–17 × 3.5–4.5 µm.

Specimen collected

On dead petiole of *Monstera* sp., Madras, Tamilnadu, Jan. 1988; On dead petiole of *Carica papya* L., Madras, Tamilnadu, Apr. 1988; On dead leaves of *Colacasia antiquorum* Schott, Madras, Tamilnadu, Mar. 1989, collected by J. Muthumary. IC No: 155.

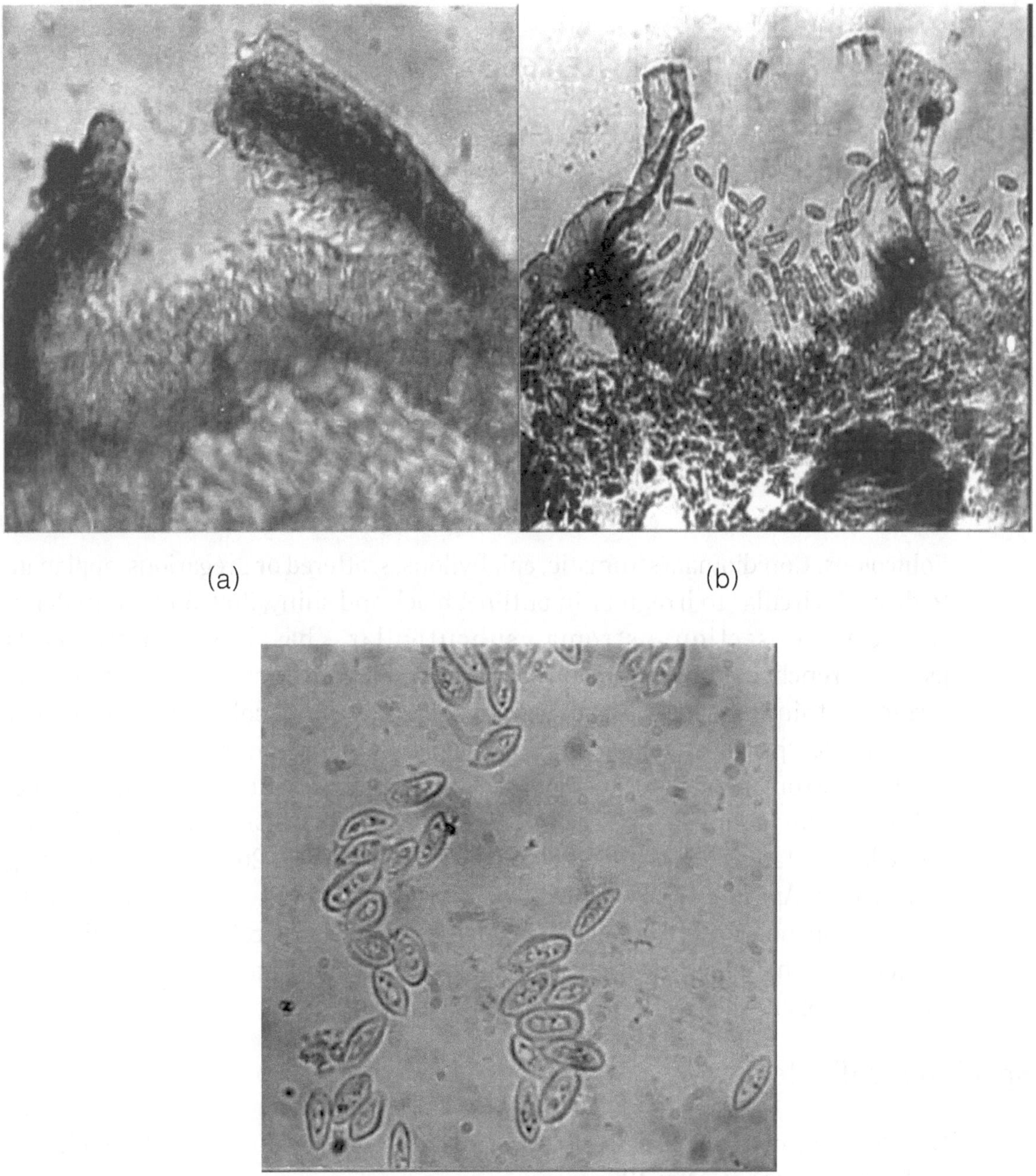

Figure 4.69 *Discogloeum veronicae*, (a–b) V.S. of conidiomata (×200); (c) Mature conidia (×900)

51. *Discosia artocreas* (Tode ex Fries) Fr.

Source

Fr., *Summa Veg. Scand.*, 423 (1849); Sacc. **III**: 153 (1884); Chona, *et al*, *Indian Phytopath.*, **10**: p. 154 (1957); R.S. Mathur, *The Coelomycetes of India*, p.91 (1979); Sutton, B.C., *The coelomycetes,* CMI, England, p. 148 (1980).

Description

Foliicolous. Conidiomata stromatic, epiphyllous, scattered or gregarious, applanate to discoid, circular to irregular in outline, black and shiny, 200–500 µm in diam. In vertical section, stroma subcuticular, basal stromatic cells pseudoparenchymatous 10–15 µm, thick, brown and thick-walled at the base, hyaline to thin-walled towards the conidiogenous region; columnar interlocular tissue made up of hyaline and elongated cells. Conidiogenous cells arising from the basal stromatic cells, interspersed with mycelium matrix, conical, hyaline, 4–7 × 3–4 µm with 2–3 annelations. Conidia sub-cylindrical, straight to slightly curved, 3-septate, cells unequal, hyaline, thin-walled, 12–20 × 2–3 µm, bearing appendages. Apical cell sub-conical with a rounded apex with an unbranched , tubular, appendage 6–10 µm long, median 2 cells cylindrical, lower median cell longer than the upper median cell, basal cell drawn into a single, unbranched, excentric appendage.

Specimen collected

On leaves of *Syzygium jambolanum*, Kodaikanal, Tamilnadu, Nov. 1987, collected by J. Muthumary. IC. No: 126.

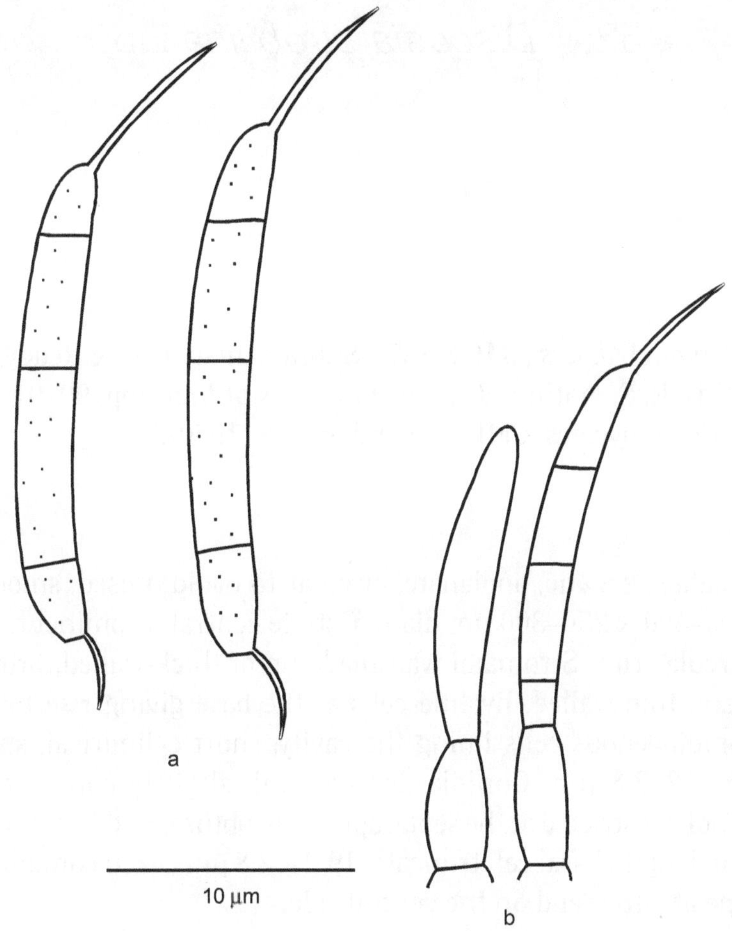

Figure 4.70 *Discosia artocreas*, (a) Mature conidia; (b) Developing conidia.

52. *Discosia strobilina* Lib.

Source

Lib., *Pl. Crypt. Ard. Exs.*, **346** (1837); Subramanian, C.V. & Reddy, K.R.C. *Kavaka* **2**: 62 (1974); R. S. Mathur, *The Coelomycetes of India*, pp. 90–91 (1979); Sutton, B.C. *The Coelomycetes*, CMI, England, p. 148 (1980).

Description

Conidiomata stromatic, applanate, circular to ovoid, raised, smooth and convex above, 300–400 × 250–300 µm diam. Ostiole central, prominent, opening with a raised circular rim. Stromatal wall made up of thick-walled, brown cells at the outside and thin-walled, hyaline cells at the base giving rise to conidiogenous cells. Conidiogenous cells lining the cavity, short cylindrical, smooth, hyaline, 9.8–12.6 × 2–2.8 µm. Conidia dorsiventral, slightly convex on dorsal side, 4-celled, not constricted at the septa, apical cell obtuse, middle two cells cylindrical, unequal in length, basal cell truncate, 19.2 × 2.8 µm., each conidium with a single appendage at either end on the ventral side.

Specimen collected

On dead leaves of *Ilex wightiana* Wall., Kodaikanal, Tamilnadu, Nov. 1987; On dead leaves of *Rhododendron* sp., Shimla, Himachala Pradesh, Oct. 1989, collected by J. Muthumary.

Recorded as endophyte in

Ocimum basilicum, Terminalia arjuna, V. Gangadevi, 2005, Ph.D. Thesis, "Studies on Fungal Endophytes from some medicinal plants and screening selected coelomycetes for taxol production". MUBL. No: 673.

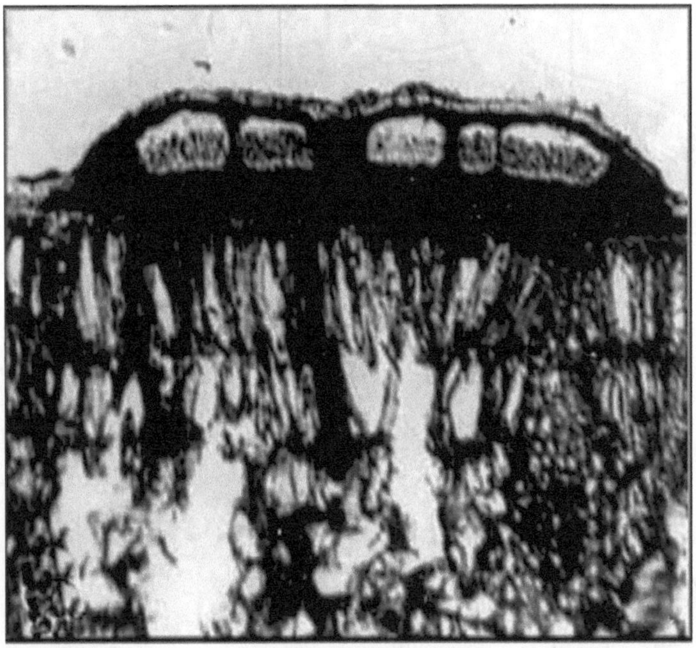

Figure 4.71 *Discosia strobilina*, (a) V.S. of conidioma (×200);

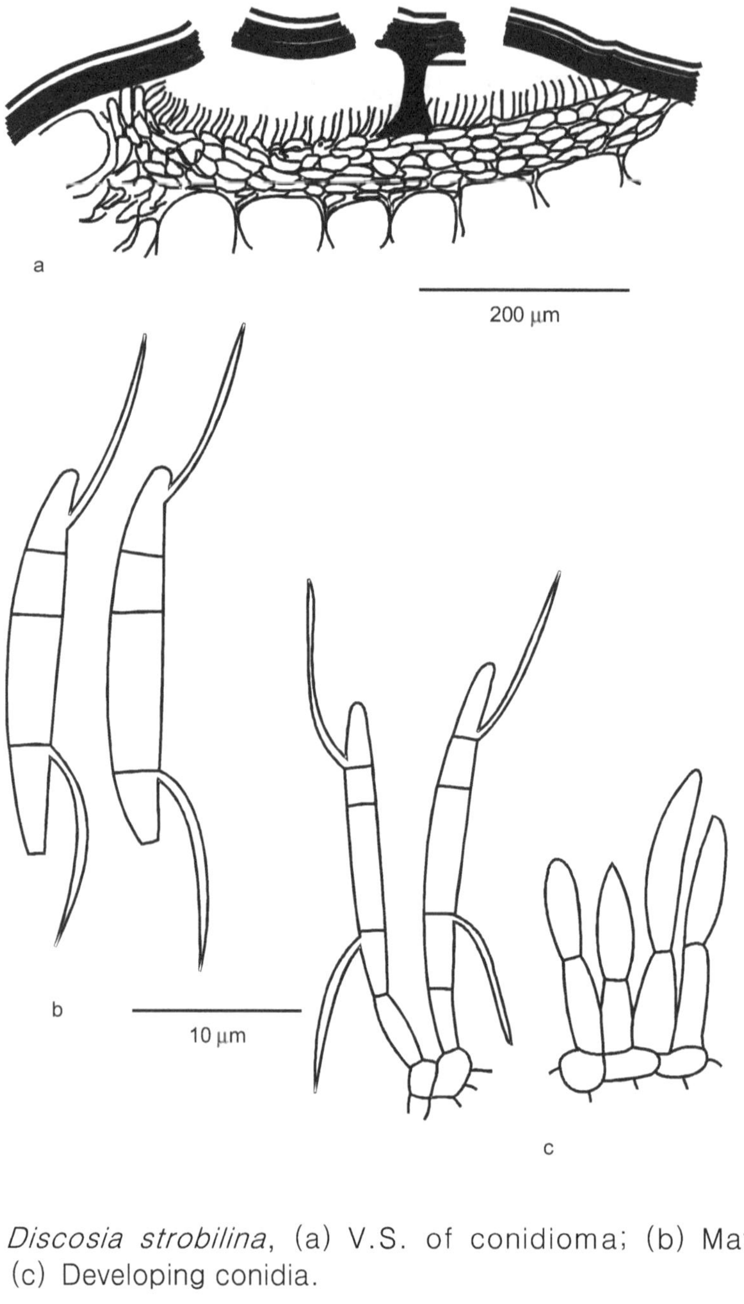

Figure 4.72 *Discosia strobilina*, (a) V.S. of conidioma; (b) Mature conidia; (c) Developing conidia.

53. *Discosia vagans* de Not.

Source

De Not., *Mem. Accad. Sci. Torino,* **2.** *Ser.* 10: 354 (1849); Subramanian, C.V. & K.R.C., Reddy, *Kavaka,* **2**: p. 69 (1974); R.S. Mathur, *The Coelomycetes of India,* p. 92 (1979); Sutton, B.C., *The Coelomycetes,* CMI, England, p. 148(1980).

Description

Conidiomata stromatic, 130–160 μm in diam., gregarious, becoming erumpent from breaking through epidermis, retaining the host cuticle above stromatal wall made up of thick-walled, brown cells at the outside and thin-walled, hyaline cells at the base giving rise to conidiogenous cells. Conidiogenous cells 8 × 2.5 μm. Conidia 27 × 3 μm.

Specimen collected

On dead leaves of *Glycosmis pentaphylla* Correa, Courtallam, Tamilnadu, Jul. 1989, collected by J. Muthumary. IC. No: 179.

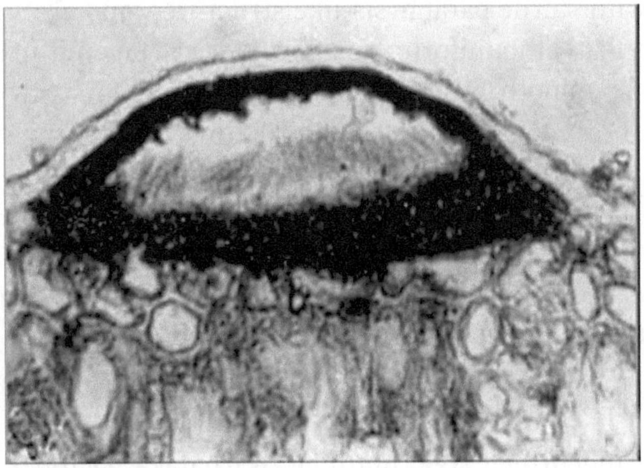

Figure 4.73 *Discosia vagans*, (b) V.S. of conidioma (×200);

54. *Dwiroopa sundara* Subramanian and Muthumary

Source

Subramanian, C.V. and J. Muthumary, *Proc. Indian. Acad. Sci.* (Plant Sci.) **96**(3): 191–197 (1986).

Description

Conidiomata, stromatic, immersed, solitary, brown to black, non-ostiolate, glabrous and 500–700 µm in diam. In vertical section the conidioma appears subspherical and somewhat flattened and divided into 1–3 incomplete compartments from the ingress of peridial tissue in different directions. The wall is pseudoparenchymatous, composed of about 5 layers of brown, spherical to angular cells, the cells progressively paler and finally hyaline towards the inner region. In the apical region of the conidioma, the wall is composed of about 8 layers of brown, vertically oriented cells forming a kind of 'lid' where the conidioma ruptures to release the conidia at maturity. Two types of conidia, here termed a- and b- conidia are produced. Both kinds are gangliar, obovoid, one-celled, truncate at the base, rounded at the apex and are produced on separate conidiogenous cells in the same conidioma. The conidiogenous cells line the cavitities of the conidiomata, intermixed with sterile paraphysis-like structures, and are simple, one-celled or rarely two-celled, ampulliform to subcylindrical tapered markedly towards the apices, hyaline, smooth, 5–12 (x=10) × 3–6 (x = 5.5) µm, each producing a solitary conidium at its tip. The paraphysislike structures are filiform to shortly cylindrical, simple or branched, many-septate, blunt at the tip, smooth, hyaline, 150–200 × 5–6 µm. The a-conidia are brown, thick-walled, guttulate, with 3–4 differentially pigmented, longitudinal striations and 25–30 (x=27) × 17–18 (x=17.5) µm. The b-conidia are brown, but paler than the a-conidia, thin-walled, smooth, eguttulate without any striations, 10–12(x=11) × 3–4 (x=3.8) µm.

Specimen collected

On twig of unidentified tree, Agumbe, Karnataka, October 1979, collected by D. Jayarama Bhat, MUBL No. 2870, IMI: 255137.

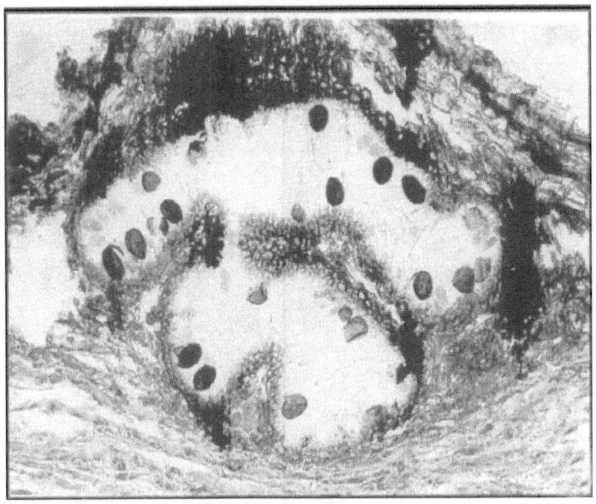

(a)

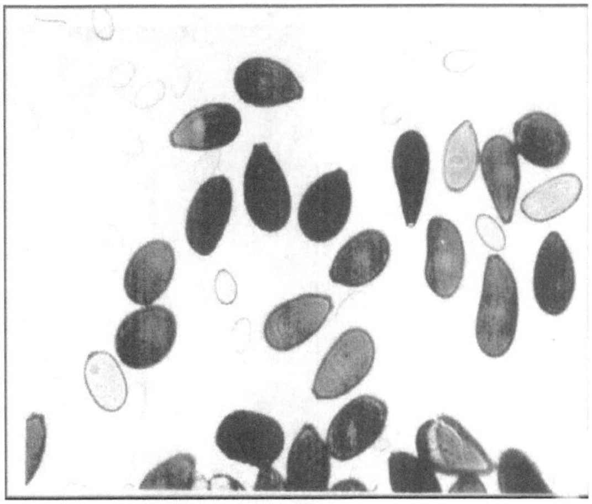

(b)

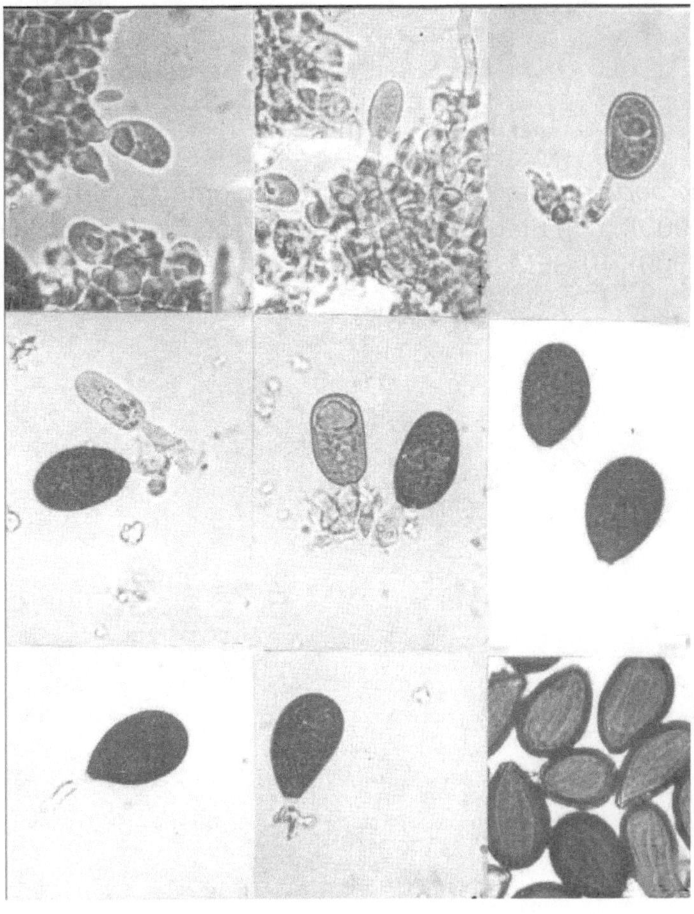

(c-k)

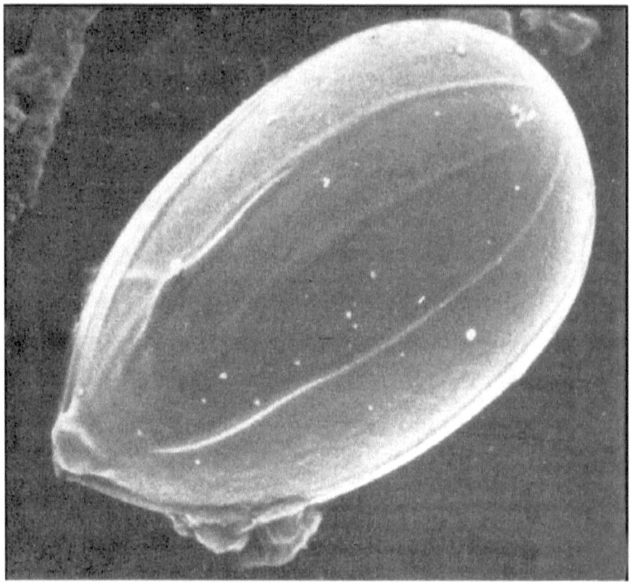

(l)

Figure 4.74 *Dwiroopa sundara*, (a) V.S. of conidioma (×260); (b) a and b conidia (×900); (c–j) Stages in the development of a-conidia; (k) Mature conidia (×900); (l) SEM of a-conidium (×8000)

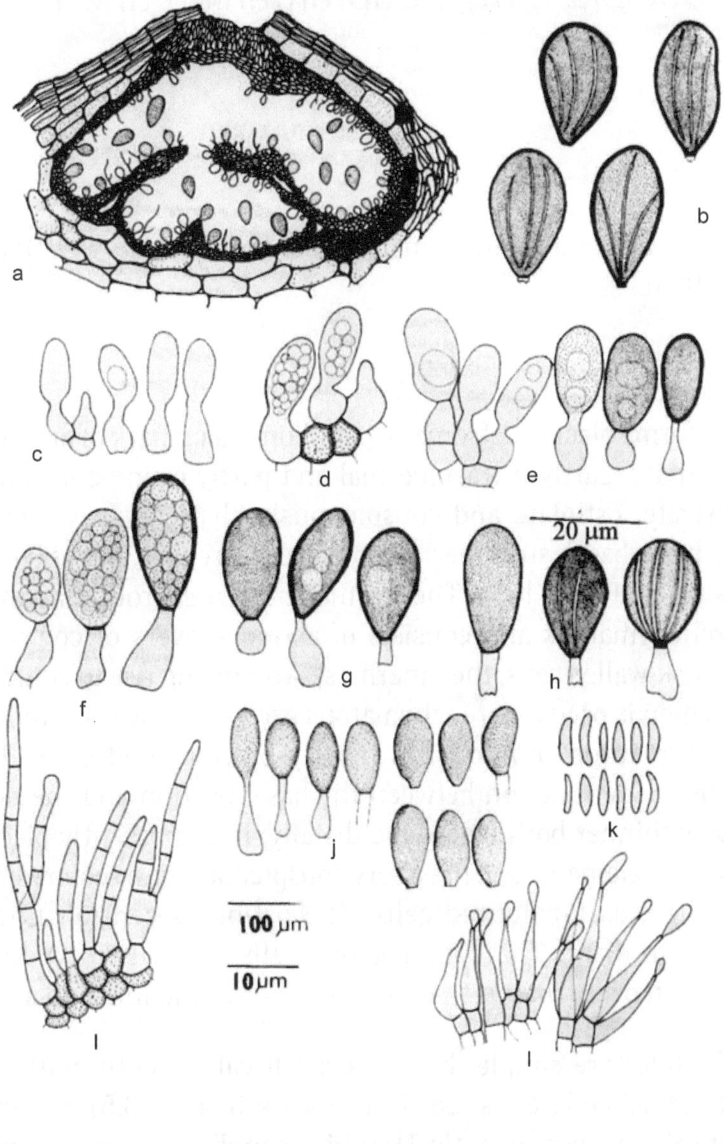

Figure 4.75 *Dwiroopa sundara*, (a) V.S. of conidioma; (b) Alpha conidia; (c–h) Stages in the develpment of alpha conidia; (i) Paraphysis-like structures; (j) Beeta conidia-developing and Mature conidia; (k) phialoconidia; (l) Phialides with developing phialoconidia.

55. *Dwiroopella ramya* Subramanian and Muthumary

Source

Subramanian, C.V. and J. Muthumary, *Proc. Indian. Acad. Sci.* (Plant Sci.) **96**(3): 199–203 (1986).

Description

The fungus forms black, eustromatic conidiomata on bark of an unidentified tree. They are solitary, partly intraepidermal and partly erumpent, unilocular, thick-walled, rostrate, ostiolate and conspicuously clypeate. In vertical section, the conidioma has a basal, subspherical to flattened region, 600–800 × 300–500 μm and a neck up to 900 μm long. The stromatic tissue surrounding the conidioma is pseudoparenchymatous and consists of about 8 layers of compactly arranged, irregular, thick-walled cells, the innermost layer giving rise to conidiogenous cells. The neck region is pseudoparenchymatous except for the innermost part where it is plectenchymatous and consists mostly of elongated cells. The clypeus is conspicuous at the junction between the basal portion and the neck, becoming progressively thinner both below and distally; it is apparently plectenchymatous with thin walled elements and has a very conspicuous, dark-colored rind, composed of thick-walled, dark-coloured cells. The ostiole is central, furfuraceous and 100–200 μm in diam. The conidiogeous cells are of two types; phialidic and annellidic, both occurring in the same conidioma and forming a palisade.

The phialides are simple, brown, cylindrical, smooth, with dark-coloured, somewhat ragged collarettes, 28–48 (×=38) × 3–4 (×=3.5) μm and proliferated enteroblastically and percurrently. The phialoconidia are elongate-oval to ellipsoidal or cylindrical or even clavate, brown, smooth, 1-septate, rarely 2–3 –septate, sometimes constricted at the septum, often mammillate at the base and 10–18 (×=16) × 5–7 (×=5.0) μm. Long neck projects beyond the substrate. A conspicuous clypeus at the junction between the base and the neck is another unique feature. Each conidioma produces within it two types of conidia, phialoconidia from simple phialides which have the ability to proliferate and also gangliar conidia from annellated conidiogenous cells. Both types of conidia are typically two-celled, the phialoconidia often being 3-celled.

Specimen collected

On unidentified tree, Quard hittow, Koppa, Karnataka, October1979, collected by J. Muthumary. MUBL. No. 2871. IMI 255129.

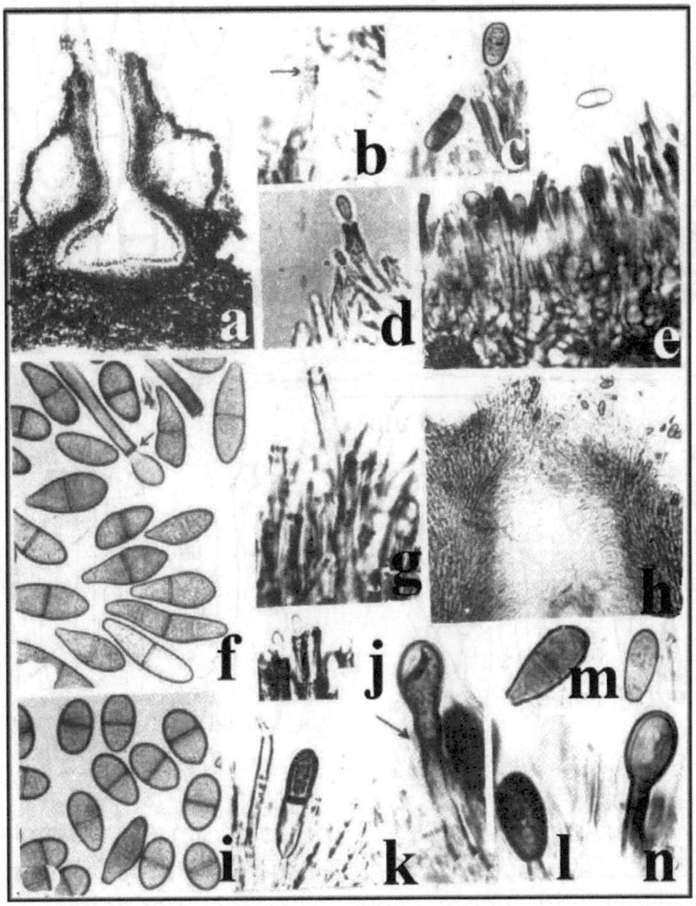

Figure 4.76 *Dwiroopella ramya*, (a) V.S. of conidioma (×200); (b–c) Developing conidia (×200); (f) Mature conidia (×900); (g) Phialides showing percurrent proliferations; (h) V.S of ostiolar region of conidioma (×200); (i) mature conidia (×200); (j–m) Phialides showing percurrent proliferations; (n) Mature conidia (×900)

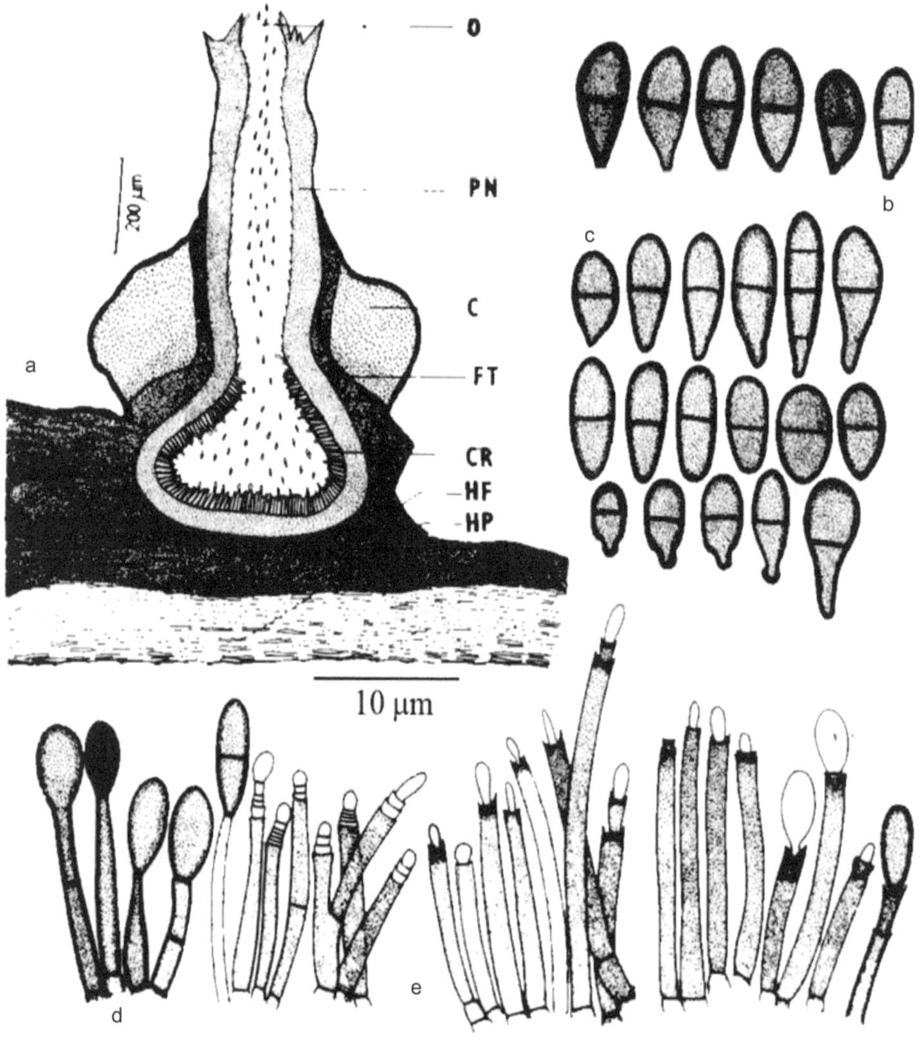

Figure 4.77 *Dwiroopella ramya*, (a) V.S. of conidioma (O–Ostiole); PN–Pycnidial neck; C–Clypeus; FT–Fungal Tissue (Stroma); CR–Conidiferous region; HF–Host and fungal tissue; HP–Host periderm); (b) Conidia from annellated condiogenous cells; (c) Phialoconidia; (d) Annellidic conidiogenous cells; (e) Phialidic conidiogenous cells.

56. *Entomosporium mespili* (Dc Ex Duby) Sacc.

Source

Saccardo, *Michelia*, **2**: 115 (1880); R.S. Mathur, *The Coelomycetes of India*, p.97 (1979); Sutton, B.C., *The Coelomycetes*, CMI, England, p.150 (1980); Muthumary, J., *Current Science*, **56**(4): pp. 195–196 (1987); Nag Raj, T.R., *Coelomycetous anamorphs with appendage bearing conidia*, Mycologue Publications, Canada, pp. 349–352 (1993)

Description

Lesions on leaves pale brown, amphigenous, irregular, 1–5 mm diam. Conidiomata acervular, subcuticular, 500–800 µm in diam., separate or aggregated, composed of hyaline to pale brown textura- angularis. Conidiogenous cells phialidic, arising from the cells lining the cavity of the acervulus, cylindrical, hyaline, smooth, with minute collarette, 10–18 × 4–8 µm. Conidia cruciform, 4–5 celled, hyaline, smooth, thin-walled, consisting of a larger upper cell and a smaller basal cell and two or more smaller lateral cells arising from the upper regions of the lower cell; apical cell globose to subglobose, 6–9 × 3–4 µm, apex obtuse, base truncate, basal cell globose to short cylindrical, 4–5 × 3.0 –3.5 µm, lateral cells globose to subglobose, 3–5 µm in diam; apical, and lateral cells are each furnished with a single cellular, unbranched, flexuous appendage, apical appendage 10–15 µm long, lateral appendage 8–11 µm long.

Specimen collected

On leaves of *Rosa* sp., Bryant Park, Kodaikanal, October 1978, collected by J. Muthumary. MUBL. No. 2915.

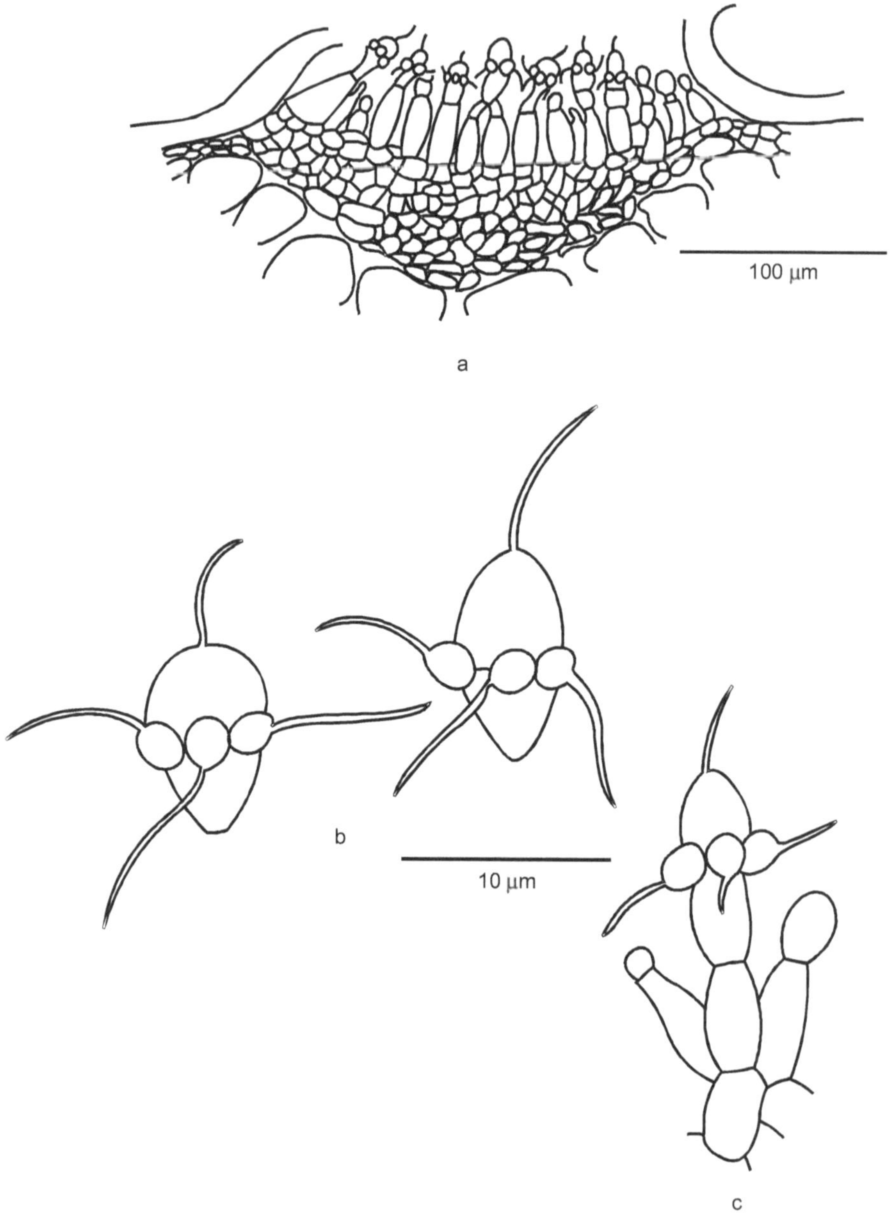

Figure 4.78 *Entomosporium mespili*, (a) V.S. of conidioma; (b) Mature conidia; (c) Developing conidia.

57. *Fusicoccum indicum* Tilak & Rokde

Source

Tilak & Rokde, *Mycopath. Mycol. appl.*, **21**: 200–204 (1964); R.S. Mathur, *The Coelomycetes of India*, p.99 (1979)

Description

Conidiomata eustromatic, 500–600 µm diam. immersed, subperidermal, separate, black, multilocular, walls made up of brown, pseudoparenchymatous cells. Conidiogenous cells lining the cavity, hyaline, smooth, cylindrical, blastic, 10–15 × 2–3 µm. Conidia aseptate, fusiform, hyaline, smooth, thin-walled, apex obtuse to blunt, base truncate, 15–20 × 2.5–3 µm.

Specimen collected

Isolated as endophyte from *Mimusops elengi*; On dead twigs of various plants, Madras Christian College campus, Tambaram and Anna University campus, Chennai, Nov. 1990, by J. Muthumary.

Recorded as endophyte in

Mimusops elengi, V. Gangadevi, 2005, Ph.D. Thesis, "Studies on Fungal Endophytes from some medicinal plants and screening selected coelomycetes for taxol production". MUBL. No: 674.

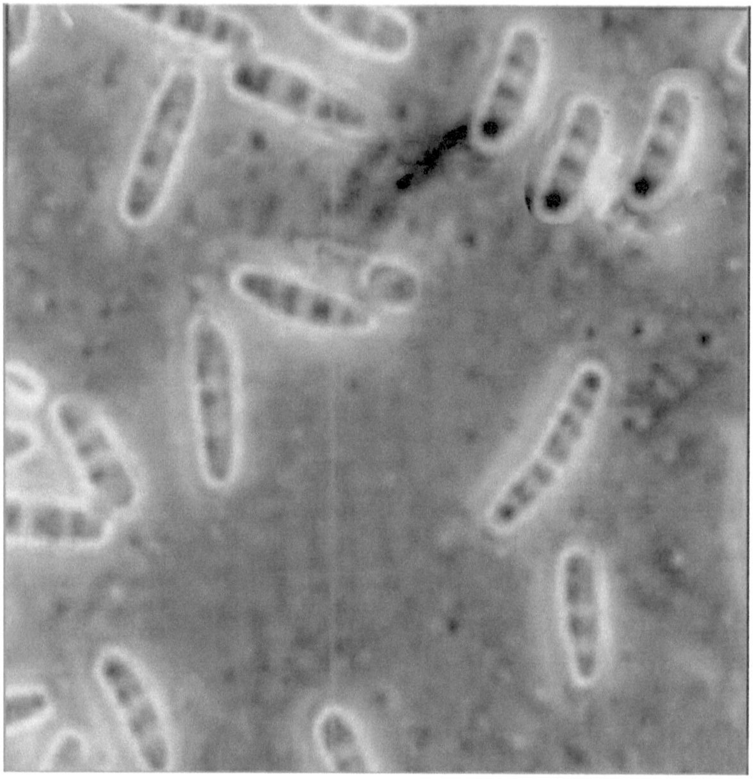

Figure 4.79 *Fusicoccum indicum*; Mature Conidia (×900)

58. *Hainesia lythri* (Desm.) Hohn.

Source

Hohn, *Ster. Akad. Wiss. Wien.*, **115**: 687 (1906); R.S. Mathur, *The Coelomycetes of India*, p.103 (1979); Sutton, B.C., *The Coelomycetes,* CMI, England, p. 552 (1980).

Description

Conidiomata initially immersed in the substrate, at maturity opening like a cup and superficial, separate, colourless, 100–200 µm, basal wall pale, thin-walled, pseudoparenchmatous, becoming free hyphae towards the sides to form the cup-like conidiomata. Conidiophores branched, hyaline, smooth, filiform. Conidiogenous cells filiform, branched septate, hyaline, smooth, 20–25 × 2.3 µm. Conidia 1-celled, hyaline, smooth, allantoid, 5–9 × 1.5–2 µm.

Specimen collected

On twigs of many plants, Kambakkam, Tamilnadu, October 1982, collected by J. Muthumary. IC. No: 180.

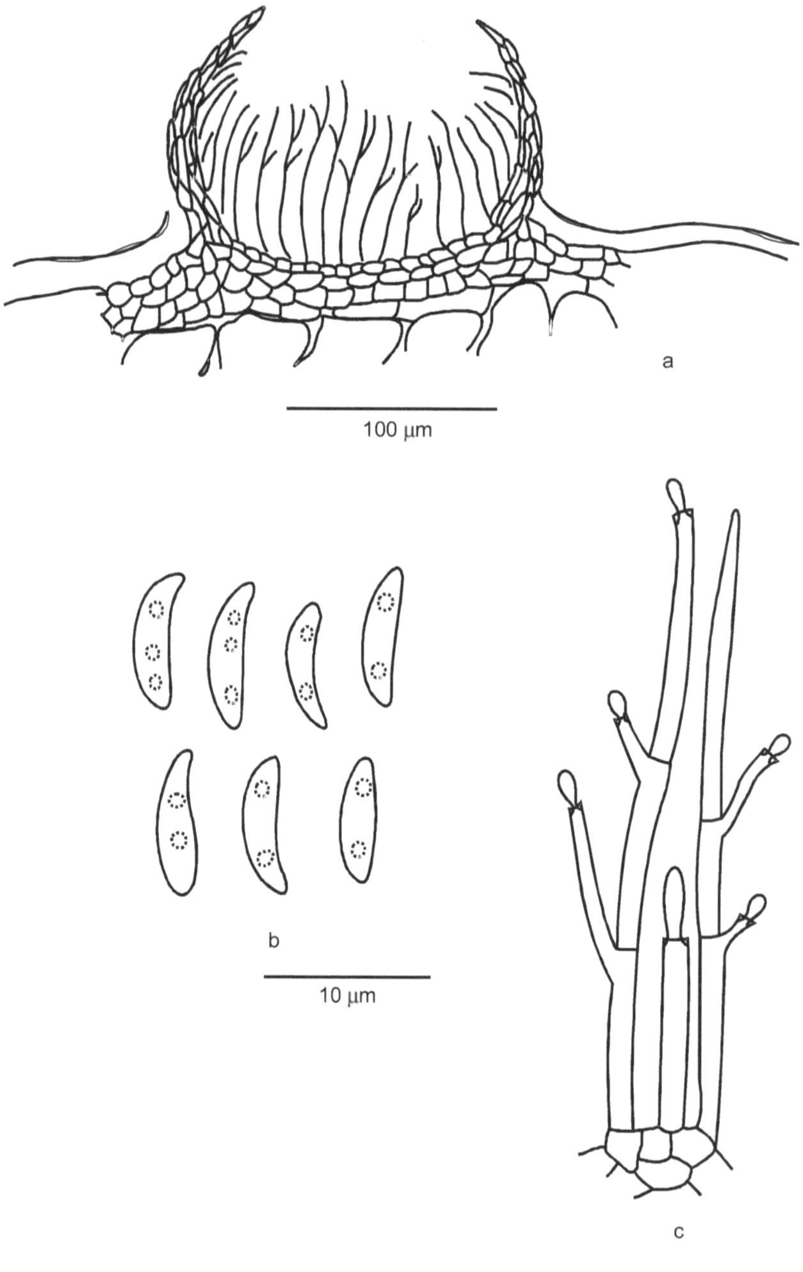

Figure 4.80 *Hainesia lythri* (a) V.S. of conidioma; (b) Mature conidia; (c) Developing conidia.

59. *Harknessia eucalypti* Cooke

Source

Cooke & Harkness, *Grevillea,* **9**: 85 (1881); Sutton, B.C., *The Coelomycetes,* CMI, England, p. 178 (1980); Nag Raj, T.R., *Coelomycetous anamorphs with appendage bearing conidia,* Mycologue Publications, Canada, p.377 (1993).

Description

Foliicolous. Conidiomata pycnidial, ampulliform, abundant, spherical, brown to black. 200–250 µm in diam. In vertical section, pycnidium subepidermal, unilocular, glabrous, made up of thick-walled tissues, pseudoparenchymatous cells on the outer side and thin-walled and hyaline cells towards the conidiogenous region. Conidiogenous cells lining the entire cavity of the pycnidium, cylindrical to lageniform, hyaline, smooth, 10–15 × 4–5 µm. Conidia broadly ventricose, straight to slightly curved, pale brown to dark brown with longitudinal striations in some conidia, basal appendage continuous with the conidiogenous cells initially, when detached become a tubular, and thin-walled structure, 10–12 × 2–4 µm.

Specimen collected

On dead leaves of *Eucalyptus* sp., Nilgiris, Tamilndu, Jul. 1990, collected by J. Muthumary. IC. No: 206.

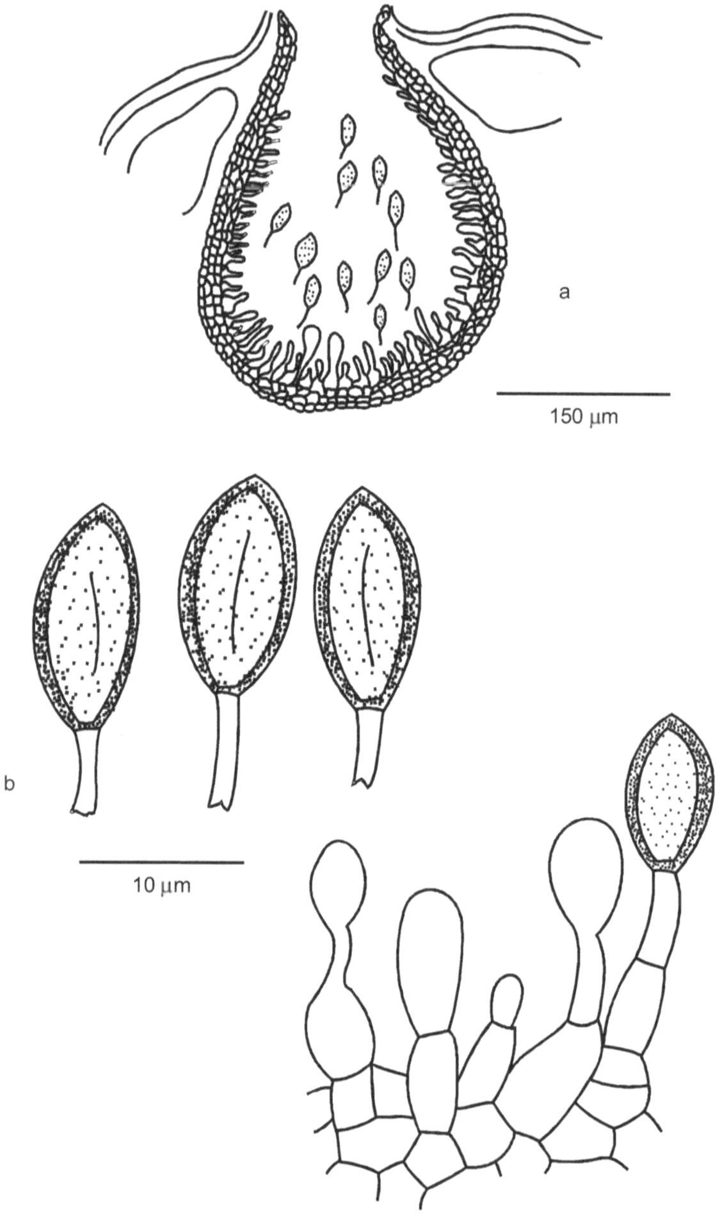

Figure 4.81 *Harknessia eucalypti*, (a) V.S. of conidioma; (b) Mature conidia; (c) Developing conidia.

60. *Harknessia podocarpi* Lindquist & Sutton

Source

Sutton, B.C., *Mycological papers,* **123**: 20–23 (1971); Sutton, B.C., *The Coelomycetes,* CMI, England, p. 180 (1980); Nag Raj, T.R., *Coelomycetous anamorphs with appendage bearing conidia,* Mycologue Publications, Canada, p. 392 (1993).

Description

Conidiomata pycnidia, ampuliform, abundant, spherical, brown to black 150–200 µm in diam, unilocular, glabrous, pseudoparenchymatous cells on the outer side and thin-walled and hyaline cells towards the conidiogenous cells lining the entire cavity of the pycnidium, cylindrical to lageniform, hyaline, smooth, 15–20 × 2–3 µm. Conidia broadly ventricose, straight to slightly curved, pale brown to dark brown, basal appendage continuous with the conidiogenous cells initially, when detached becomes a tubular and thin-walled structure, 18–22 × 10–3 µm.

Specimen collected

On leaves of *Podocarpus parlatorei*, Kodaikanal, TamilNadu, Nov. 1990, collected by J. Muthumary. IC. No: 207.

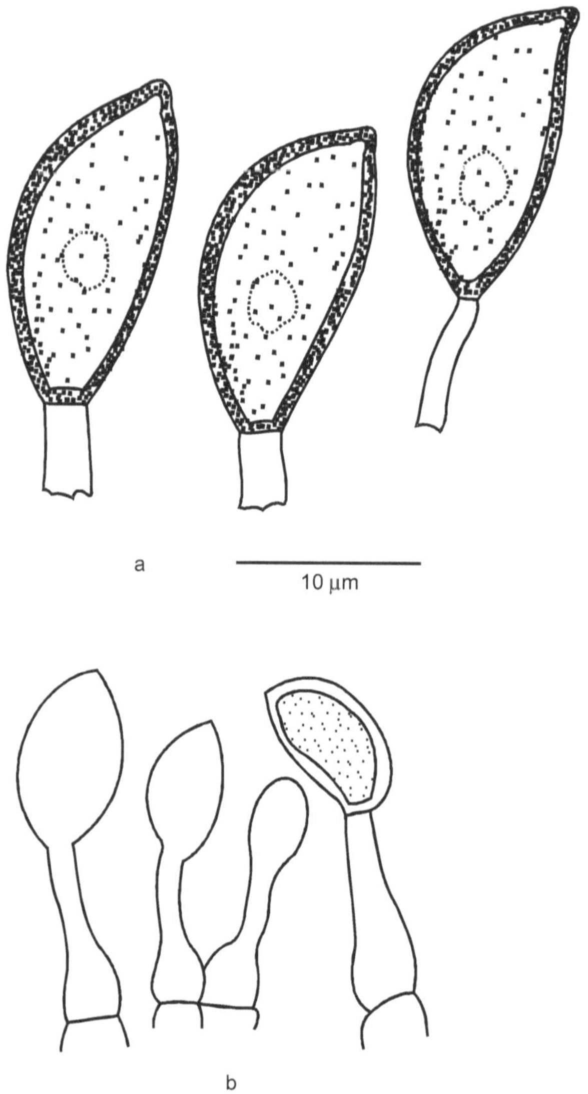

Figure 4.82 *Harknessia podocarpi*, (a) Mature conidia; (c) Developing conidia.

61. *Harknessia shearii* Petrak

Source

Petrak, *Sydowia,* **7**: 404 (1953); Sutton, B.C., *The Coelomycetes,* CMI, England, p. 180 (1980); Nag Raj, T.R., *Coelomycetous anamorphs with appendage bearing conidia,* Mycologue Publications, Canada, p. 396 (1993).

Description

Conidiomata pycnidial, immersed, globose to subglobose, unilocular, pale brown, pycnidial wall made up of thick walled, brown cells at the outside and thin-walled pseudoparenchyma, hyaline cells at the innerside giving rise to conidiogenous cells. The upper part of the stroma opens widely to release the conidia, furfuraceous. Conidiogenous cells formed from the inner cells of the pycnidial wall, lageniform, hyaline, smooth, unbranched, 7–23 µm long. Conidia holoblastic, elliptical, aseptate, apiculate at the apex, pale brown, globose to irregular, guttulate, smooth, 11–13 (12) × 6–7.5 (6.5) µm, with the conidiogenous cell persisting as a cylindrical, hyaline, basal appendage devoid of cytoplasm.

Specimen collected

On dead leaves of *Betula ulva,* Kodaikanal, Tamilnadu, Jul. 1990, collected by J. Muthumary. IC. No: 208.

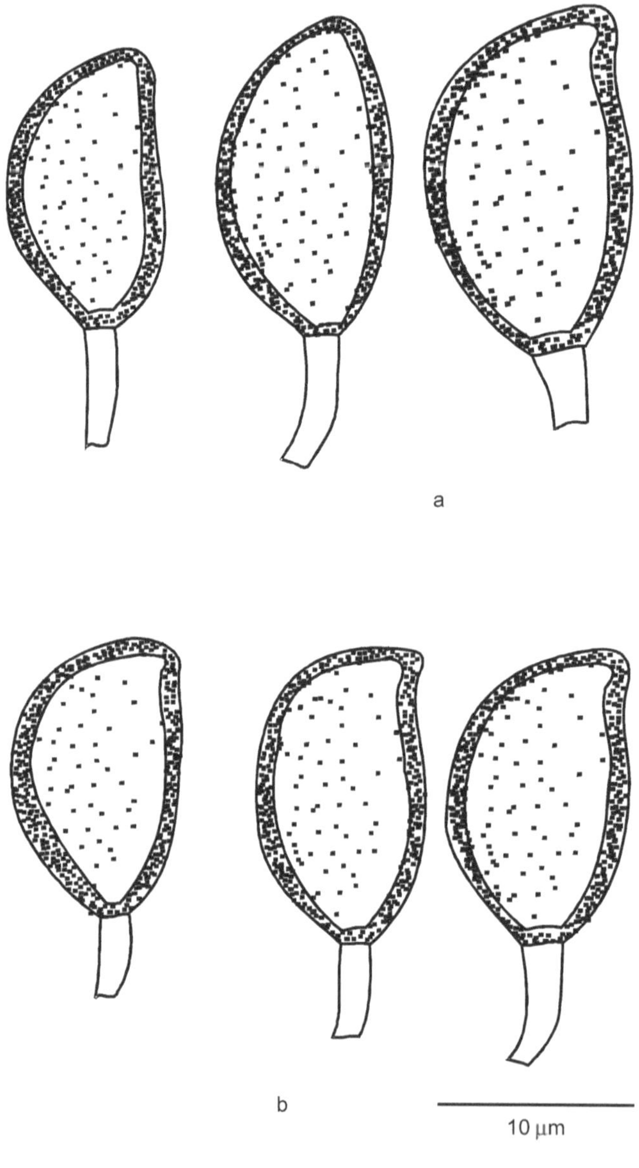

Figure 4.82 *Harknessia shearii*, (a–b) Mature conidia;

62. *Harknessia thujina* Ell. and Ev.

Source

Ell. & Ev., *Proc. Acad. Nat. Sci. Philad.*, 459 (1893); Sutton, B.C., *The Coelomycetes*, CMI, England, p. 184 (1980); Masilamani, S. and J. Muthumary, *J. Indian Bot. Soc.*, **72**: 311–312 (1993). Nag Raj, T.R., *Coelomycetous anamorphs with appendage bearing conidia*, Mycologue Publications, Canada, pp.399-400 (1993).

Description

Conidiomata eustromatic, unilocular, pale brown. Pycnidial wall made up of thick-walled, brown cells at the outside and thin-walled, hyaline cells at the innerside giving rise to conidiogenous cells. Ostiole central, unspecialized, furfuraceous. Conidiogenous cells lining the pycnidial cavity, holoblastic, discrete, formed from the wall lining the locule, often carried away by conidia at secession. Conidia acrogenous, fusiform, brown, with a cellular unbranched basal appendage resulting from the persistent conidiogenous cell, 19–22 × 9–10 µm; basal appendage 19–30 µm long.

Specimen collected

On dead leaves of *Elaeocarpus ferrugineus* W., Kodaikanal, Tamilnadu, Nov. 1987, collected by J. Muthumary. IC. No: 127.

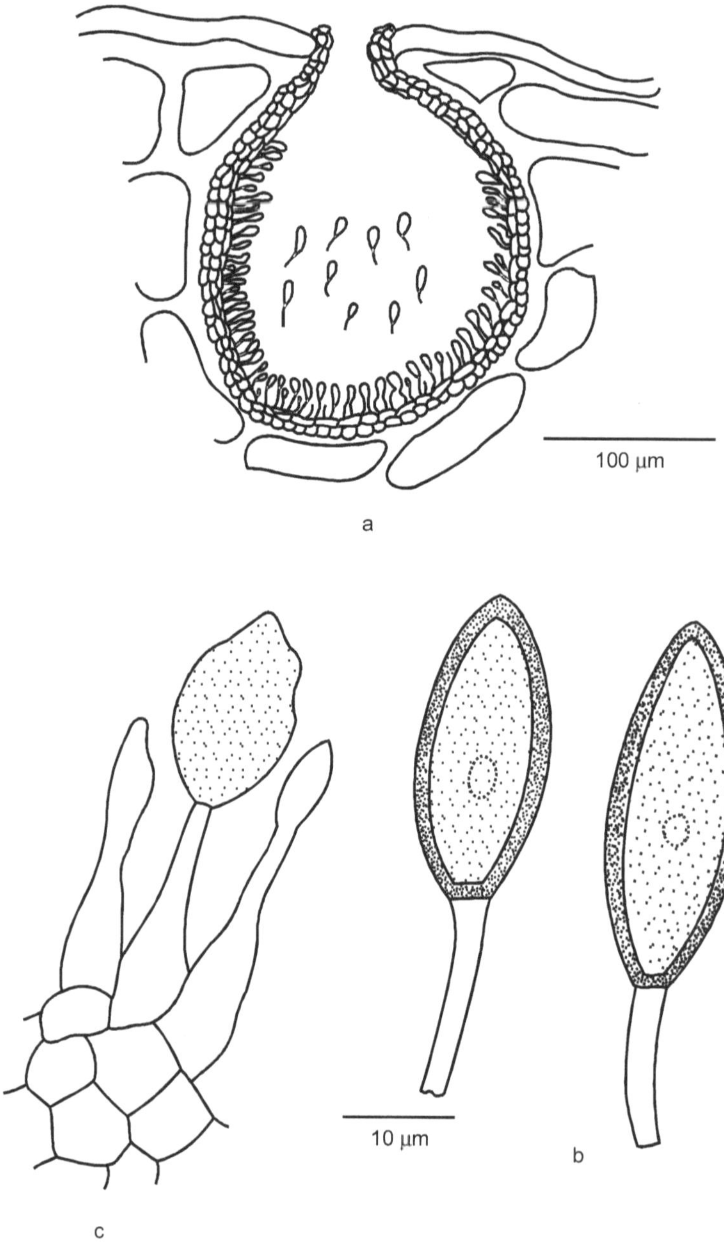

Figure 4.83 *Harknessia thujina*, (a) V.S. of conidioma; (b) Mature conidia; (c) Developing conidia.

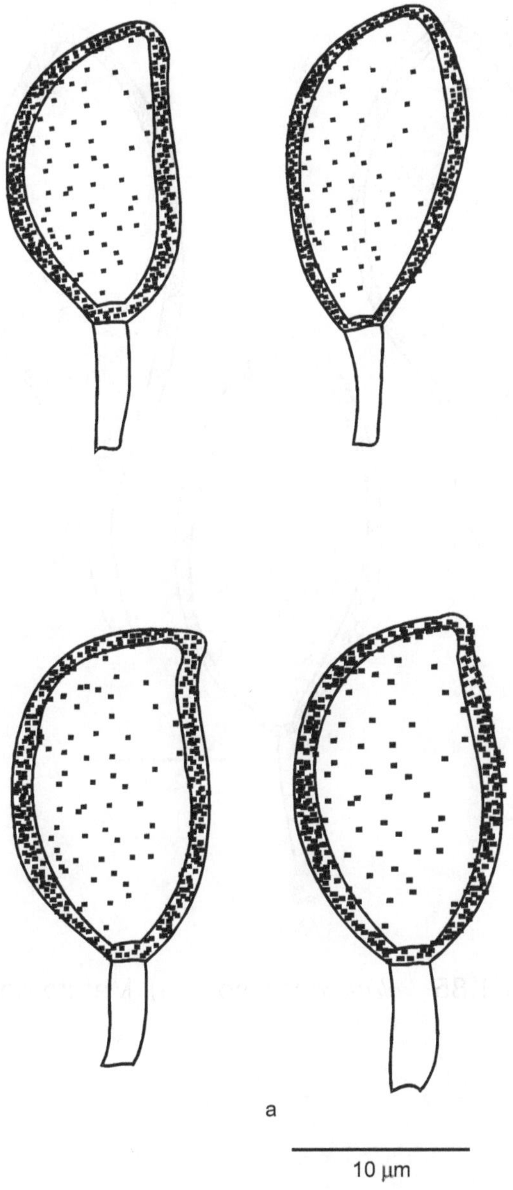

Figure 4.84 *Harknessia* sp., (a) Mature conidia;

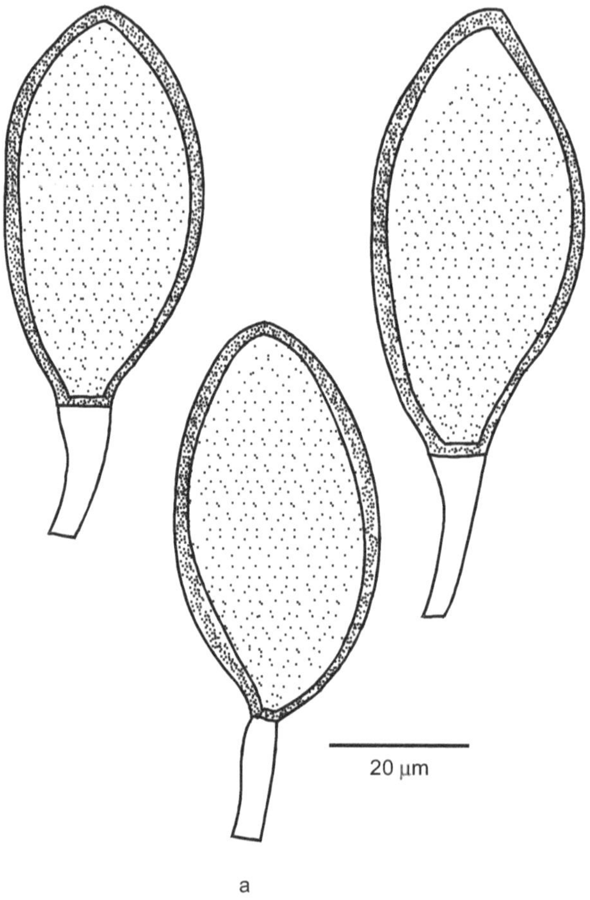

Figure 4.85 *Harknessia* sp., (a) Mature conidia

63. *Hendersonia* sp.

Source

R.S. Mathur, *The Coelomycetes of India*, p.108 (1979).

Description

Conidiomata eustromatic, immersed, subperidermal, dark brown to black, multilocular, 200–300 μm in diam. outer layer of dark brown thick-walled, pseudoparenchymatous, inner layer of thin-walled, hyaline cells. Ostiole papillate. Conidiogenous cells holoblastic, single or annellidic, cylindrical to ampulliform, hyaline smooth. Conidia obovoid, cylindrical to clavate, 1-septate, apex obtuse, base truncate, sometimes both thickened. 15–20 × 5–6 μm in diam.

Specimen collected

On twigs of *Betula ulva*, Kodaikanal, Tamilnadu, Nov. 1990, collected by J. Muthumary. IC. No: 209.

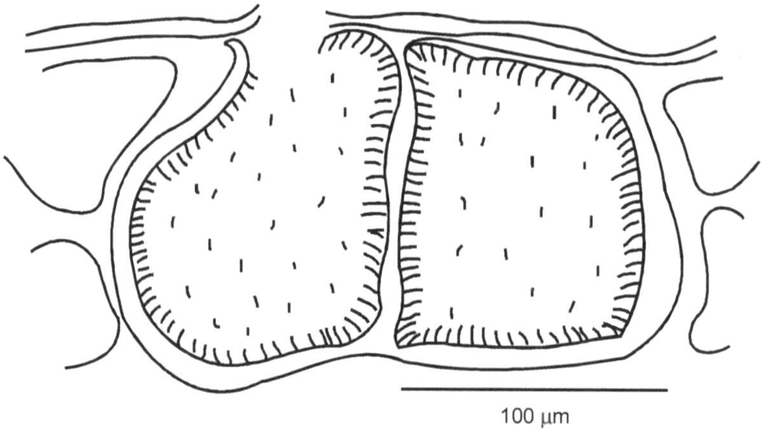

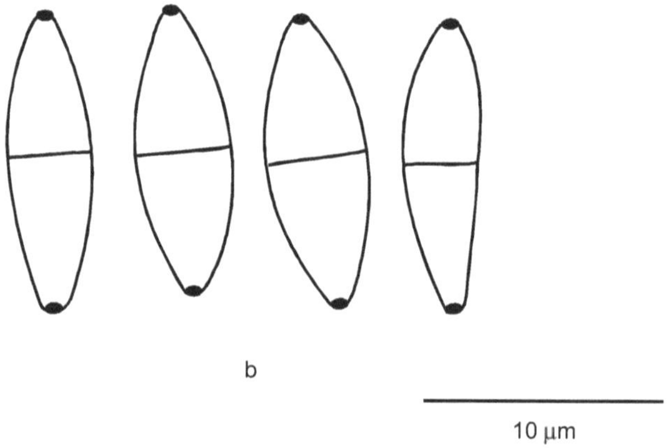

Figure 4.86 *Hendersonia* sp., (a) V.S. of conidioma; (b) Mature conidia.

64. *Heteropatella indica* Muthumary

Source

Muthumary J., *Current Science*, **55**(5): 250–251 (1986).

Description

The fungus forms black, cupulate conidiomata on dead twigs. The conidiomata are solitary, at first closed, later opening to become cupulate, 250–400 μm in diam. In vertical section, the conidioma is peridermal; the basal part is immersed in the peridermal tissue and the cupulate part is erumpent. The wall of the conidioma is multilayered, consisting of thin-walled, hyaline to pale brown, textura-angularis at the base, at the sides of similar darker tissue which develops into textura-porrecta at the periphery. Setae are present on the outer wall of the conidioma. They are brown, rigid, septate, smooth, tapering, paler towards the subacute apex, 150–200 × 3.75–4.0 μm, confined to the sides of the conidioma.

The conidiogenous cells line the cavity of the conidioma; arranged in a palisade, blastic, hyaline, smooth, cylindrical, 25–30 × 3.5–4.0 μm. The conidia are fusiform, 3-septate, very rarely 4-septate, hyaline and fusiform, 3-septate, very rarely 4-septate, hyaline to slightly curved, often guttulate, apex obtuse, base truncate, 45–50 (×=46.0) × 5.0–5.5 (×=4.5) μm apical appendage terminal, unbranched, 11-5 – 14.0 μm long, basal appendage excentric, filiform, unbranched, shorter than the apical appendage, 7–8 μm.

Specimen collected

On fallen twigs of *Eugenia jambolana*, Madras, Madras Christian College Campus, Tambaram, Chennai, March 1984, collected by J. Muthumary, MUBL No: 2873.

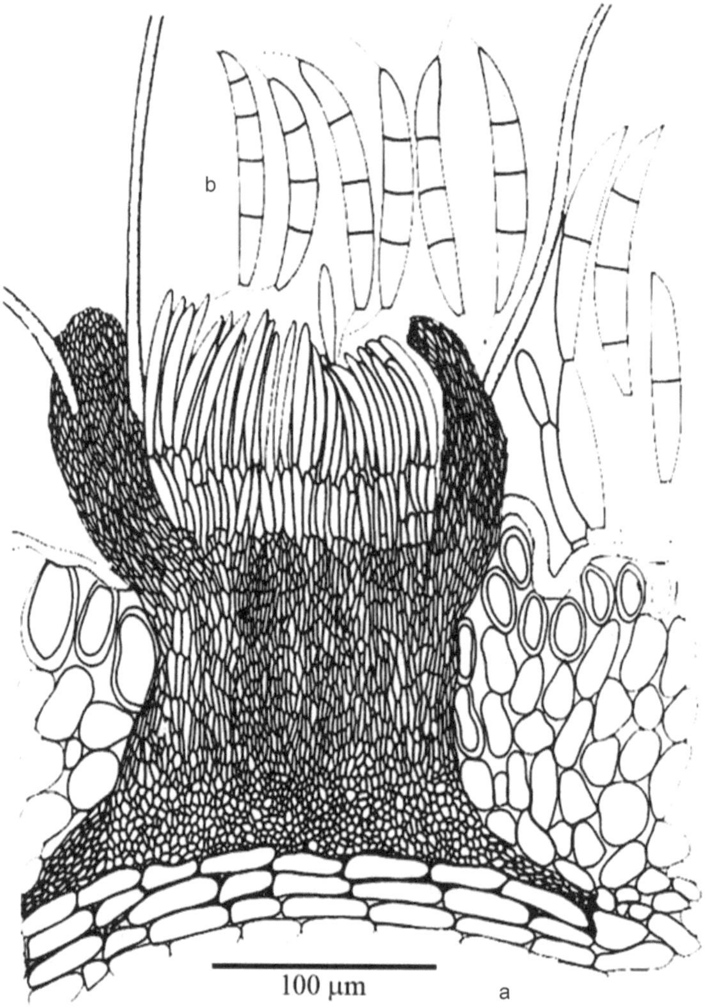

Figure 4.87 *Heteropatella indica*, (a) V.S. of conidioma; (b) Mature conidia;

65. *Hyalotiella transvalensis* Papendorf

Source

Papendorf *Trans, Br. Mycol. Soc.,* **50**: 69 (1967); Mathur, *The Coelomycetes of India,* pp. 112–113 (1979); Sutton, B.C., *The Coelomycetes,* CMI, England, 70–71 (1980); Nag Raj, T.R., *Coelomycetous anamorphs with appendage bearing conidia.* Mycologue Publications, Canada, pp. 422–425 (1993).

Description

Conidiomata pycnidial, immersed, globose, black, unilocular, 200–300 µm in diam. Dehiscence irregular. Pycnidial wall composed of outer layer of thick-walled cells and inner layer of thin-walled cells, conidiogenous cells lining the pycnidial cavity, holoblastic, filiform to cylindrical, hyaline, smooth, 8–15 × 1.5–2.0 µm. Conidia 3- euseptate, 4-celled, apical cells hyaline with 2–3 appendages; appendage 10–15 µm long; subepical cells pale brown, upper two cells longer than the basal cell, smooth, 15–20 × 2.0–2.5 µm.

Specimen collected

On dead leaves of *Eucalyptus* sp., Kambakkam, Nov. 1984, collected by J. Muthumary. IC. No: 107.

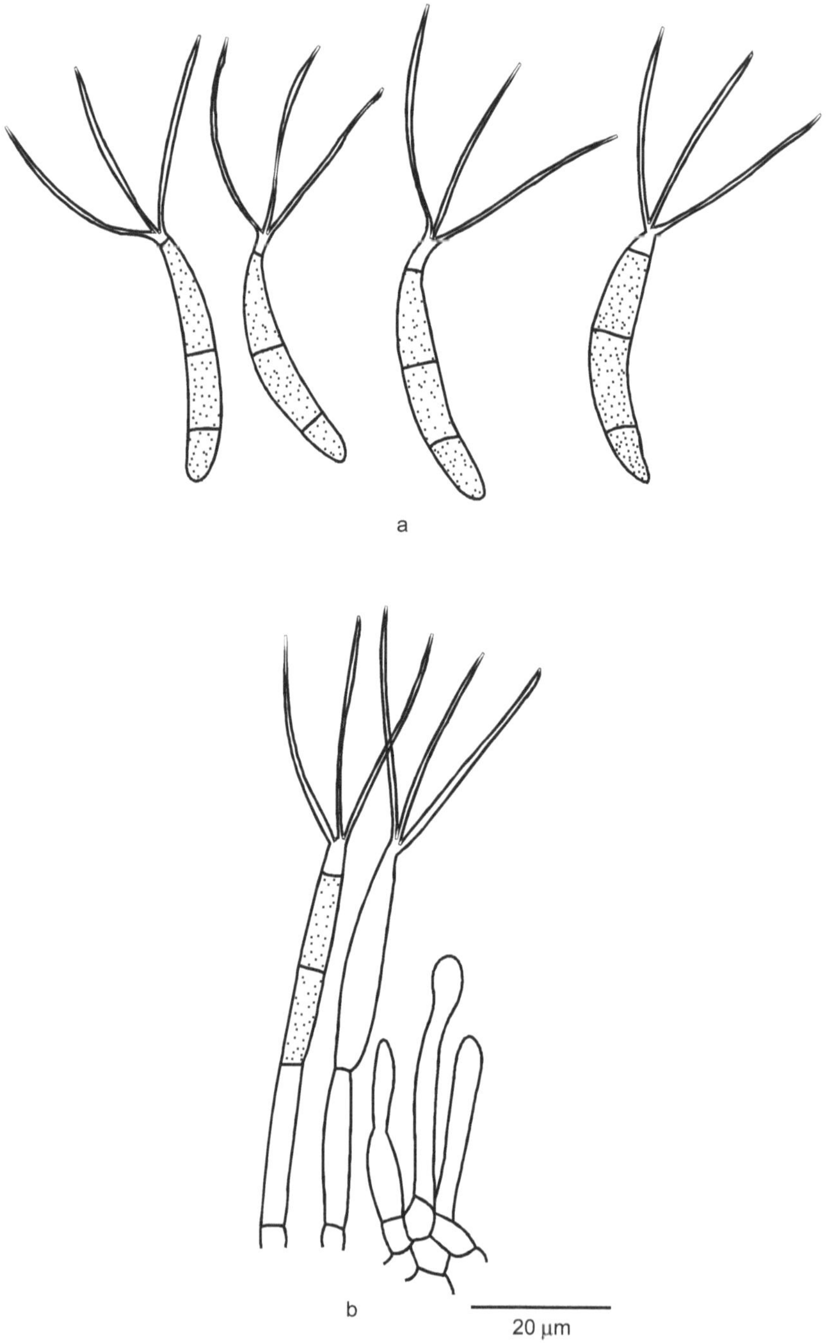

Figure 4.88 *Hyalotiella transvalensis*, (a) Mature conidia; (b) Developing conidia.

66. *Kendrickomyces indicus* Sutton, Rao and Mhaskar

Source

Sutton, Rao & Mhaskar, *Trans. Br. Mycol. Soc.*, **67**: 243 (1976); R.S. Mathur, *The Coelomycetes of India*, pp.114–115 (1979); Sutton, B.C. *The Coelomycetes*, CMI, England, p. 569 (1980).

Description

Conidiomata eustromatic, multilocular, locules arranged vertically, each locule made up of thin-walled pseudoparenchyma from which the conidiogenous cells arise, 800 µm diam. The ostiolar channels of each locule merge to form a common ostiole. Conidiogenous cells phialidic, integrated, cylindrical, hyaline, smooth, 11–15 × 1.5–2 µm. Conidia hyaline, aseptate, smooth, fusiform, curved, 4.5–6.5 × 1.5–2 µm.

Specimen collected

On dead fruits of *Eucalyptus globulus* Labill., Nilgiris, Tamilnadu, Feb. 1988, collected by J. Muthumary. IC. No: 156.

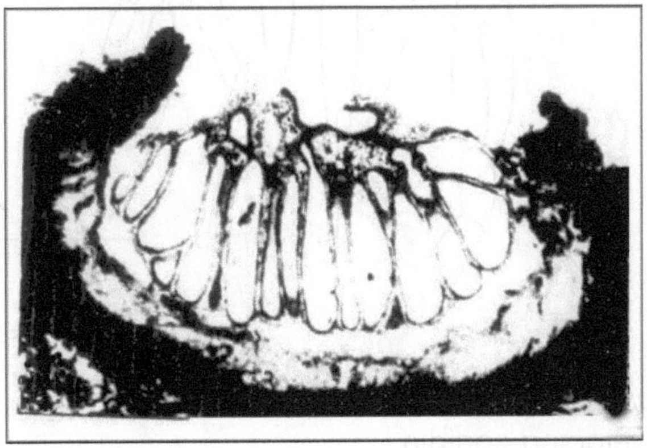

Figure 4.89 *Kendrickomyces indicus*; V.S. of conidioma (×260)

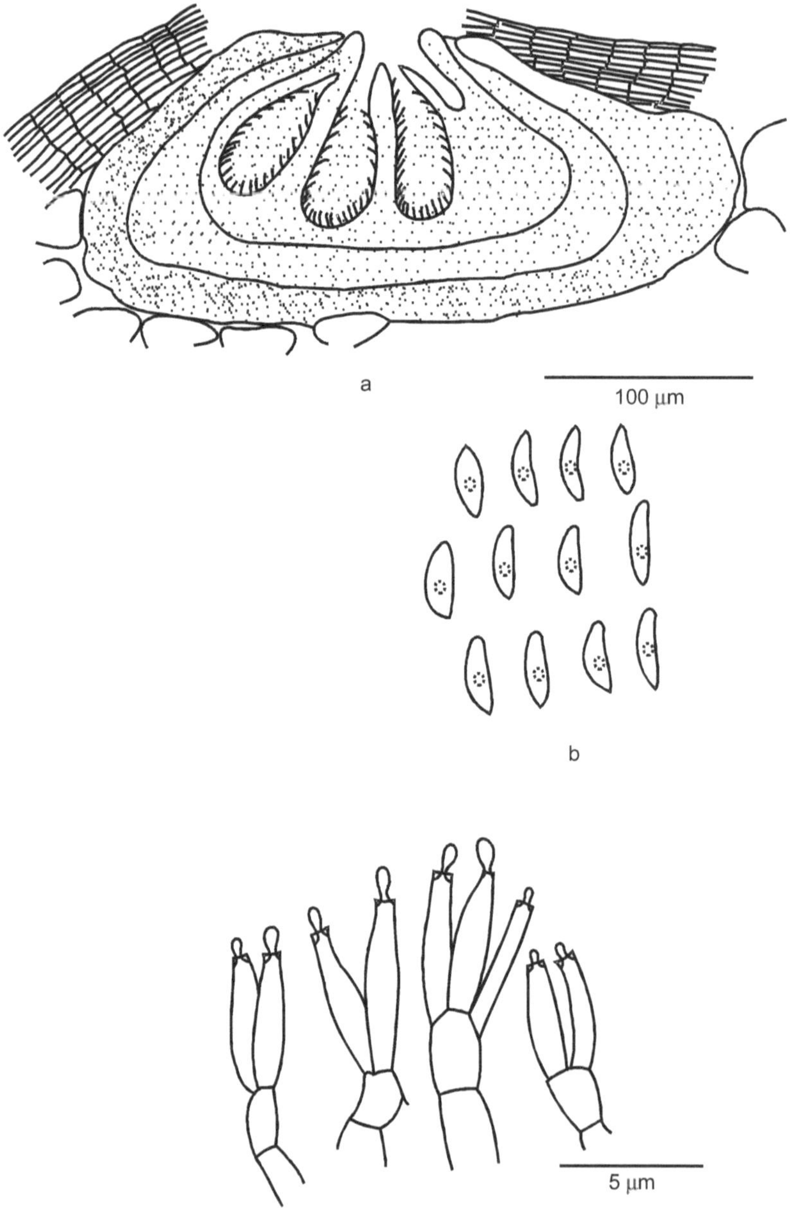

Figure 4.90 *Kendrickomyces indicus*, (a) V.S. of conidioma; (b) Mature conidia; (c) Developing conidia.

67. *Marssonina rosae* (Lib.) Died.

Source

Died., *Kryot. Fl. MK. Brandenb,* **9**: 830 (1915); Sutton, B.C., *The Coelomycetes*, CMI, England, p. 300 (1980); Muthumary, J. and S. Masilamani, *J. Indian bot. Soc.*, **68**: 131–133 (1989).

Description

Conidiomata acervular, brown. Conidiogenous cells annellidic, doliiform, 5–7 × 3–4 μm. Conidia hyaline, 1-septate, septum separating two unequal cells, slightly curved, base truncate, apex obtuse, constricted at the septum, guttulate, 15–18 × 5–5.5 μm.

Specimen collected

On living leaves of *Rosa* sp. (Var. Queen Elizebeth, Summer Holiday, Tapestry, Golden master piece, Princes Elizebeth, Sombruil, Laleenabert and Eclipse), Kodaikanal, Tamilnadu, Nov. 1987, collected by J. Muthumary. IC. No: 136, 137, 138, 139, 140, 141, 142, 143.

Figure 4.91 *Marssonina rosae*; V.S. of conidioma (×200)

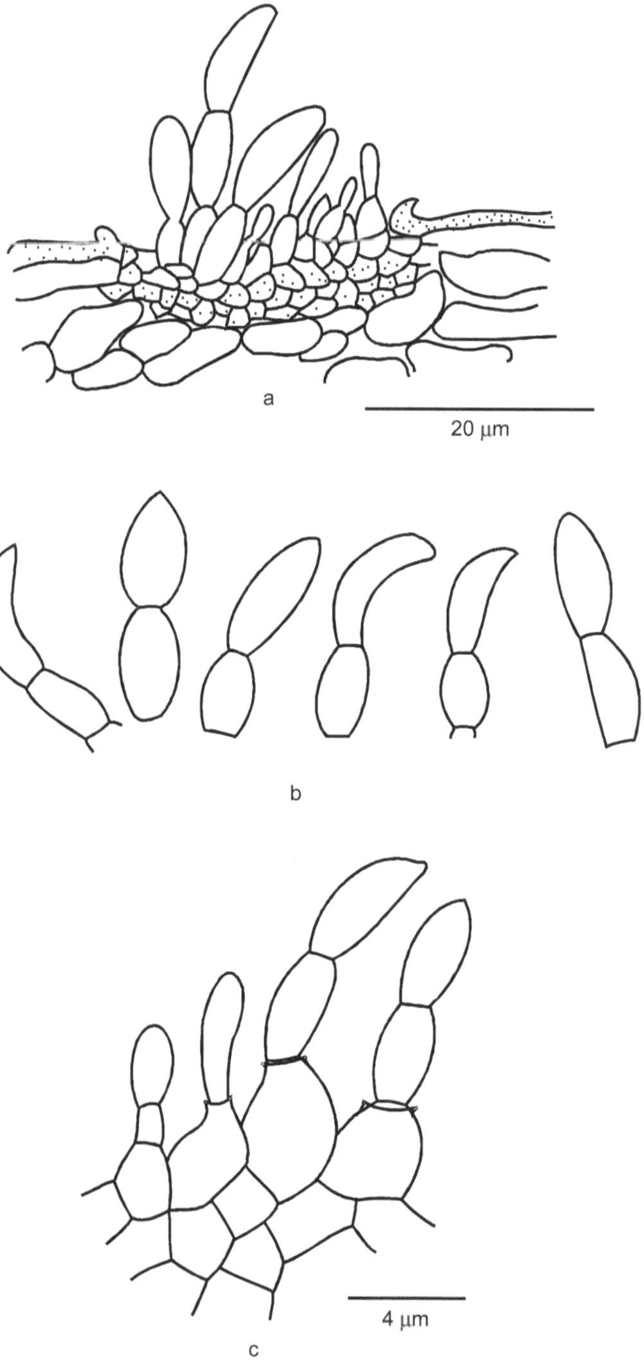

Figure 4.92 *Marssonina rosae*, (a) V.S. of conidioma; (b) Mature conidia; (c) Developing conidia.

68. *Melanconium* sp.

Source

Link in Maga 2. Ges. *Nature,* Freande, *Berlin* **3**: 9, (1809); Sutton, B.C., *Persoona,* **3** (1): 193–198, (1964); R.S. Mathur, *The Coelomycetes of India,* p.127 (1979); Sutton, B.C., *The Coelomycetes,* CMI, England, p. 330 (1980).

Description

Acervuli scattered, initially immersed, breaking irregularly at maturity, 500–900 µm in diam. Stroma composed of smooth, brown pseudoparenchymatous cells forming the base of the acervular conidioma. Conidiogenous cells formed from the upper cells of the stroma, straight, hyaline, cylindrical, smooth, with 1–4 annellations, 15–20 × 2–3 µm. Conidia unicellular, pale brown, smooth, base truncate with a marginal frill, apex obtuse, 10–15 × 6–7 µm.

Specimen collected

On dead bark of *Quercus alba,* Kodaikanal, Tamilnadu, October 1989, collected by J. Muthumary. IC. No: 181.

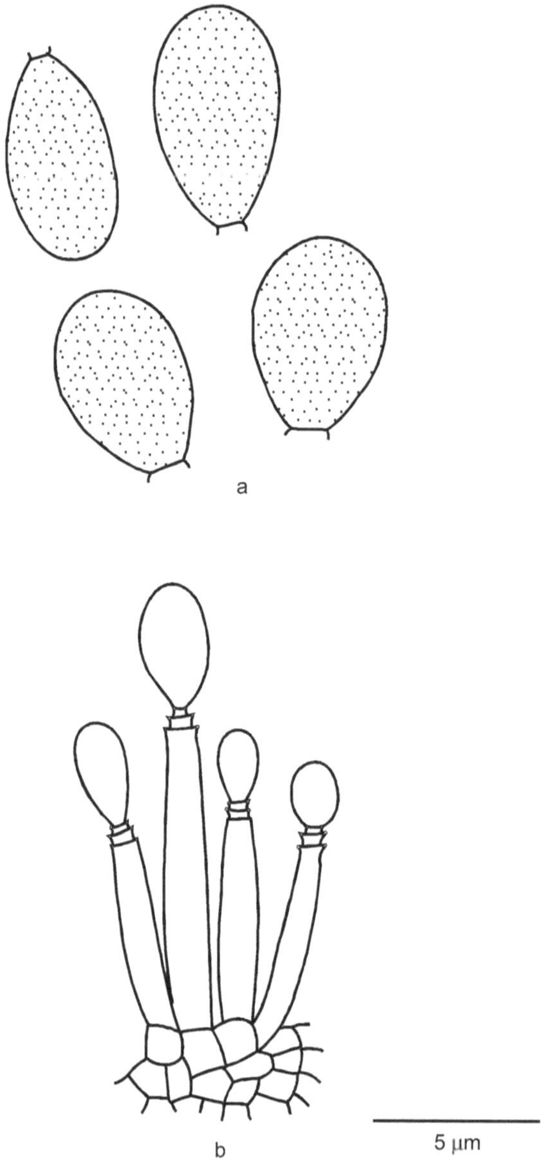

Figure 4.93 *Melanconium* sp, (a) Mature conidia; (b) Developing conidia.

69. *Microsphaeropsis* sp.

Source

Morgan-Jones, *Can. J. Bot.,* **52**: 2576 (1974); R. S. Mathur, *The Coelomycetes of India,* p.130 (1979); Sutton, B.C., *The Coelomycetes,* CMI, England, p. 422 (1980).

Description

Conidiomata pycnidial, globose, separate, brown, unilocular, erumpent, 350–420 μm diam. Pycnidial wall made up of thick-walled, brown cells at the outside and thin-walled, hyaline cells at the innerside giving rise to conidiogenous cells. Conidiogenous cells lining the pycnidial cavity, enteroblastic, phialidic, ampulliform, hyaline, smooth, 3–6 × 2.5–3 μm. Conidia brown, thick-walled, aseptate, ornamented, globose to subglobose, 5–6 × 4.5–5 μm.

Specimen collected

On dead twigs of *Rosa* sp. and several unidentified plants, Nilgiris, Tamilnadu, Feb. 1988, collected by J. Muthumary. IC. No: 157.

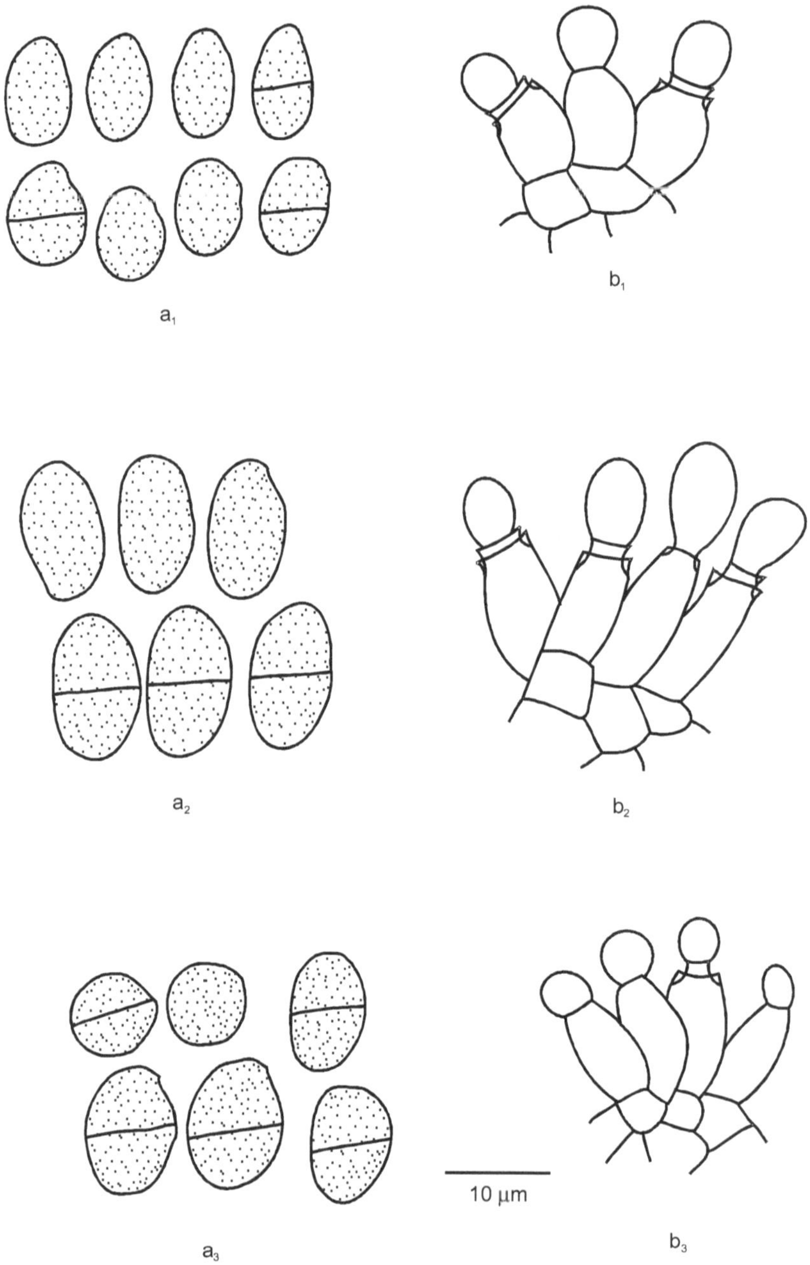

Figure 4.94 *Microsphaeropsis* sp, (a1–a3) Mature conidia; (b1–b3) Developing conidia.

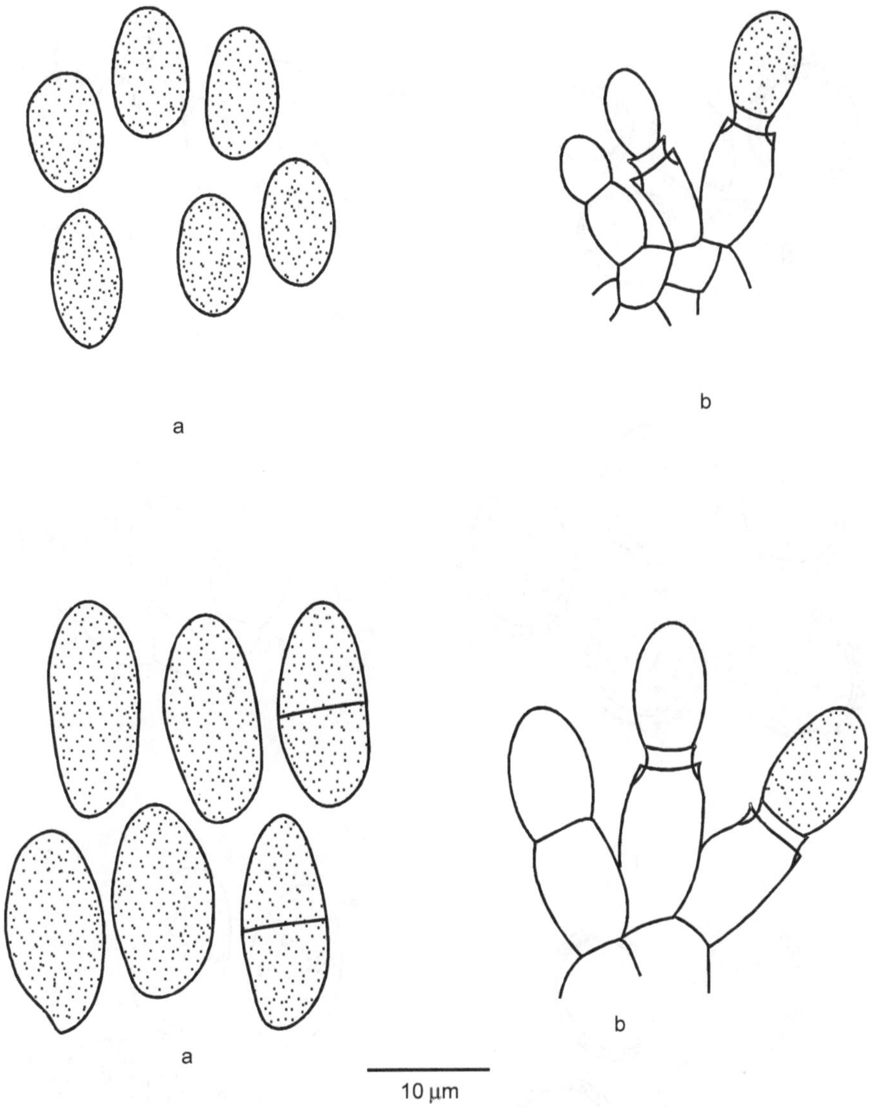

Figure 4.95 *Microsphaeropsis* sp, (a) Mature conidia; (b) Developing conidia.

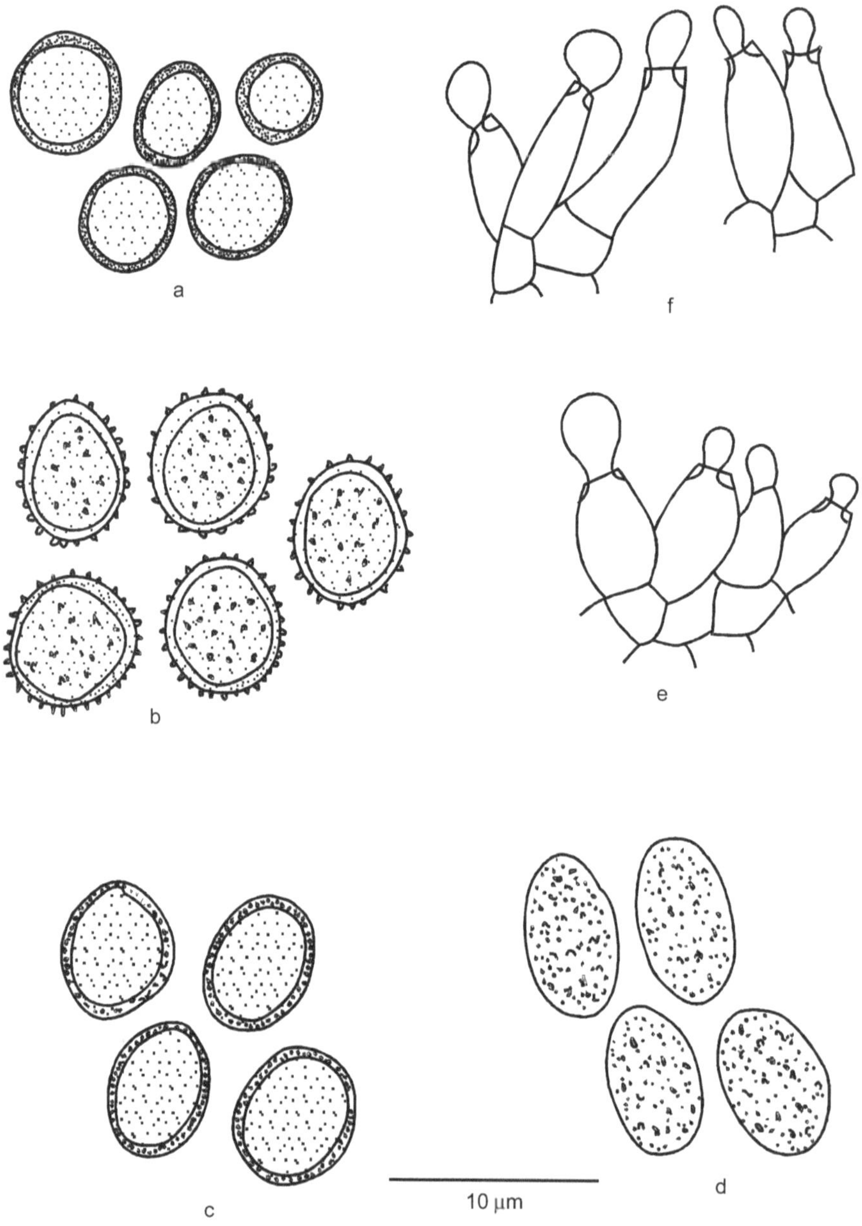

Figure 4.96 *Microsphaeropsis* sp, (a–d) Mature conidia; (e–f) Developing conidia.

70. *Mirimyces pulcher* Nag Raj

Source

Nag Raj, T.R., *Coelomycetous anamorphs with appendage bearing conidia*, Mycologue Publications, Canada, p. 478 (1993).

Description

Conidiomata pycnidial, unilocular, 300 μm. Conidiophores arising from the upper cells of the stroma and part way up the side walls. Conidiogenous cells colourless, smooth, immersed in mucus, discrete, ampulliform to lageniform, cylindrical, holoblastic. Conidia 12–16 × 3–5 μm, fusiform, straight or slightly curved, eguttulate, lageniform, aseptate with truncate base.

Specimen collected

On fresh leaves of *Philodendron* sp. (Araceae), Theosophical Society, Adayar, Chennai, May 2000, collected by Senthil Priya, IC. No: 261.

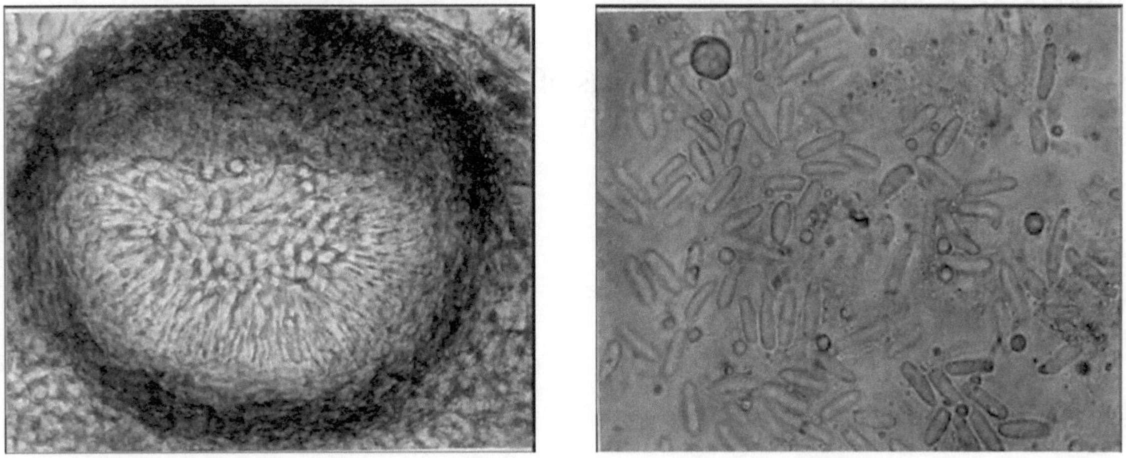

Figure 4.97 *Mirimyces pulcher*; (a) V.S. of conidioma (×200); (b) Mature conidia (×700)

71. Monochaetia kansensis (Ellis and Bartholomew) Saccardo

Source

Saccardo, *Syll. Fung.*, **18**: 485 (1906); Sutton, B.C., *The Coelomycetes*, CMI, England, p. 272 (1980); Nag Raj, T.R., *Coelomycetous anamorphs with appendage bearing conidia,* Mycologue Publications, Canada, pp. 490–493 (1993).

Description

Conidiomata acervular, isolated as endophyte, irregular to subglobose, 250–300 μm in diam; wall composed of brown, psuedoparenchymatous cells. Conidiogenous cells lining the conidiomatal cavity, subcylindrical to irregular, hyaline, smooth, with 1–2 percurrent proliferations. Conidia fusiform, 4-septate, 20–25 × 5–6 μm with three median cells and an apical and basal cell, basal and apical cells provided with apical and basal appendage, 10–15 μm long.

Specimen examined

Isolated as endophyte from twigs of *Taxodium mucronatum,* Ootacamund, TamilNadu, Jun. 2007, by S. Kamalraj, IC. No: 314.

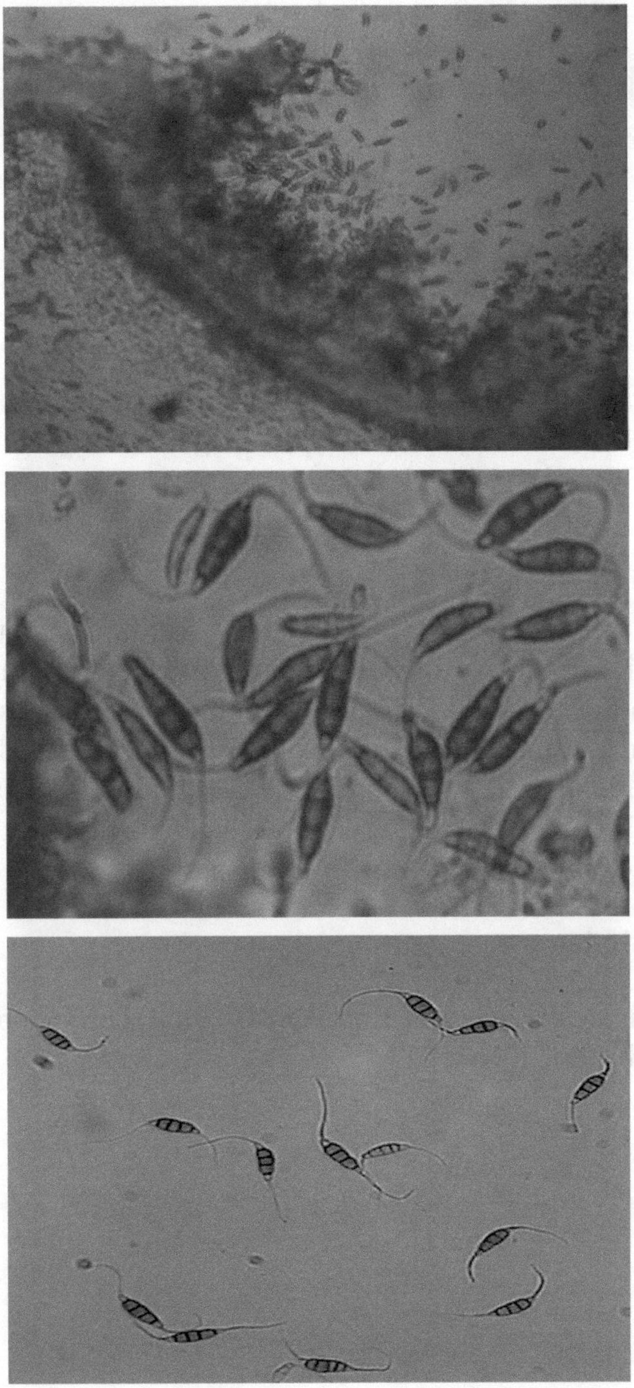

a b c

Figure 4.98 *Monochaetia kansensis*; (a) V.S. of conidioma (×200); (b–c) Mature conidia b. ×900, c. ×400

72. *Monochaetia lutea* Swart and Griffiths

Source

Griffths, *Trans. Br. Mycol. Soc.*, **62**(1): 151–161 (1974); Nag Raj, T.R., *Coelomycetous anamorphs with appendage bearing conidia*, Mycologue Publications, Canada, p. 493 (1993).

Description

Conidiomata acervular, globose, irregular, 90–125 µm, Stroma at first closed, fertile all over the inner surface. Conidiogenous cells cylindrical, annellidic. Conidia 4-celled, occasionally 2-celled, due to degeneration of end cells, slightly constricted at the median septum, 14 (12–16) × 6.5 (5.5–7) µm, two middle cells with thick walls, brown, 10–5 (9–12) µm long, apical cell thin-walled, hyaline, with a short apical appendage, basal cell thin-walled, hyaline, sometimes with a minute central appendage, both hyaline cells sometimes collapsing.

Specimen collected

On litter of unidentified leaves, Kodaikanal, TamilNadu, October 1990, collected by J. A. Jayachandra, IC. No: 210.

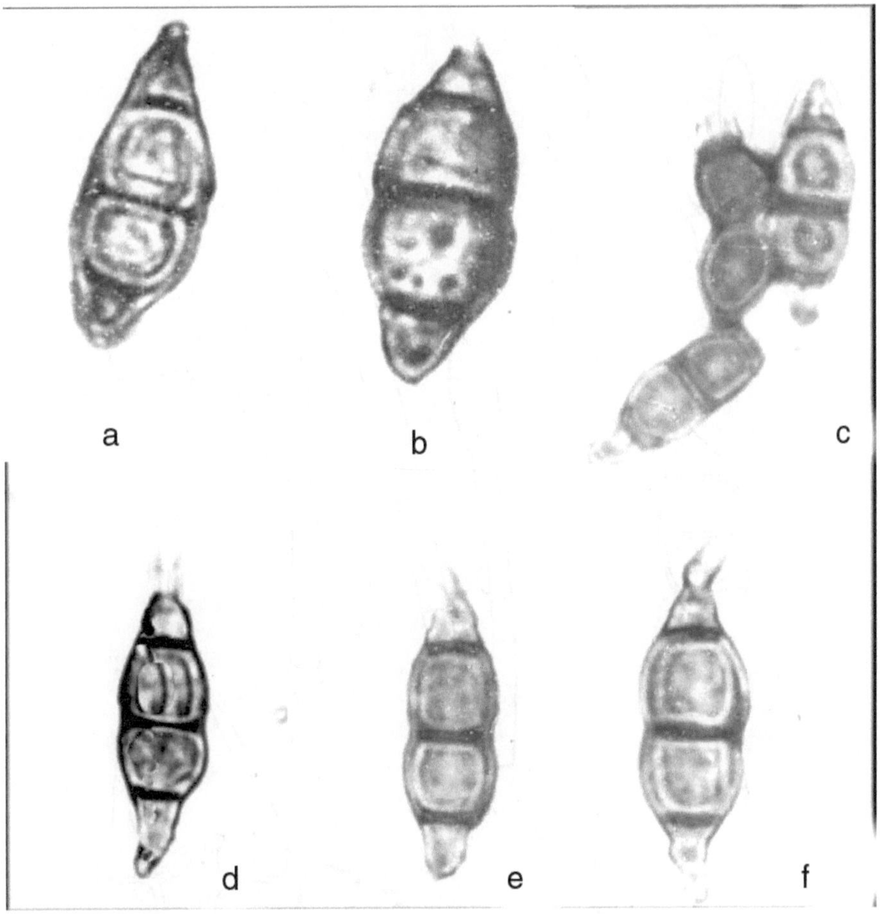

Figure 4.99 *Monochaetia iutea*; (a–f) Mature conidia (×900)

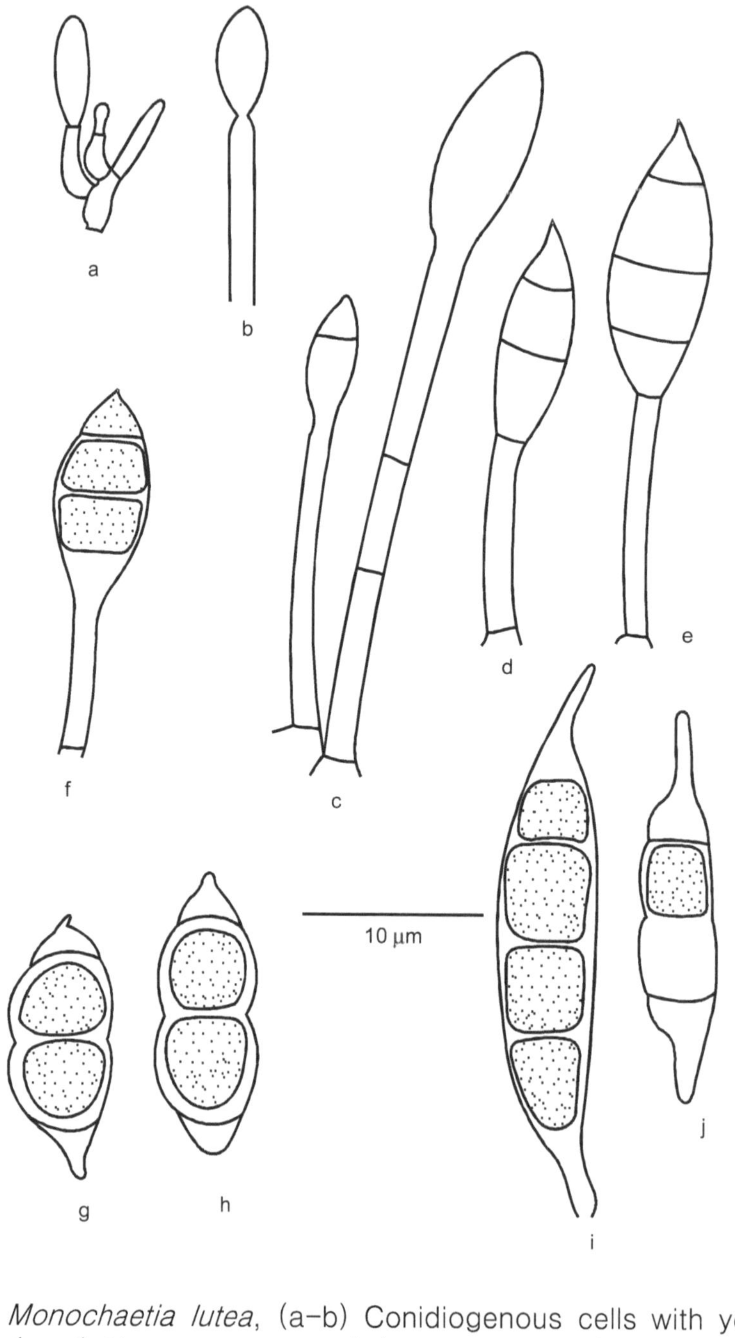

Figure 4.100 *Monochaetia lutea*, (a–b) Conidiogenous cells with young conidia; (c–d) Young conidia; (e) Annellides; (f–h) Young conidia; (i–j) Mature conidia.

73. *Monostichella indica* Sutton

Source

Sutton, B.C., *The coelomycetes,* CMI, England, p. 517 (1980).

Description

Conidiomata acervular, 180 µm, subcuticular, separate, thin-walled. Conidiogenous cells enteroblastic, phialidic, determinate, cylindrical, hyaline. Conidia 9–13 × 2–3 µm, hyaline, aseptate, guttulate, obtuse at both ends, cylindrical.

Specimen collected

On fresh leaves of *Hibiscus rosasinensis* (Malvaceae) Theosophical society, Adayar, Chennai, June 2000, collected by Senthil Priya, IC. No: 262.

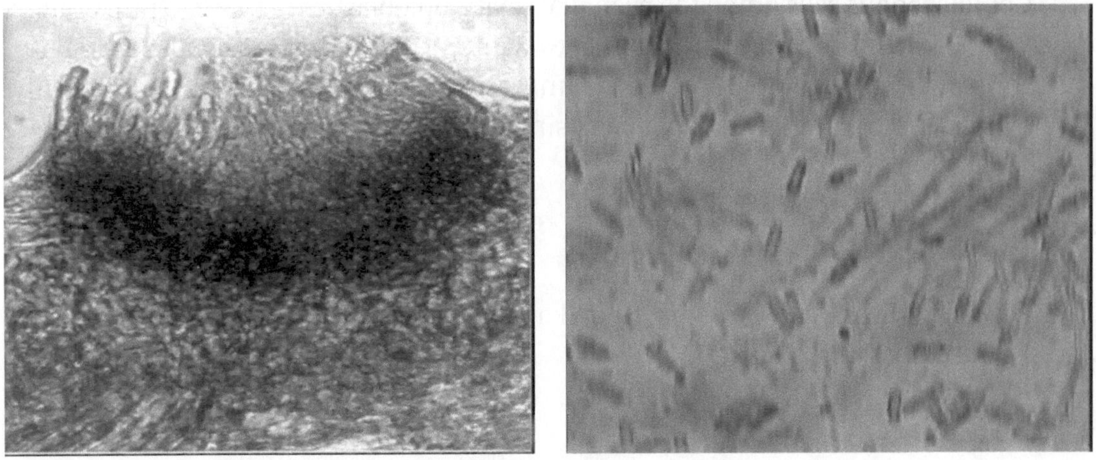

Figure 4.101 *Monostichella indica*; (a) V.S. of conidioma (×200); (b) Mature conidia (×700)

74. *Mycotribulus mirabilis* Nag Raj and Kendrick

Source

Nag Raj & Kendrick, *Can. J. Bot.*, **48**: 2219 (1970); R.S. Mathur, *The Coelomycetes of India*, p-137 (1979); Sutton, B.C., *The coelomycetes,* CMI, England, p. 66 (1980); Nag Raj, T.R., *Coelomycetous anamorphs with appendage bearing conidia*, Mycologue Publications, Canada, pp. 530–532 (1993).

Description

Conidiomata pycnidial, immersed, unilocular, globose, thick-walled, 200–600 µm diam. Paraphyses hyaline, septate, smooth, deeply lobed and irregular at the apices, 30–50 × 1.5–2 µm. Pycnidial wall 20–25 µm, thick, outer layers dark brown, thick-walled, inner layer hyaline and thin-walled giving rise to the conidiogenous cells. Conidiogenous cells lining the pycnidial cavity, holoblastic, 7–15 × 2–5 µm. Conidia aseptate, fusiform, base truncate, apex acute, with a single appendage, smooth, thin-walled, guttulate, 2–3 basal, unbranched appendages inserted laterally, 8–15 × 2–4.5 µm. Apical appendage single, polar, 12 µm long. *Mycotribulus* is a monotypic genus.

Specimen collected

On living leaves of *Rosa* sp., Kodaikanal, Tamilnadu, Nov. 1987, collected by J. Muthumary, IC. No: 129.

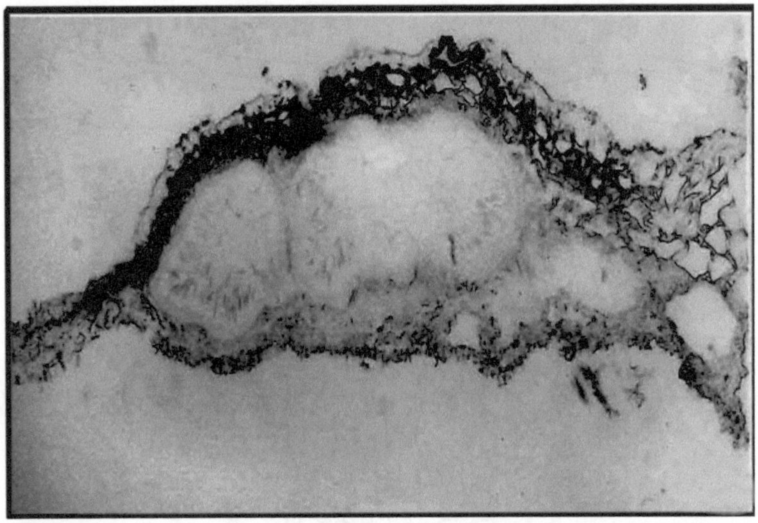

a

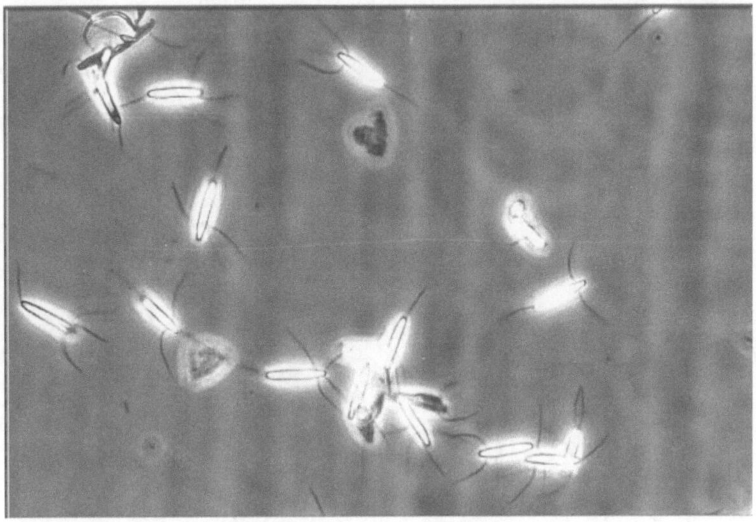

b

Figure 4.102 *Mycotribulus mirabilis*; (a) V.S. of conidioma (×260); (b) Mature conidia (×900)

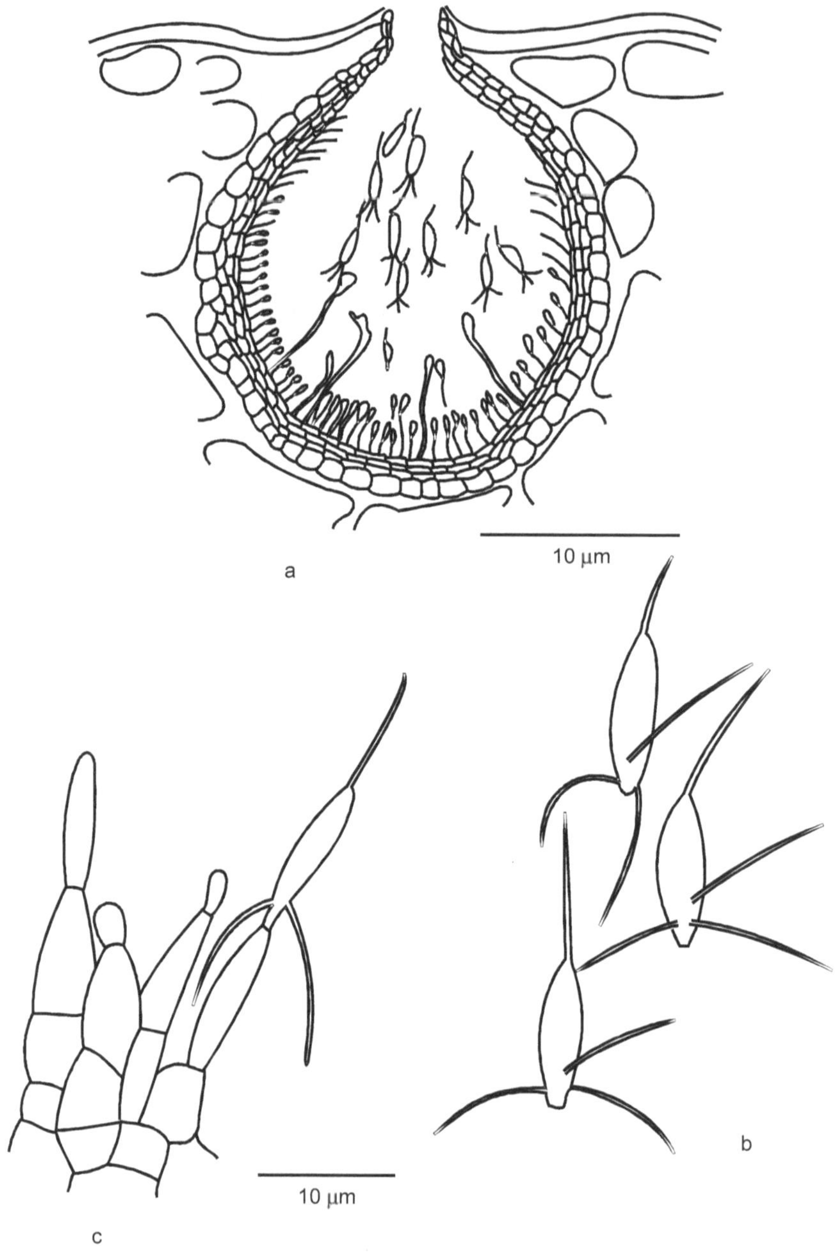

Figure 4.103 *Mycotribulus mirabilis*; (a) V.S. of conidioma; (b) Mature conidia; (c) Developing conidia.

75. *Paramassariothea falcata* Subram. and Muthumary

Source

Subramanian, C.V. and J. Muthumary, *Can. J. Bot*, **57**: 2422–2427 (1979).

Description

The fungus was collected on fallen twigs of *Ficus asperima* in the Coorg forests, India. It forms black, stromatic pycnidia immersed in the substrate, and slightly raised, mostly solitary, rarely in groups of two to three, glabrous, and 600–900 µm × 200–300 µm with a papillate ostiole of diameter about 25–35 µm. The peridium is pseudoparenchymatous, 30–45 µm wide and is composed of an outer layer of spherical to irregular, thick-walled cells and an inner layer of elongated, thin-walled and light-coloured cells. In three-loculate pycnidia, the central locule is 300–400 µm × 200–300 µm, with an ostiole 25–35 µm in diameter; the other locules are smaller and 150–200 µm in diameter and apparently without ostiole. The conidiogenous cells line the cavity of the pycnidium and are simple, annellidic with one or four annellations, hyaline, smooth, ampulliform to cylindrical, 10–12 µm × 3–4 µm; they are intermixed with sterile, paraphysis-like structures. The paraphysis-like structures are hyaline, smooth, simple or branched, septate and 180–200 µm × 5–7 µm. The conidia are extruded through the ostiole in the form of long, thin, white cirrhi and are hyaline, smooth, straight to falcate, solitary, 10–29 (X = 24) transversely distoseptate with a truncate base and tapering end, 90–140 µm (X = 110 µm) × 10–14 µm (X = 13 µm) with a length:width ratio of 10:1. In some pycnidia, spermatia are produced from hyaline, smooth, branched, septate hyphae 20–40 µm × 2–3 µm; the spermatia are hyaline, smooth, filiform, straight to curved, continuous, and 18–30 µm × 0.5–1.0 µm.

Specimen collected

On fallen twigs of *Ficus asperima*, Coorg forests, Karnataka, November 1977, collected by D. Jayarama Bhat. MUBL No: 2368, IMI No: 238249.

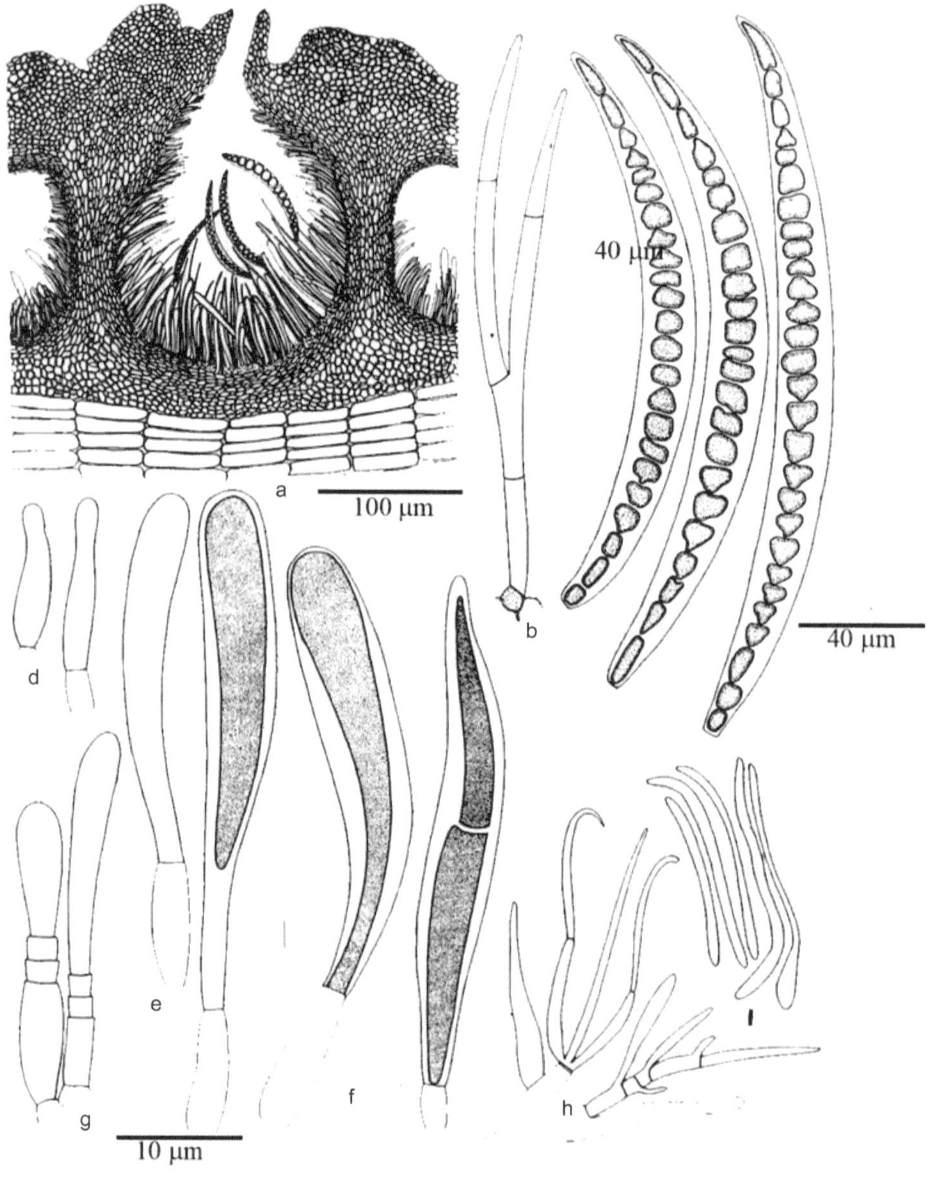

Figure 4.104 *Paramassariothea falcata*, (a) V.S. of conidioma; (b) Paraphysis like structure; (c) Mature conidia; (d–g); In (f), note the formation of the first septum in the middle of the young conidium and in, (g) annellations; (h) Hyphae bearing spermatia; (i) Spermatia.

76. *Pestalotiopsis adusta* (Ell & Ev.) Steyaert

Source

Stayaert, R.L., *Trans. Br. Mycol. Soc.,* **36**: 82 (1953); Ponnappa, K.M. *Proc. Indian. Acad. Sci.,* **71B**: 66-67 (1970); Prasad, N and R.K. Verma, *Sci. and Cult.,* **32**: 558 (1966); Sohi, H.S., *et al. Indian Phytopath.,* **25**: 441(1972); Sharma, *Botanique,* **6**: 100 (1975); Singh, S.M., *Indian Phytopath.,* **29**: 19 (1977); R.S. Mathur, *The Coelomycetes of India,* p.149 (1979).

Description

Pustules black, minute, 70–120 µm diam. Conidiomata pycnioid, unilocular, glabrous, brown. Conidiogenous cells holoblastic, annellidic. Conidia 5-celled, oblong or elliptic-fusoid, erect, usually 16–22 × 5–6 µm; intermediate colored cells olivaceous, concolorous, 10–14 µm long, only slightly constricted at septa; end cells hyaline, apical cells short, conic, bearing a crest of 3, rarely 2 appendages, erect, divergent at right angles, usually 5–12 µm, sometimes up to 15 µm long; basal cells short, obtuse; pedicels 2–7 µm long.

Recorded as endophyte in

Isolated as endophyte in *Erythroxylum monogynum,* K.P. Kannan, 2002, Ph.D. Thesis, "Biodiversity of Endophytic Mycobiota from selected Medicinal plants and some Gymnosperms". MUBL No: 4166.

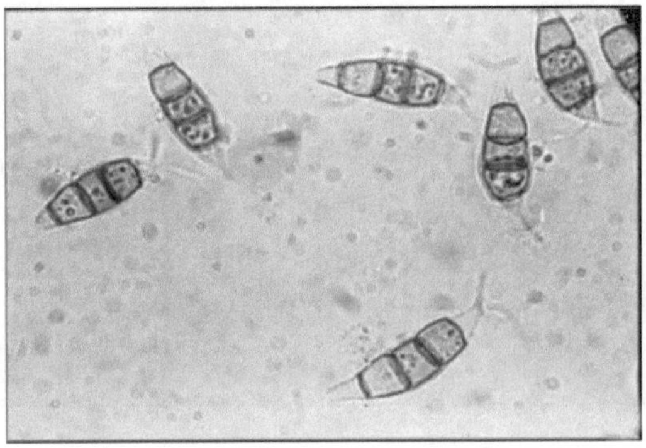

Figure 4.105 *Pestalotiopsis adusta;* Mature conidia (×900)

77. *Pestalotiopsis breviseta* (Sacc.) Stey.

Source

Steyaert, R.L., *Bull. Jard. Bot. Brux.*, **19**: 316 (1909); Guba, E.F., *Monographs of Monochaetia and Pestalotia*, Harvard Univesity Press, Cambridge, Massachusetts, p. 134 (1961); R.S. Mathur, *The Coelomycetes of India*, p.149 (1979).

Description

Pustules punctiform, gregarious, subglobose, 120–180 µm diam. Conidiomata acervular. Conidiogenous cells holoblastic, annellidic. Conidia 5-celled, elliptic fusiform, 19.95–28.5 × 6–10 µm, hardly constricted at septa, intermediate coloured cells olivaceous, concolorous, 12–16 µm, apical appendages 1–3, 5–6.5 µm long, divergent, basal appendage up to 3.9 µm long.

Recorded as endophyte in

Isolated as endophyte in *Madhuca indica, Pinus kesiya,* K.P. Kannan, 2002, Ph.D. Thesis, "Biodiversity of Endophytic Mycobiota from selected Medicinal plants and some Gymnosperms". MUBL No: 540.

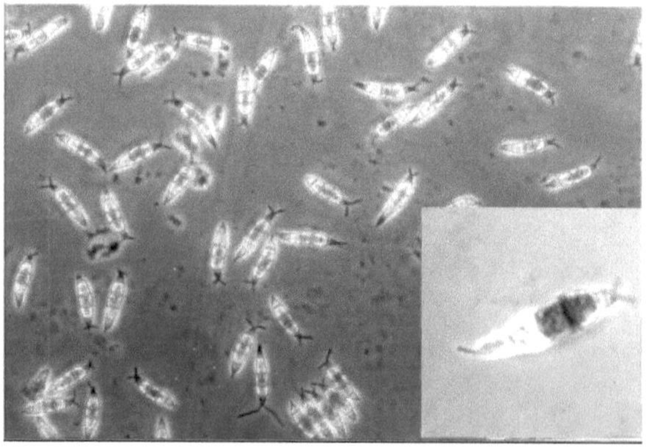

Figure 4.106 *Pestalotiopsis breviseta;* Mature conidia (×900)

78. *Pestalotiopsis cruenta* (Syd.) Stey.

Source

Steyaert, R.L., *Bull. Jard. Bot. Brux,* **19**: 318 (1949); Guba, E.F. *Monograph of Monochaetia and Pestalotia,* Harvard University Press, Cambridge, Massachusetts, p. 193 (1961); Tandon, R.N. and M.P. Srivastava, *Curr. Sci.,* **33**: 86–87 (1964); R.S. Mathur, *The Coelomycetes of India,* p-150 (1979).

Description

Pustules black, 120–180 µm diam. gregarious. Conidiomata acervular. Conidiogenous cells holoblastic, annellidic. Conidia fusiform, 5-celled, 20–24(23) × 6–7 µm, hardly constricted at septa; Coloured cells 12–17 µm, central or upper 2 umber, lowest olivaceous, guttulate; end cells hyaline, conical; apical appendages 1–4, usually 2–3, 8–8.5 (8.4) µm long, divergent, basal appendage up to 8 µm long, usually 5–5.3 (5.2) µm long.

Recorded as endophyte in

Isolated as endophyte in *Canthium dicoccum,* K.P. Kannan, 2002, Ph.D. Thesis, "Biodiversity of Endophytic Mycobiota from selected Medicinal plants and some Gymnosperms". MUBL No: 541; MTCC No: 3441.

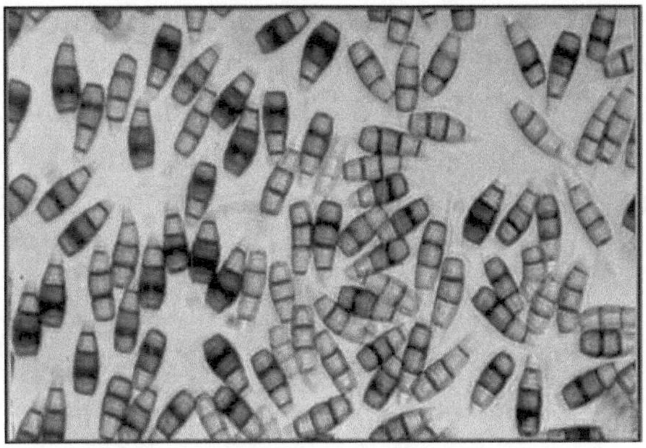

Figure 4.107 *Pestalotiopsis cruenta*; Mature conidia (×900)

79. *Pestalotiopsis disseminata* (Thuem.) Stey.

Source

Steyeart, R.L., *Bull. Jard. Bot. Brux.,* **19:** 319 (1949); Sharma, M.P., *Botanique.,* **6**: 101(1975); Guba, E.F., *Monograph of Monochaetia and Pestalotia,* Harvard University Press, Cambridge, Massachusetts, p. 342 (1961); Sharma, Botanique, **6**: 101 (1975); R.S. Mathur, *The Coelomycetes of India,* pp.150-151 (1979).

Description

Conidiomata acervular, black, 75–250 µm in diam. Conidiogenous cells holoblastic, annellidic. Conidia 5-celled, elliptic to clavate, fusiform, tapering to the base, slightly curved, 19.8 × 5.08 µm long, median coloured cells unequally colored, sometimes two upper cells slightly darker than the lower most cell. The basal cell hyaline with a single unbranched appendage, 2–5 (5) µm. Apical cell hyaline with 2–3 apical appendages, 18–22 (20) µm. The apical and basal cells conic to cylindrical, 4–5 (5) µm.

Recorded as endophyte in

Isaolated as endophyte in *Terminalia arjuna,* K.P. Kannan, 2002, Ph.D. Thesis, "Biodiversity of Endophytic Mycobiota from selected Medicinal plants and some Gymnosperms", MUBL–542; MTCC-3411.*Psidium guajava, Terminalia arjuna,* V. Gangadevi, 2005, Ph.D. Thesis, Studies on Fungal Endophytes from some medicinal plants and screening selected coelomycetes for taxol production. MUBL No: 676.

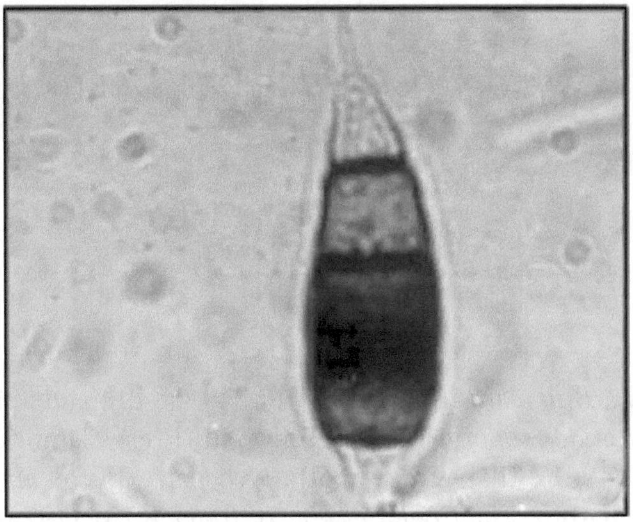

a

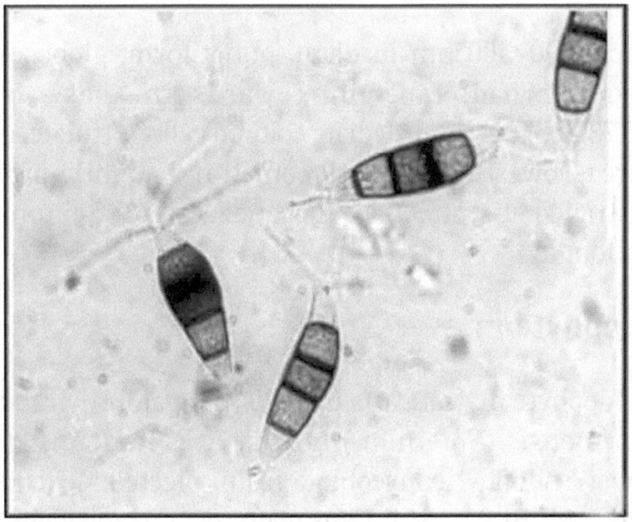

b

Figure 4.108 *Pestalotiopsis disseminata*; (a–b) Mature conidia a. (×900), b. (×400)

80. *Pestalotiopsis funerea* (Desm.) Stey.

Source

Steyaert, R.L., *Bull. Jard. Bot. Brux.*, **19**: 340 (1949); Guba, E.F., *Monograph of Monochaetia and Pestalotia*, Harvard University Press, Cambridge, Massachusetts, pp. 154–155 (1961); Tilak & Viswanath, *Mycopath. Mycol. appl.*, **13**: pp. 100–106 (1960); R.S. Mathur, *The Coelomycetes of India*, p.151 (1979);

Description

Pustules black, 100–300 µm in diam. punctiform, globose-lenticular, Conidia clavate–fusoid, broad, tapering towards the base, 5-celled, straight, 15.69–29 × 6.73–9.5 µm; intermediate colored cells guttulate, umber or olivaceous, equally colored, lowest colored cell sometimes slightly paler, 15–20 µm long. Slightly constricted at septa. Apical appendages 1–2; 5-6 µm, basal appendage 1, 2.92–4.5 µm long.

Recorded as endophyte in

Isolated as endophyte in *Araucaria bidwilli* (Bangalore), *Araucaria cunninghamii, Juniperus chinensis, Solanum nigrum,* K.P. Kannan, 2002, Ph.D. Thesis, "Biodiversity of Endophytic Mycobiota from selected Medicinal plants and some Gymnosperms", MUBL No: 544, *Vitex negundo,* V. Gangadevi, 2005, Ph.D. Thesis, "Studies on Fungal Endophytes from some medicinal plants and screening selected coelomycetes for taxol production". MUBL No: 677.

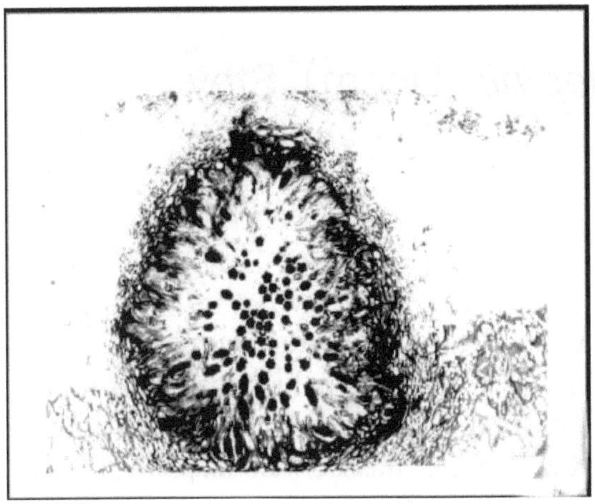

(a)

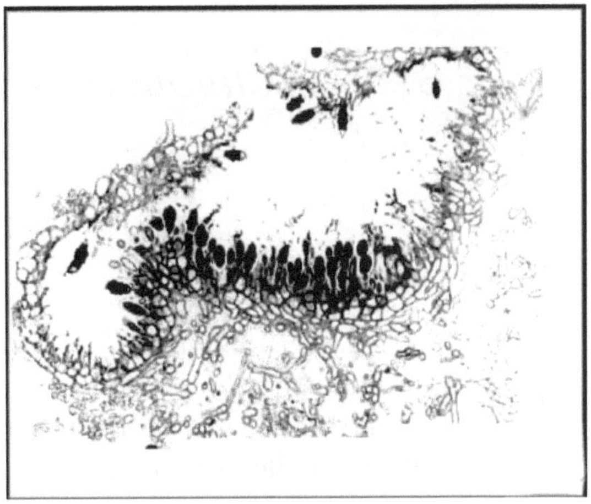

(b)

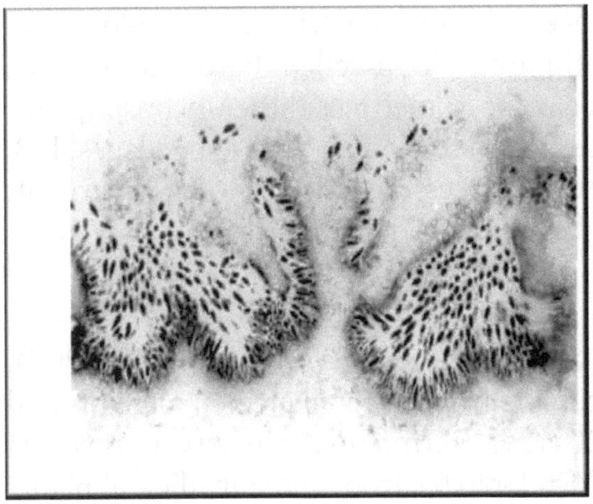

(c)

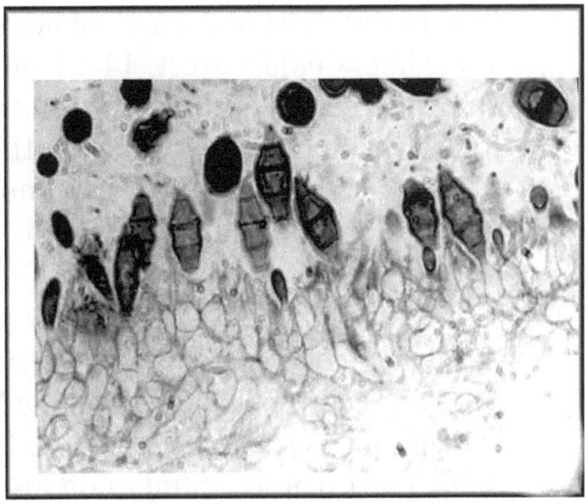

(d)

Figure 4.109 *Pestalotiopsis funerea*; (a–c) V.S. of conidiomata (×200); (d) Conidiogenous cells with Mature conidia (×900)

81. *Pestalotiopsis guepinii* (Desm) Stey.

Source

Steyaert, R.L., *Bull. Jard. Bot. Brux.*, **19**: 312 (1949); Ramakrishnan, T.S. and K. Subramaniam, *J. Madras. Univ.*, **22B**: 33 (1962); Ramakrishnan, T.S. *et al.*, *Proc. Indian Acad. Sci.*, **36B**: 85-95 (1952); Guba, E.F. *Monograph of Monochaetia and Pestalotia*, Harvard University Press, Cambridge, Massachusetts, p. 342 (1961); R.S. Mathur, *The Coelomycetes of India*, p.152 (1979).

Description

Pustules 90–170 µm diam. Conidiomata acervular. Conidiogenous cells holoblastic, annellidic. Conidia erect, 14–21 µm, elliptic-fusoid, 5-celled, hardly constricted at septa; end cells hyaline, short and broad conic, sometimes rather inconspicuous; intermediate cells olivaceous, guttulate, 12–15 × 5.5–6.6 µm; appendages 1–3, rarely 4, sometimes minutely knobbed at apices, 10–24 µm long, divergent, commonly branched, pedicels short, straight, often deciduous.

Recorded as endophyte in

Isolated as endophyte in *Phyllanthus maderaspatensis*, K.P. Kannan, 2002, Ph.D. Thesis, "Biodiversity of Endophytic Mycobiota from selected Medicinal plants and some Gymnosperms", MUBL-547; MTCC-4242. *Terminalia arjuna*, V. Gangadevi, 2005, Ph.D. Thesis, "Studies on Fungal Endophytes from some medicinal plants and screening selected coelomycetes for taxol production". MUBL No: 679.

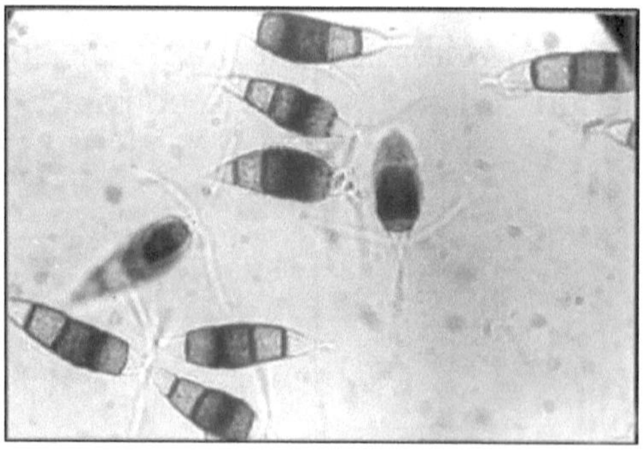

Figure 4.110 *Pestalotiopsis guepinii*; Mature conidia (×900)

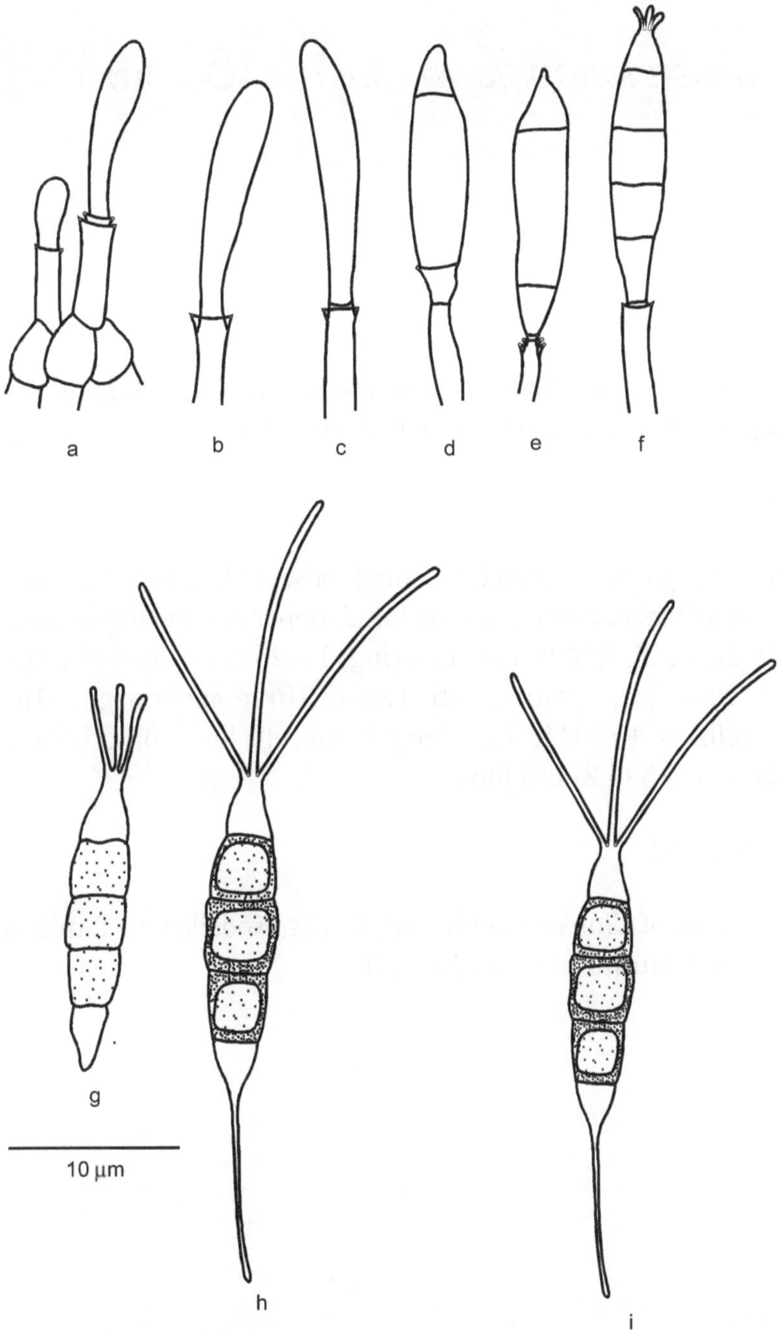

Figure 4.111 *Pestalotiopsis guepinii*, (a–c) Conidiogenous cells with young conidia; (d–e) Septa forming young conidia; (f) Annellides; (g) Young Conidium; (h–i) Mature conidia.

82. *Pestalotiopsis maculans* (Corda) Nag Raj

Source

Nag Raj, T.R., *Coelomycetous anamorphs with appendage bearing conidia*, Mycologue Publications, Waterloo, Ontario, Canada, pp. 639–641(1993).

Description

Conidiomatal pustules, punctate, subglobose, black, 100–120 µm diam. conidia 5-celled, fusiform to clavate, versicolored, upper two median cells dark brown and lower median cell light brown, tapering, 19–25 (24) × 6–8 (7.2) µm. Apical cells conic to cylindric, hyaline, with two to three appendages, 16–17 µm long. Median cells 10–15.5 (14.5) µm long. Basal cells long conic, hyaline, with a basal appendage, 4.2–5 (4.2) µm long.

Specimen collected

On dead leaves of *Cinnammomum wightii* L., Kodaikanal, Tamilnadu, Jan. 1998, collected by J. Muthumary, IC. No: 181.

83. *Pestalotiopsis palmarum* (Cooke) Stey.

Source

Stey, *Bull. Jard. Bot. Etat. Bruxelle,* **19**: 322 (1949); Ramakrishnan, K. and C.V., Subramanian, *The fungi of India*–Second supplement *J. Madras Univ.,* **22 B** : 33 (1952); R.S. Mathur, *The Coelomycetes of India,* pp.154–155 (1979).

Description

Conidiomata acervular, 95–150 µm diam. Conidiogenous cells holoblastic, annellidic. Conidia fusiform, 4-euseptate, 23–27 × 6.5–8 µm; apical cell conic, with 3-apical appendages 26–32 µm long; basal cell truncate, with single appendage 10–14 µm long.

Specimen collected

On dead leaves of *Areca catechu* L., Madras, Tamilnadu, Jan. 1989, collected by J. Muthumary, IC. No: 182.

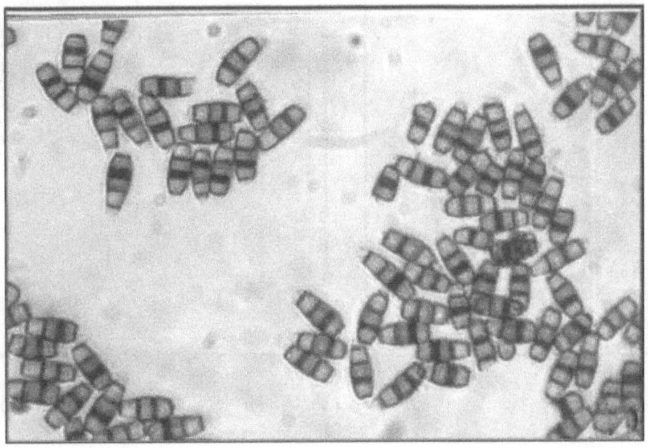

Figure 4.112 *Pestalotiopsis palmarum*; Mature conidia (×700)

84. *Pestalotiopsis sydowiana* Bres.

Source

Bres., *Hedwigia*, 35: 32 (1896).

Description

Conidiomata 150–300 µm diam. Conidia 5-celled, 23–28 × 8–9 µm, the lower coloured cells olivaceous, the upper two darker; apical 3-simple appendages 23–40 µm long; basal single unbranched appendage 6–12 µm long.

Specimen collected

On dead leaves of *Araucaria bidwillii* Hook., Nilgiris, Tamilnadu, Jan. 1988; On dead leaves of *Mimusops elengi* L., Madras, Tamilnadu, Nov. 1989, collected by J. Muthumary. IC No: 158.

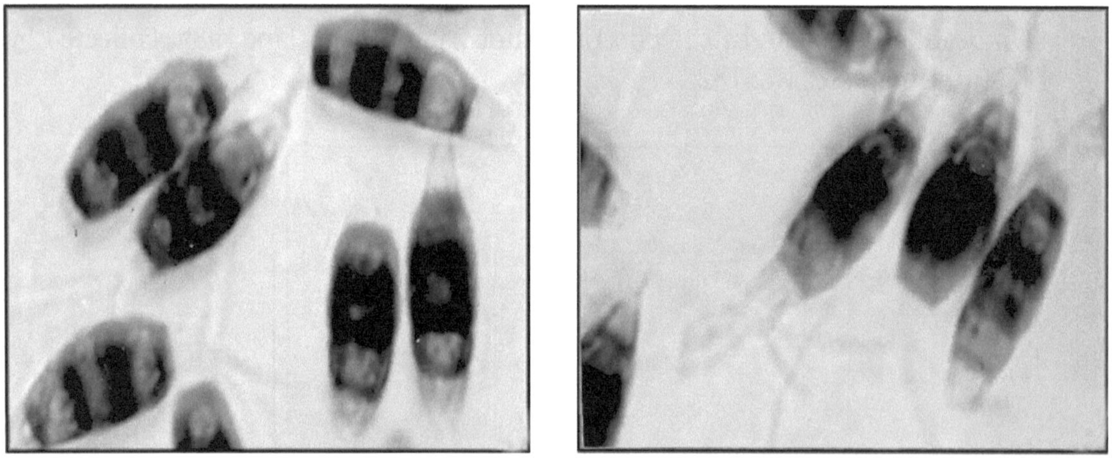

(a) (b)

Figure 4.113 *Pestalotiopsis sydowiana*; (a–b) Mature conidia (×900)

85. *Pestalotiopsis theae* (Sawada) Stey.

Source

Sawada, Sepceial Rep., *Agr. Exp. Sta.,* Taiwan (Formosa) **11**: 113 (1915); Stey., *Bull. Jord. Bot. Brux.*, **19**:327 (1949): Dube, H.K. & K.S. Bilgrami, *Mycopath. Mycol. Appl.,* **29**: 30-34 (1966).

Description

Conidiomata acervular, separate, 88–176 μm diam. made up of thick-walled psudoparenchyma. Conidiogenous cells holoblastic, annellidic, cylindrical, hyaline, smooth, with 1–3 percurrent proliferations, 10–15 × 1.3 μm. Conidia fusiform, straight, 5-celled dark brown, 23–30 × 7–8 μm, base truncate, with an endogenous, simple appendage, apical cell hyaline conic, with 3 spathulate appendages, 35–45 μm long.

Specimen collected

On dead leaves of *Eugenia jambolana* Lam., Madras, Tamilnadu, Nov. 1987, collected by J. Muthumary. IC No: 130.

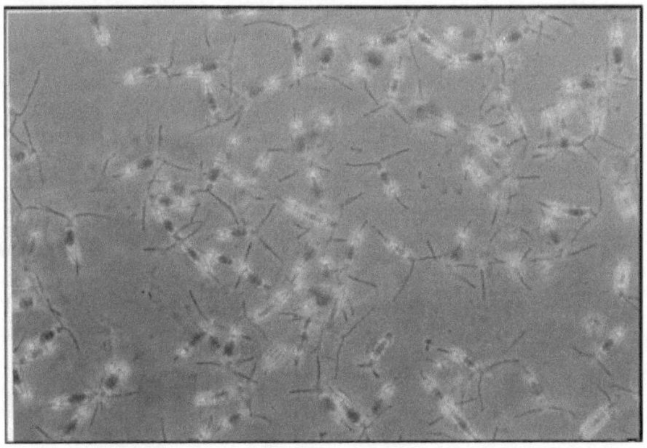

Figure 4.114 *Pestalotiopsis theae*; Mature conidia (× 400)

86. *Pestalotiopsis uvicola* (Spegazzini) Bissett

Source

Bissett., *Can. J. Bot.* **60**: 2752 (1982); Nag Raj, T.R., *Coelomycetous anamorphs with appendage bearing conidia,* Mycologue Publications, Waterloo, Ontario, Canada, pp. 662–663 (1993).

Description

Spots, round, blackish, 0.5–0.8 cm in diam, acervuli globose-lenticular, black, erumpent, 150–300 µm in diam. Conidia 5-celled, slender, fusiform to elliptic, 18-25 × 5.5–7.5 µm, erect or slightly curved, hardly constricted at septa; apical cells long conic to cylindric, bearing a crest of 3 rarely 2, filiform setulae or apical appendages usually up to 13 µm, sometimes up to 22 µm long, often distinct at base, divergent pedicels or basal appendages up to 7 µm long.

Specimen collected

On Leaves of *Cinnamomum wightii,* Kodaikanal, Tamilnadu, India, August 1995, collected by Murugan, IC No: 232.

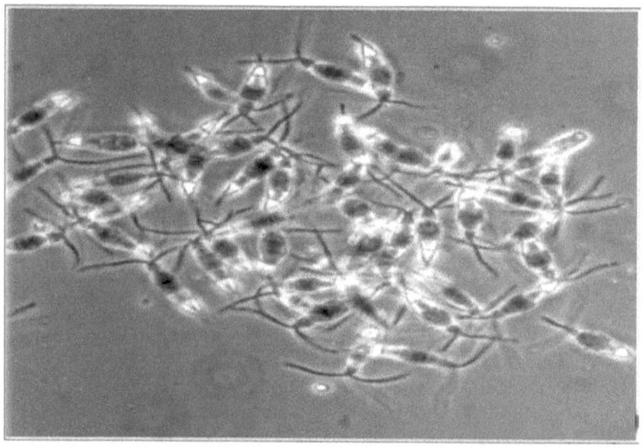

Figure 4.115 *Pestalotiopsis uvicola*; Mature conidia (×400)

87. *Pestalozziella parva* Nag Raj

Source

Nag Raj, *Trans. Br. Mycol. Soc.,* **52:** 205 (1969); Nag Raj, T.R., *Coelomycetous anamorphs with appendage bearing conidia,* Mycologue Publications, Canada, pp.673–676 (1993); R.S. Mathur, *The Coelomycetes of India,* p.159 (1979).

Description

Conidiomata pycnidial, separate, immersed, unilocular, 180–250 µm diam. Ostiole single, circular. Pycnidial wall made up of thick-walled, brown cells at the outside and thin-walled, hyaline cells at the innerside giving rise to conidiogenous cells. Conidiogenous cells lining the pycnidial cavity, holoblastic, sympodial, ampulliform, hyaline, smooth, 6–10 × 2–3 µm. Conidia hyaline to very pale brown, 5–15 × 1.5–3.5 µm, smooth, straight with 3–4 subapical, dichotomously branched, cellular appendages of 18 µm long.

Specimen collected

On dead leaves of *Terminalia arjuna* W. & A., Theosphical society, Madras, Tamilnadu, Jul. 1988, collected by J. Muthumary, IC No: 159.

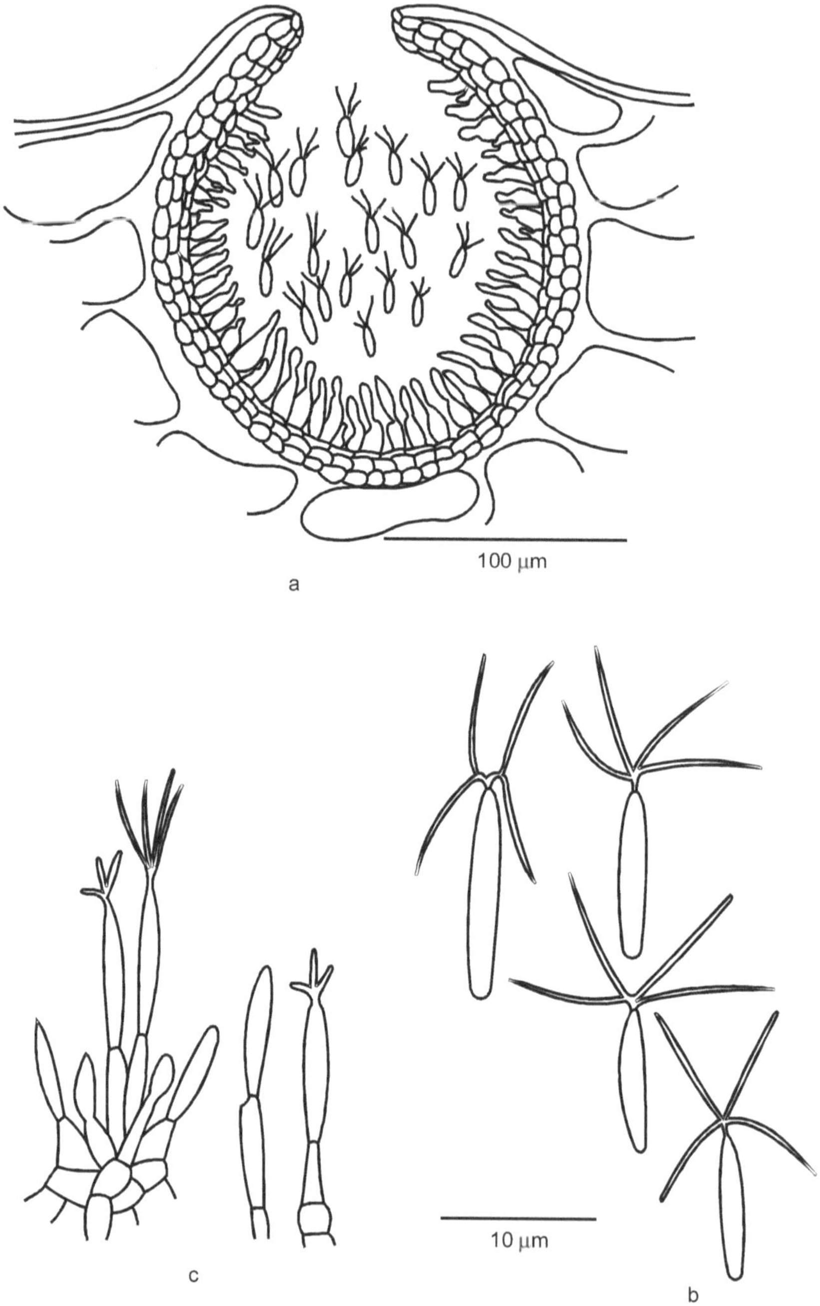

Figure 4.116 *Pestalozziella parva*, (a) V.S. of conidioma; (b) Mature conidia; (c) Developing conidia.

88. *Phaeocytostroma sacchari* (Ell. & Ev.) Sutton

Source

Sutton, B.C., *Mycol. Pap.*, **97**: 26 (1964). Sutton, B.C., *The Coelomycetes*, England, p. 631 (1980).

Description

Conidiomata pycnidial, 350–650 µm diam. Pycnidial wall made up of thick-walled, brown cells at the outside and thin-walled, hyaline cells at the innerside giving rise to conidiogenous cells. Conidiogenous cells formed all around the pycnidial cavity along with the paraphyses –like structures lining the pycnidial wall, smooth, cylindrical, 20–24 × 2–3 µm. Conidia cylindrical, single celled, pale brown, apex obtuse, base truncate, 10–15 × 3–5 µm. Paraphyses–like structures hyaline, smooth, branched, septate, 15–40 × 2.4 µm.

Specimen collected

On dead twigs of *Bambusa* sp., Kodaikanal, Tamilnadu, Nov. 1987, collected by J. Muthumary, IC. No: 133.

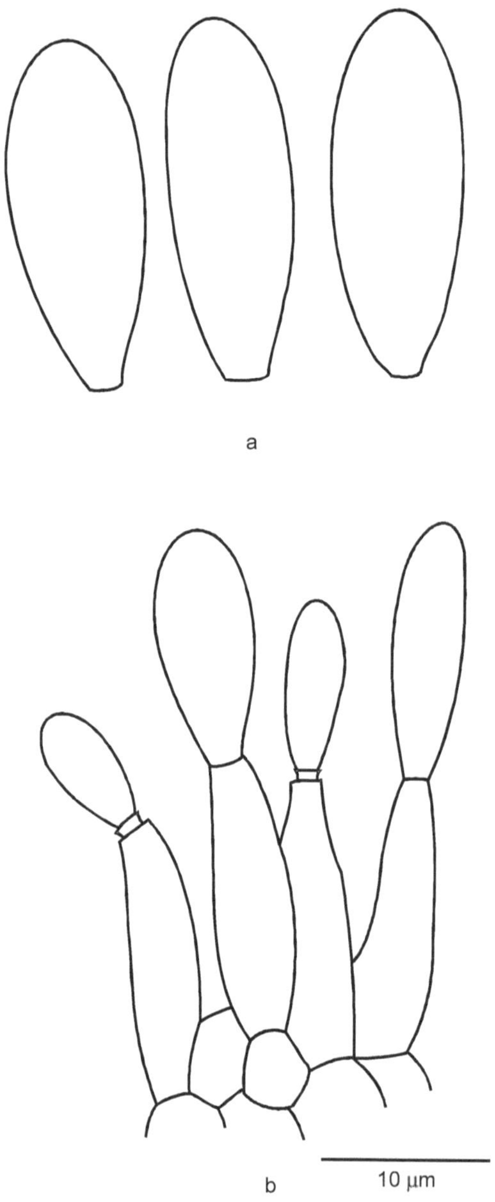

Figure 4.117 *Phaeocytostroma sacchari*, (a) Mature conidia; (b) Developing conidia.

89. *Phloeospora maculans* (Berg.) Allesch

Source

Allesch., In Rabenh., *Krypt. Fl.,* **6**: 935 (1900); Sutton, B.C., *The Coelomycetes,* CMI, England, pp. 309–311 (1980).

Description

Lesions on leaves circular, white with brown mycelia and later becoming irregular, 2–3 cm in diam. Conidiomata acervular, circular, subepidermal, discrete, composed of pale brown to hyaline, smooth pseudoparenchymatous cells, 300–350 µm in diam. Conidiogenous cells lining the cavity of the acervular, ampuliform to cylindrical, with 1–4 apical annellations, hyaline, smooth, 10–12 × 2–3 µm, Conidia straight to slightly curved, 3-septate, cylindrical, hyaline to pale brown, smooth, tapered towards the obtuse apex, 20–30 × 3–4 µm.

Specimen collected

On living leaves of *Ficus benghalensis,* Guindy campus, University of Madras, Chennai, November 2004, collected by J. Muthumary, IC. No: 290.

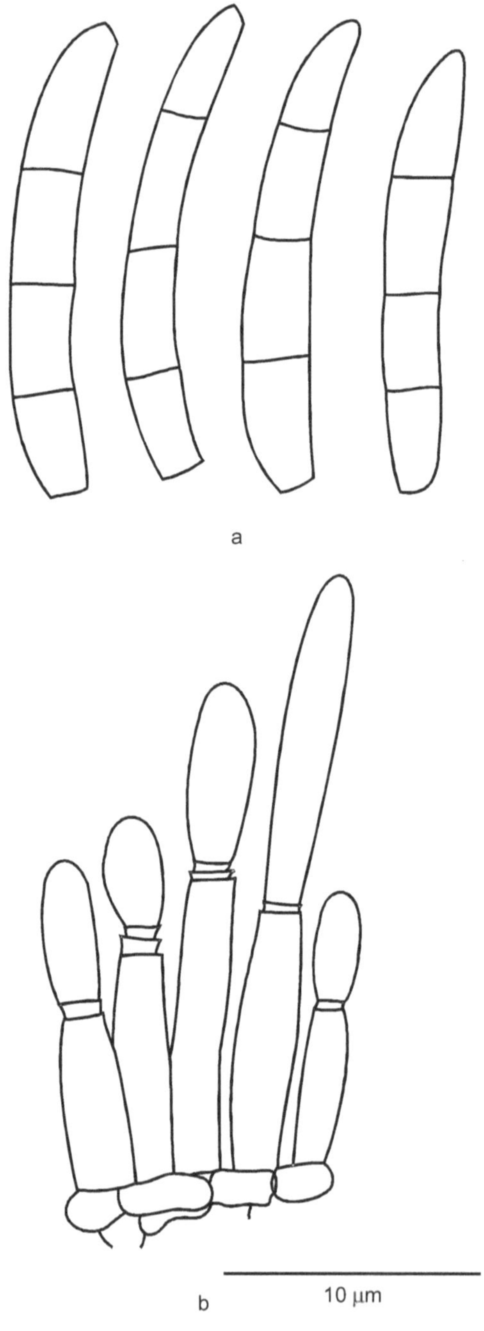

Figure 4.118 *Phloeospora maculans*, (a) Mature conidia; (b) Developing conidia.

90. *Phloeosporella leucosceptri* (Keissl.) Sutton

Source

Sutton, B.C., *The Coelomycetes,* CMI, England, pp. 241–246 (1980).

Description

Lesions on leaf brown, circular, epiphyllous, up to 1 cm in diam. Conidiomata acervular, pale white, amphigenous to epiphyllous, 100–200 μm in diam. dehiscence by irregular tears. Conidiogenous cells holoblastic, sympodial, hyaline, smooth, lageniform to subcylindrical, 25–30 × 3–3.5 μm. Conidia hyaline, scolecospores, thin walled, smooth, straight to slightly curved, 3–7–septate, tapered gradually with truncate base, 40–55 × 3–4 μm.

Specimen collected

Type specimen of *Septoria gurezae.* Mirakhur, Tahir IMI No. 205047; collected by R.K. Mirakhur on leaves of *Acer caesium,* June 1976. Identified as *Septoria gurezae.*

The above fungus is repositioned in the genus *Phloeosporella* because of the acervular conidiomata with irregular dehiscence.

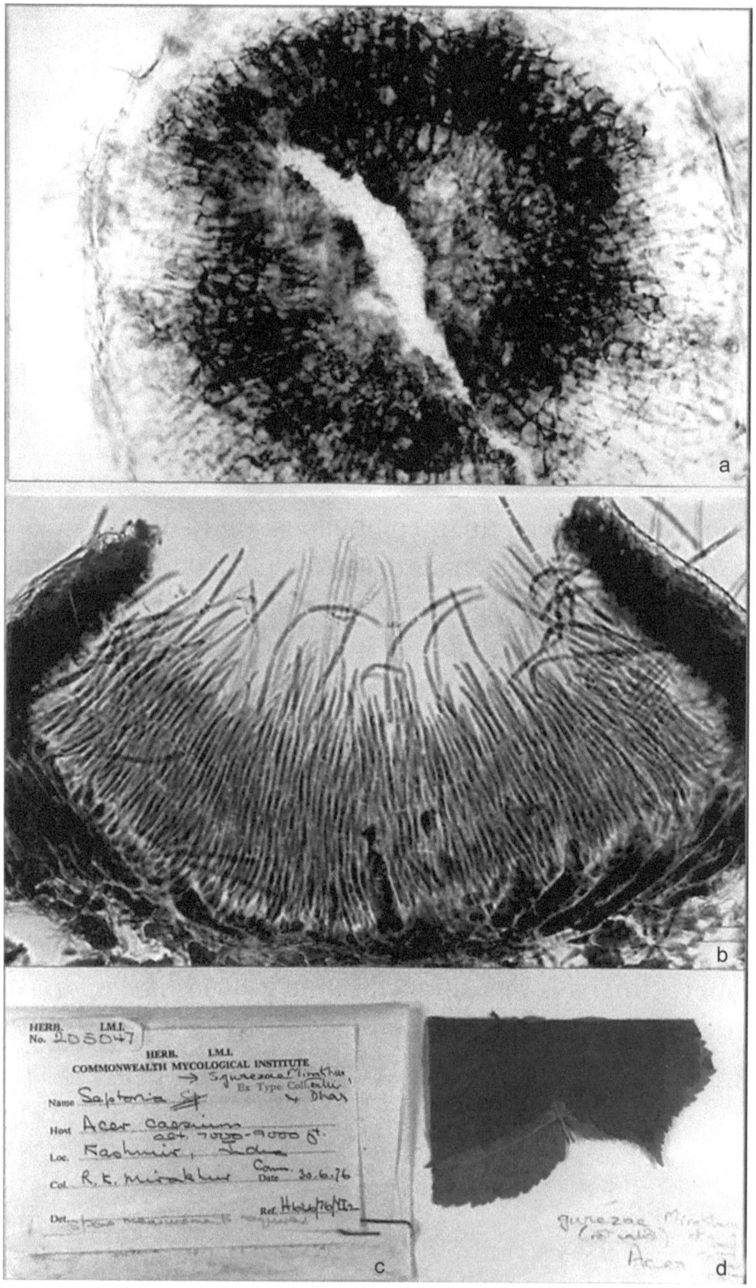

Figure 4.119 *Phloeosporella leucosceptri*; (a) Surface view of conidioma (×260); (b) V.S. of conidioma (×260); (c) Herbarium of septoria gurezae; (d) Acer leaf

91. *Phoma glomerata* (Cda) Wollenw. & Hochapt.

Source

Wollenw & Hochapt. Z., *Parasitkade,* **8:** 592 (1936); Padma & Mukerji, *Indian Phytopath,* **25:** 104–197 (1972); Sutton, B.C., *The Coelomycetes,* CMI, England, p. 385 (1980).

Description

Conidiomata pycnidial, immersed or semi-immersed, globose, brown, separate or aggregated, ostiolate, 200–250 µm diam. In vertical section pycnidium unilocular, wall composed of thin-walled, pale brown pseudoparenchma. Conidiogenous cells formed from the inner cells of the pycnidium, smooth, phialidic, hyaline, smooth, ampulliform to subcylindircal, 5–8 × 2–3 µm with well-defined collarette. Conidia hyaline, smooth, one-celled, ellipsoid to cylindrical, often gutulate, 7–8 × 3–4 µm.

Recorded as endophtye in

Isolated as endophyte in *Gloriosa superba, Araucaria columanaris,* K.P. Kannan, 2002, Ph.D. Thesis, "Biodiversity of Endophytic Mycobiota from selected Medicinal plants and some Gymnosperms", MUBL–553. *Justicia gendarussa, Ocimum basilicum,* V. Gangadevi, 2005, Ph.D. Thesis, "Studies on Fungal Endophytes from some medicinal plants and screening selected coelomycetes for taxol production". MUBL No: 687.

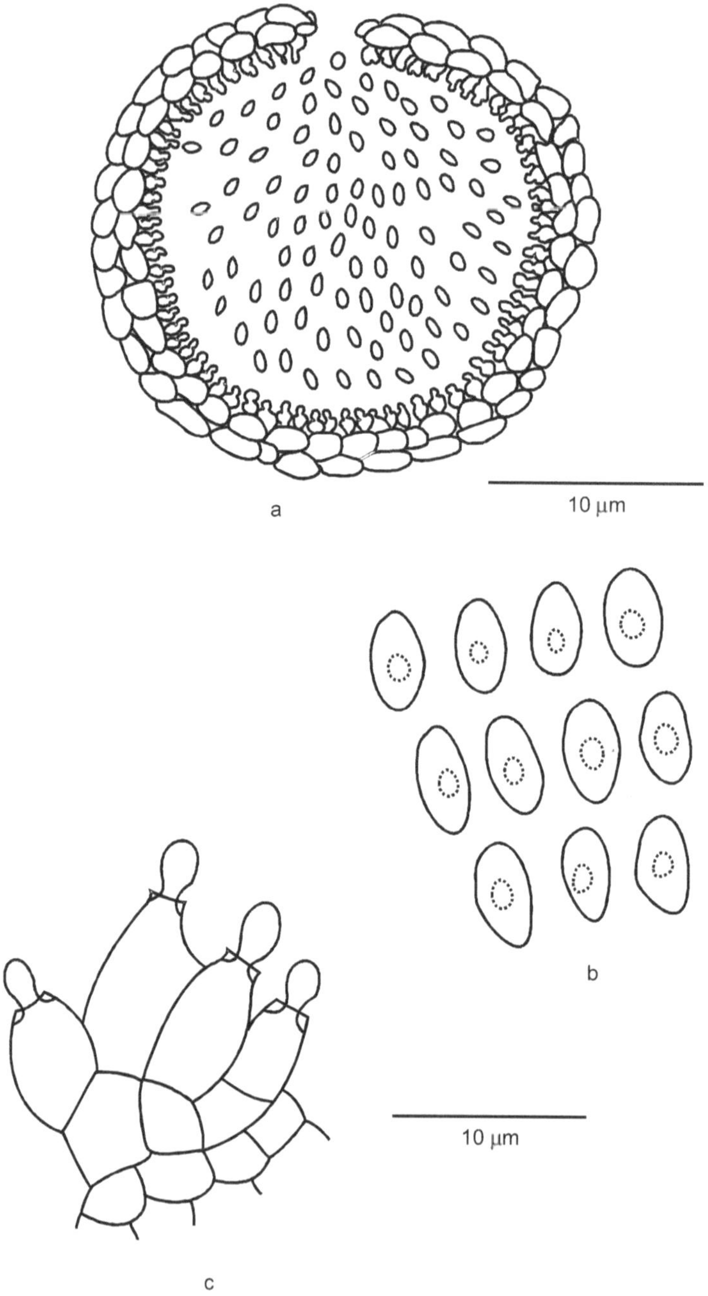

Figure 4.120 *Phoma glomerata*, (a) V.S. of conidioma; (b) Mature conidia; (c) Developing conidia.

92. *Phoma* sp.

Source

Frank, Z., *Ver. Rubenz. Indus. Dtsch. Reichs.*, **42**: 904 (1892); Agarwal, V.K. & O.V. Singh, *Indian Phytopath.*, **26:** p. 741(1973); R.S. Mathur, *The Coelomycetes of India*, pp.170–173 (1979); Sutton, B.C., *The Coelomycetes*, CMI, England, p. 380 (1980).

Description

Colonies olivaceous, aerial mycelium immersed, branched, septate, hyaline, pale brown, felty. Conidiomata pycnidial, immersed, erumpent, unilocular, brown, globose 100–200 µm in diam. Ostiole single. Pycniidal wall comprising an outer layer of brown, thick-walled cells and inner layer of hyaline, thin-walled cells. Conidiogenous cells enteroblastic, Phialidic, hyaline, ampulliform, smooth 5–7 × 2–3 µm. Conidia single-celled, hyaline, smooth, globose to allantoid, straight, slightly curved, ellipsoid with two large guttules.

Recorded as endophyte in

Isolated as endophyte in *Santalum album,* K.P. Kannan, 2002, Ph.D. Thesis, "Biodiversity of Endophytic Mycobiota from selected Medicinal plants and some Gymnosperms", MUBL – 550; MTCC – 4272; *Mimusops elengi, Ocimum basilicum, Vitex negundo,* V. Gangadevi, 2005, Ph.D. Thesis, "Studies on Fungal Endophytes from some medicinal plants and screening selected coelomycetes for taxol production". MUBL No: 696.

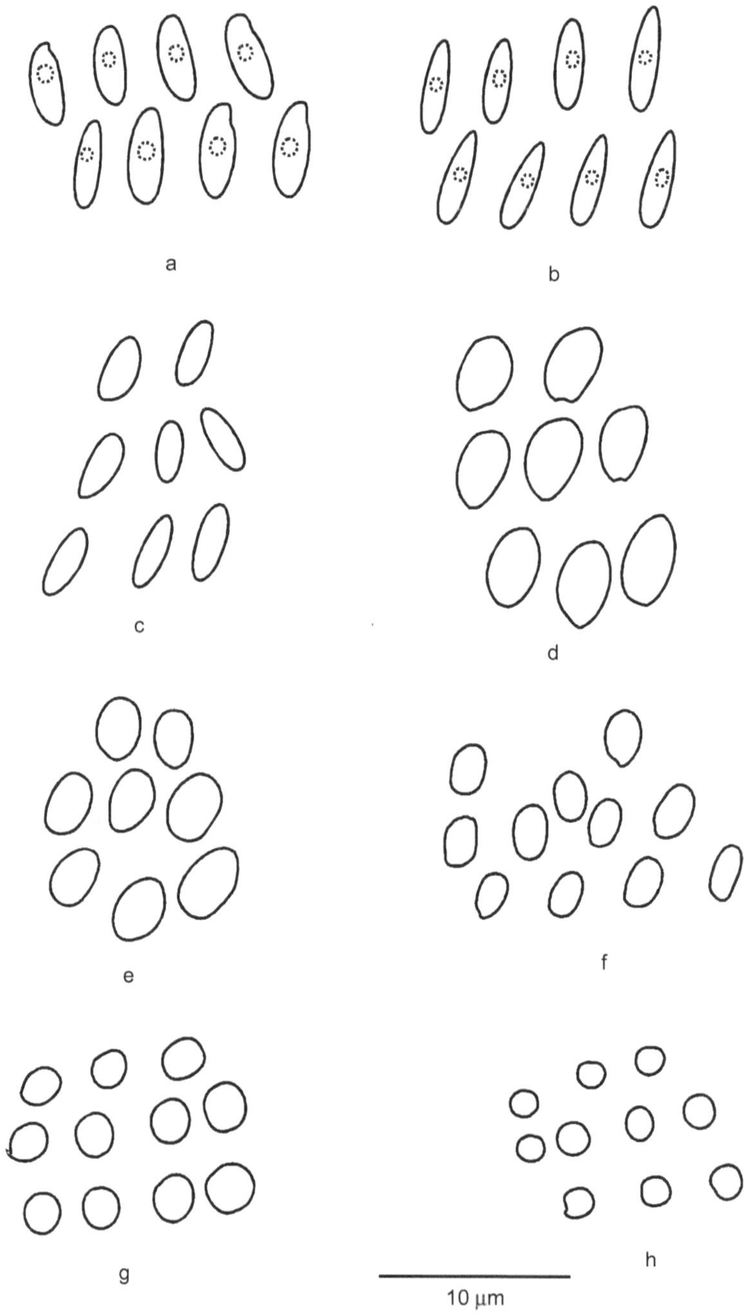

Figure 4.121 *Phoma* sp., (a–h) Mature conidia.

93. *Phomopsis abdita* Sacc.

Source

Sacc., *Trav., F. It. Crypt.* (Pyrenomycetes) **2**(1): 237 (1906); Sutton, B.C., *The Coelomycetes,* CMI, England, pp. 570–571 (1980).

Description

Conidiomata eustromatic, immersed, dark brown, separate, globose, unilocular, thick-walled, 248–320 μm diam. Ostiole single, circular. Conidiogenous cells enteroblastic, phialidic, determinate, 28–30 × 2–3 μm. α–Conidia hyaline, fusiform, biguttulate, 8–10 × 1.6–2 μm. β–Conidia absent.

Specimen collected

On phyllode of *Cactus indicus* Roxb. On dead twigs of *Caesalpenia pulcherrima* Sw., Madras, Tamilnadu, May 1988; on dead leaves of *Guazuma tomentosa* kunth., Madras, Tamilnadu, Sep. 1989, Collected by J. Muthumary.

Recorded as endophyte in

Isolated as enophyte in *Madhuca indica, Melia azedarach,* K.P. Kannan, 2002, Ph.D. Thesis, "Biodiversity of Endophytic Mycobiota from selected Medicinal plants and some Gymnosperms", MUBL–558; MTCC–4275. *Psidium guajava,* V. Gangadevi, 2005, Ph.D. Thesis, "Studies on Fungal Endophytes from some medicinal plants and screening selected coelomycetes for taxol production". MUBL No: 690.

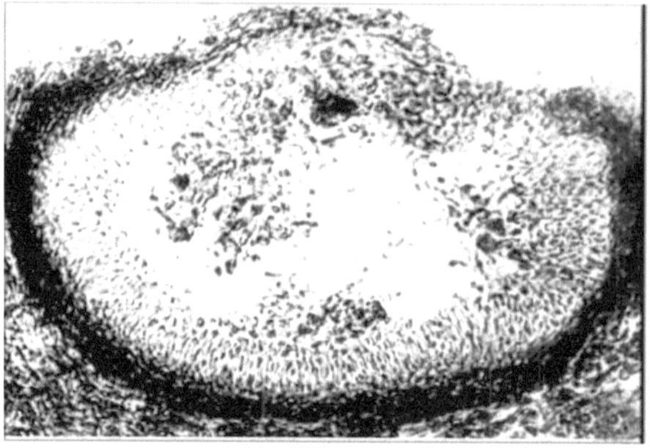

Figure 4.122 *homopsis abdita*; (a) V.S. of conidioma (×200);

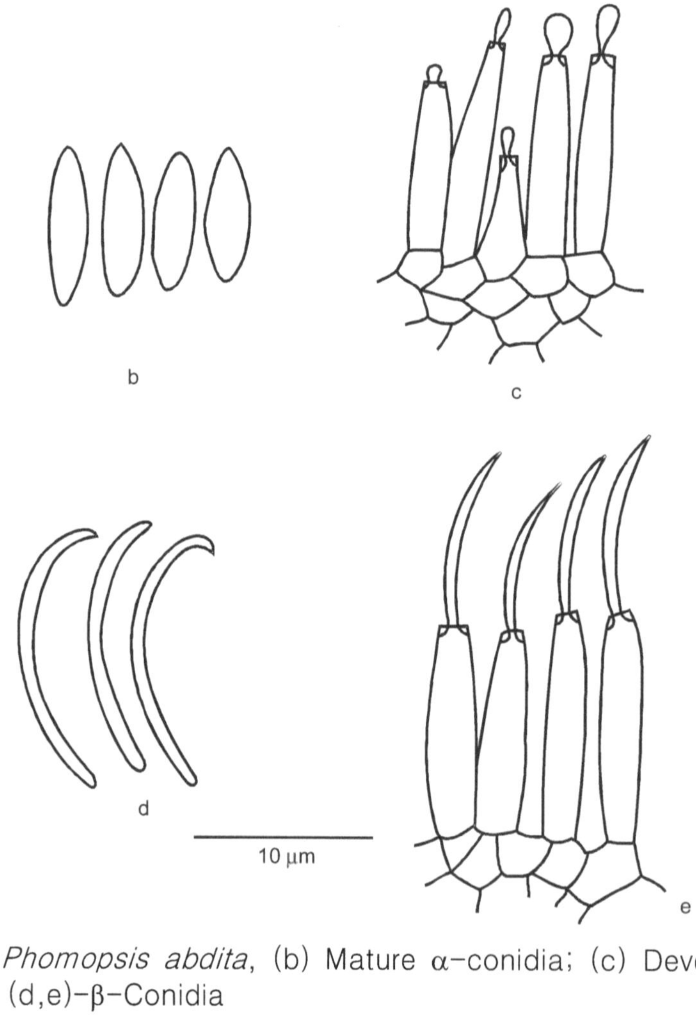

Figure 4.123 *Phomopsis abdita*, (b) Mature α-conidia; (c) Developing conidia; (d,e)-β-Conidia

94. *Phomopsis archeri* Sutton

Source

Sutton, B.C., *The Coelomycetes*, CMI, England, p. 571 (1980).

Description

Conidiomata 112–160 μm. Conidiogenous cells 12–15 × 2–3 μm. α– 6–7 × 1.5–2 μm; β–conidia 12.5–19.5 × 1.6–2 μm.

Specimen collected

On dead twigs of *Eucalyptus teretecornis* Sm. Madras, Tamilnadu, Mar. 1990, collected by J. Muthumary.

Recorded as endophyte in

Isolated as endophyte in *Withania somnifera, Wrightia tinctoria,* V. Gangadevi, 2005, Ph.D. Thesis, "Studies on Fungal Endophytes from some medicinal plants and screening selected coelomycetes for taxol production". MUBL No: 559; MTCC N0: 4276.

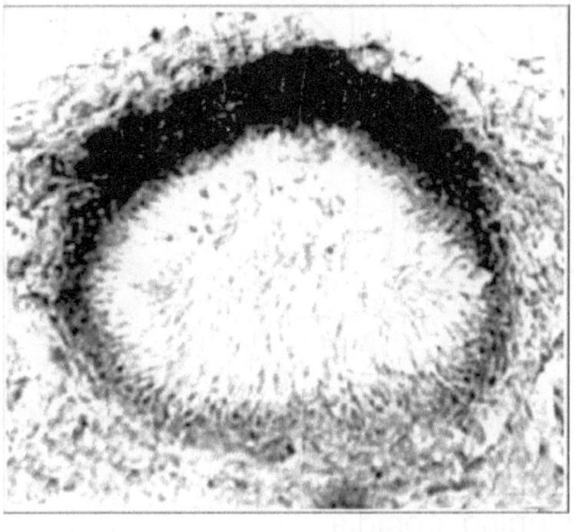

Figure 4.124 *Phomopsis archeri*; V.S. of conidioma (× 200).

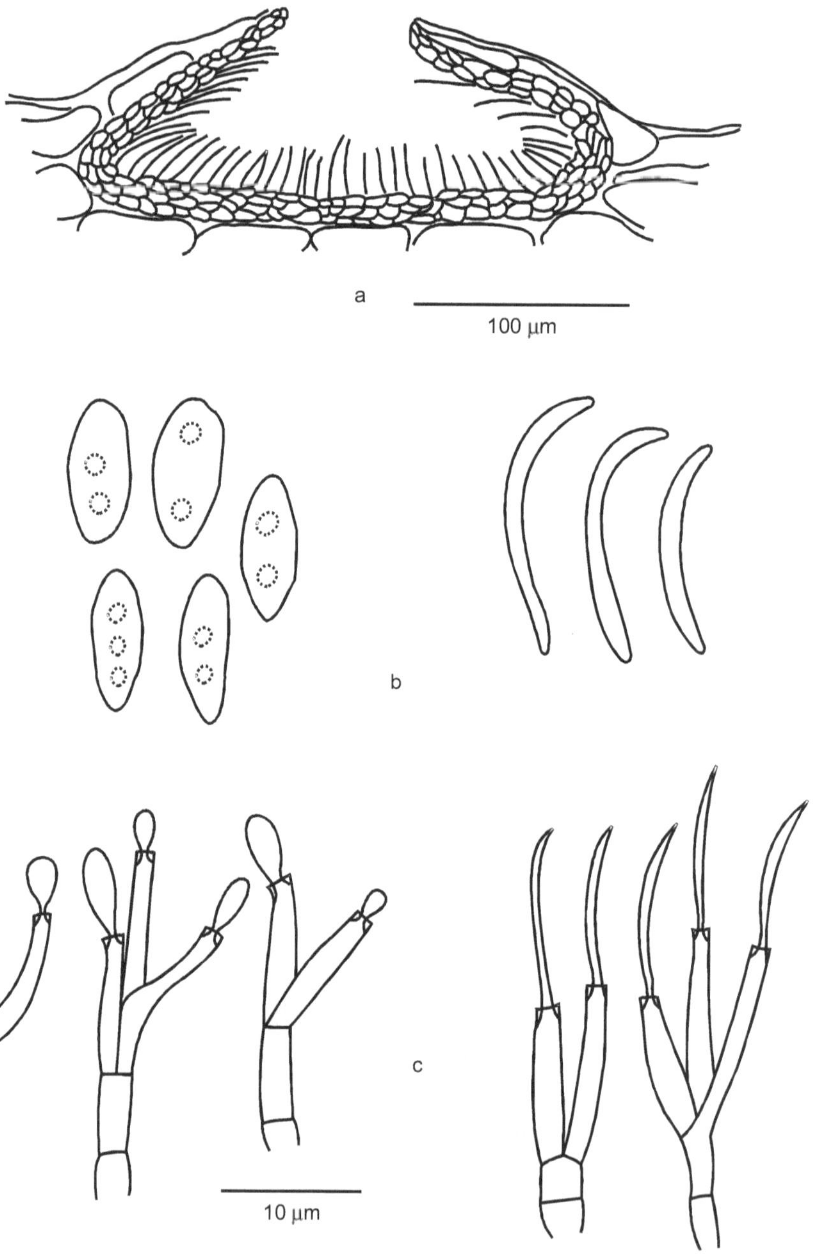

Figure 4.125 *Phomopsis archeri*, (a) V.S. of conidioma; (b) Mature conidia; (c) Developing conidia.

95. *Phomopsis dracaenae* Sahni

Source

Sahni, *Mycopath. Mycol. Appl.*, **36**: pp. 267–288 (1965); R.S. Mathur, *The Coelomycetes of India*, p.196 (1979).

Description

Conidiomata 216 µm diam. α–conidia aseptate, 9.8 × 2–2.5 µm. β–conidia 10–25 × 1.5–2 µm.

Specimen collected

On litter of *Dracaena terminalis,* West Madras, Tamilnadu, Jan. 1989, collected by J. Muthumary, IC. No: 183.

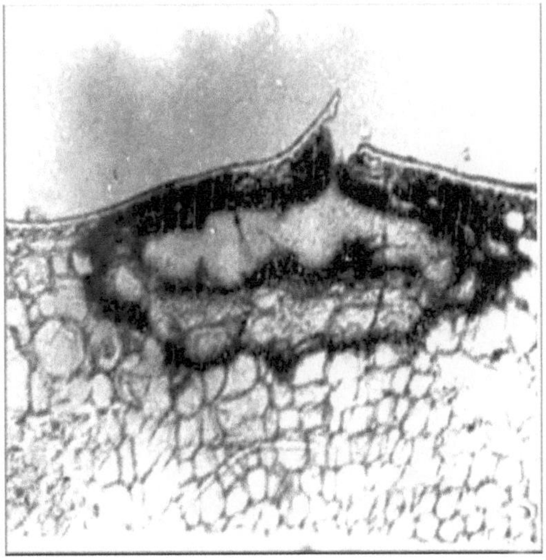

Figure 4.126 *Phomopsis dracaenae*; V.S. of conidioma (× 200).

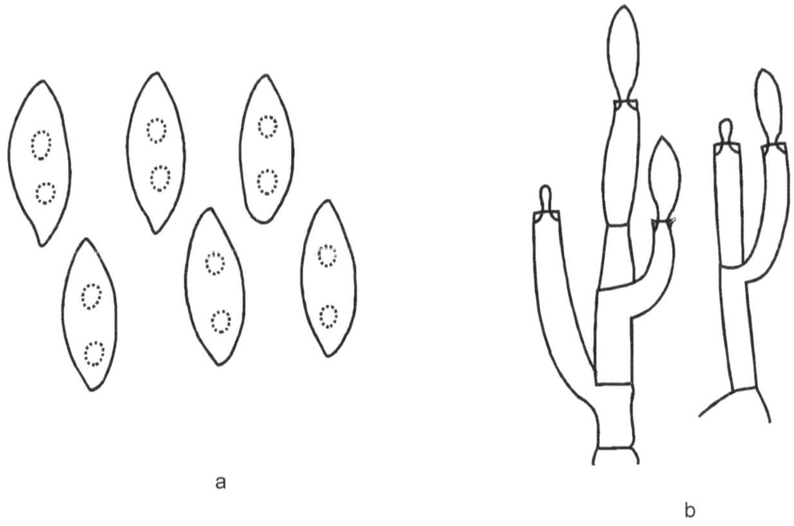

Figure 4.127 (a) *Phomopsis dracaenae*, Mature conidia; (b) Developing conidia.

96. *Phomopsis eugeniae* Punithalingam

Source

Punithalingam, E., *Trans. Br. Mycol. Soc.*, **63** (2): pp. 229–236 (1974); Shreemali, J.L., *Curr. Sci.*, **41**: p. 39 (1972); R.S. Mathur, *The Coelomycetes of India*, p.196 (1979).

Description

Conidiomata eustromatic. α–conidia 6 × 2–3.5 μm. β–conidia 15–20 × 0.5–1 μm.

Specimen collected

On dead twigs of *Eugenia jambolana* Lam., Madras, Tamilnadu, Nov. 1988, collected by J. Muthumary, IC No: 160.

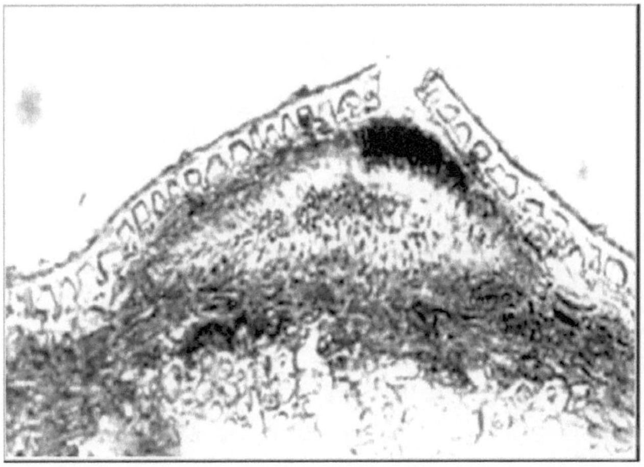

Figure 4.128 *Phomopsis eugeniae*; V.S. of conidiomata (×200).

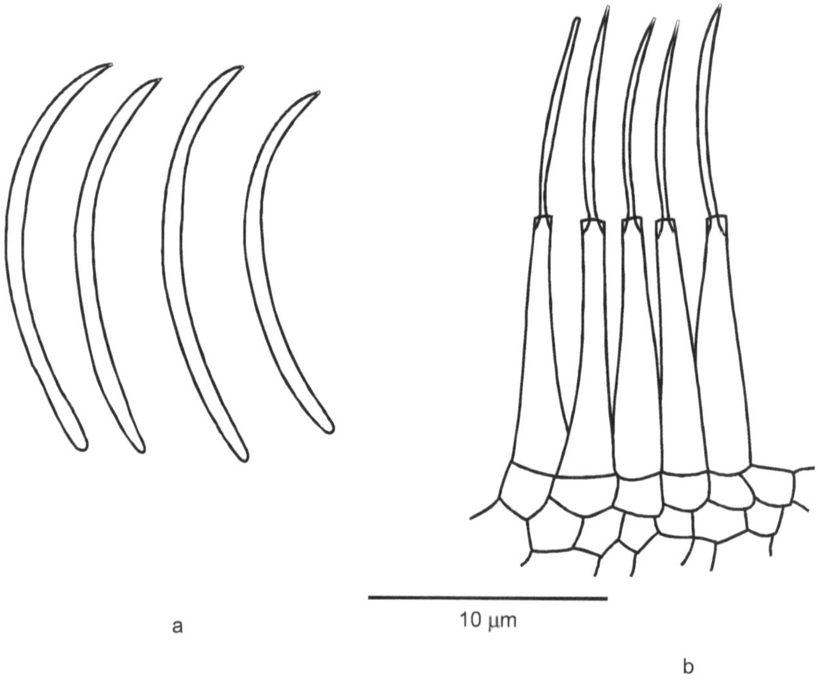

Figure 4.129 *Phomopsis eugeniae;* (a) Mature conidia; (b) Developing conidia

97. *Phomopsis lantanae* (Costa & Camara) Sutton

Source

Sutton, B.C., *The Coelomycetes*, CMI, England, p. 571 (1980).

Description

Conidiomata 160–180 µm diam. Conidiogenous cells 15 × 3 µm. α–conidia 7–10 × 1.6–3.2 µm. β–conidia absent.

Specimen collected

On dead twigs of *Pongamia glabra* Vent., Madras, Tamilnadu, Jan. 1989; On dead leaves of *Ficus religiosa* L., Madras, Tamilnadu, Sep. 1990, collected by J. Muthumary, IC No: 184.

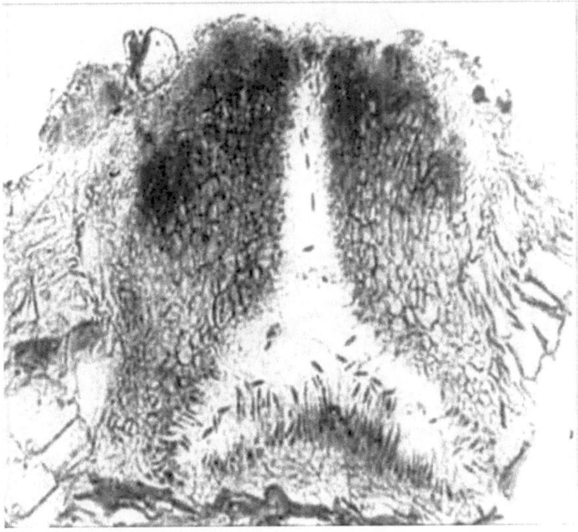

Figure 4.130 *Phomopsis lantanae*; V.S. of conidioma (× 200).

98. *Phomopsis mangiferae* Ahmad

Source

Petrak, F. and Ahmad, *Sydowia,* **8**: p. 183 (1953).

Description

Conidiomata 95–128 μm diam. Conidiogenous cells 20–25 × 2–3 μm. α–conidia 5–7 × 1.5–2.5 μm. β–conidia absent.

Specimen collected

On dead leaves of *Mangifera indica* L., Madras, Tamilnadu, Apr. 1991, collected by J. Muthumary, IC No: 224.

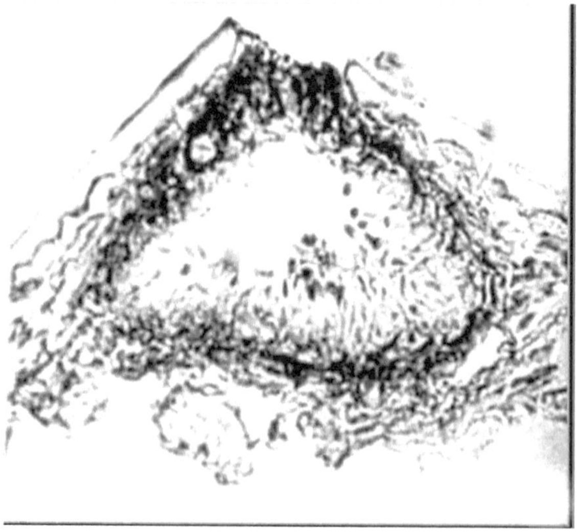

Figure 4.131 *Phomopsis mangiferae*; V.S. of conidioma (× 200).

99. *Phomopsis obscurans* (Ell. & Ev.) Sutton

Source

Sutton, *Trans. Br. Mycol. Soc.*, **48**: 615 (1965); Sutton, B.C., *The Coelomycetes*, CMI, England, p. 571 (1980).

Description

Conidiomata 140–200 µm diam. Condiogenous cells 10–15 × 2–3 µm. α-conidia 6.5–10 × 1.5–3.2 µm. β–conidia absent.

Specimen collected

On dead leaves of *Lagerstroemia microcarpa* West Madras, Tamilnadu, Sep. 1988; On dead leaves of *Tamarindus indica* L. Madras, Tamilnadu, Apr. 1991, collected by J. Muthumary, IC No: 420.

Recorded as endophyte in

Isolated as endophyte in *Curcuma longa, Vitex negundo,* K.P. Kannan, 2002, Ph.D. Thesis, "Biodiversity of Endophytic Mycobiota from selected Medicinal plants and some Gymnosperms", MUBL- 562; *Curcuma longa, Vitex negundo,* V. Gangadevi, 2005, Ph.D. Thesis, "Studies on Fungal Endophytes from some medicinal plants and screening selected coelomycetes for taxol production".

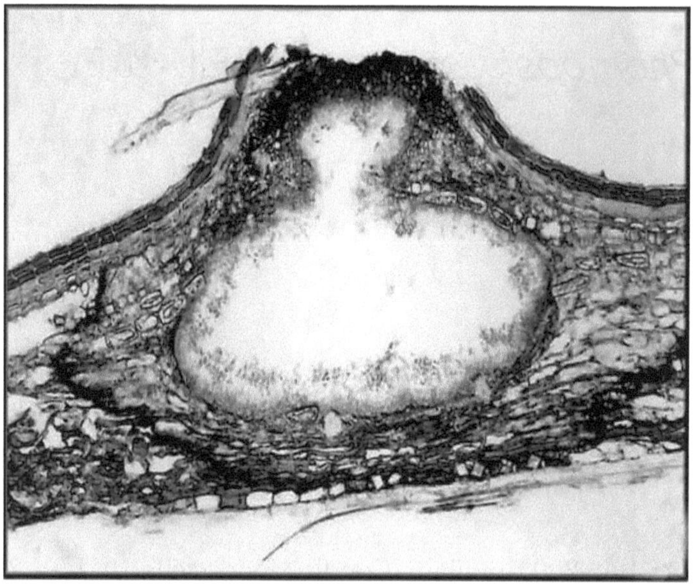

(a)

Figure 4.132 *Phomopsis obscurans*; (a) V.S. of conidioma (× 200).

100. *Phomopsis terminaliae* (P. Henn) Sutton

Source

Sutton, B.C., *The coelomycetes*, CMI, England, p. 573 (1980).

Description

Conidiomata 80–190 μm diam. Conidiogenous cells 20 μm long. α–conidia 8–10 × 3–3.2 μm. β–conidia absent.

Specimen collected

On dead twigs of *Caesalpenia pulcherrima* Sw., Madras, Tamilnadu, Sep. 1988; On dead twigs of *Cassia fistula* L., Madras, Tamilnadu, Dec. 1990, collected by J. Muthumary, IC No: 421.

Recorded as endophyte in

Isolated as endophyte in *Terminalia arjuna,* K.P. Kannan, 2002, Ph.D. Thesis, "Biodiversity of Endophytic Mycobiota from selected Medicinal plants and some Gymnosperms", MUBL–565; MTCC–4277; *Cardiospermum helicacabum, Terminalia arjuna,* V. Gangadevi, 2005, Ph.D. Thesis, "Studies on Fungal Endophytes from some medicinal plants and screening selected coelomycetes for taxol production". MUBL No: 694.

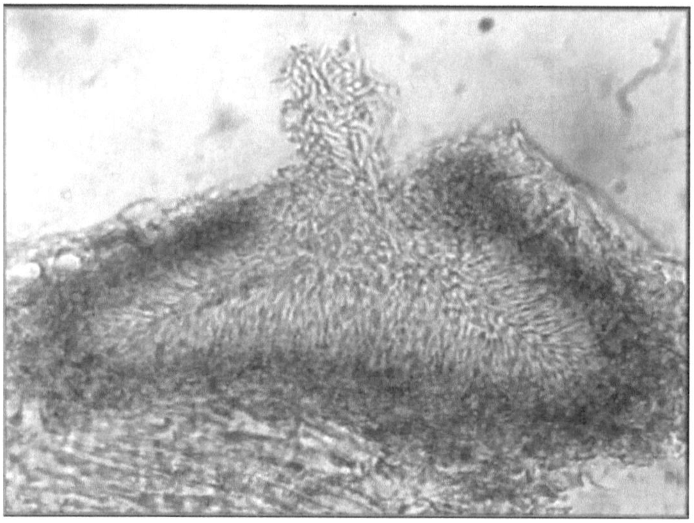

(a)

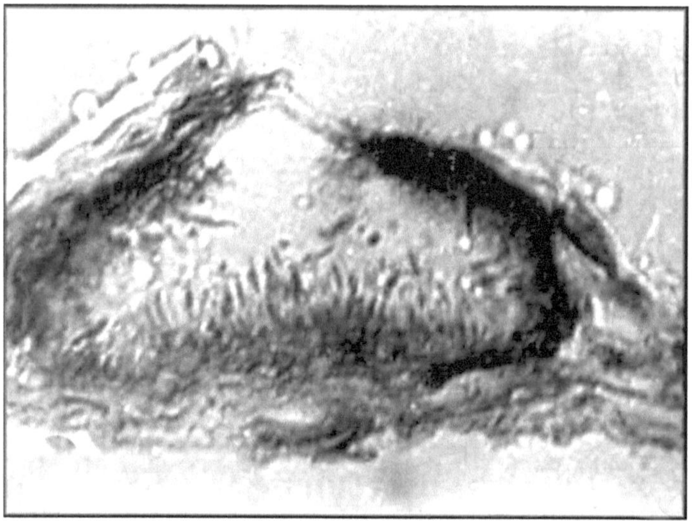

(b)

Figure 4.133 *Phomopsis terminaliae*; (a–b) V.S. of conidiomata (×200).

101. *Phomopsis tersa* (Sacc.) Sutton

Source

Sutton, B.C., *The Coelomycetes*, CMI, England, p.573 (1980).

Description

Mycelium immersed, branched, septate, hyaline, pale brown. Conidiomata up to 650 µm diam., flat, multilocular. Conidiophores filiform to subulate, branched and septate only at the base, up to 15 µm long. Conidia: a - conidia hyaline straight, cylindrical, obtuse at each end, sometimes slightly constricted in the middle, prominently biguttulate, 6.5 – 7.5 x 2.5 µm; β-conidia absent.

Recorded as endophyte in

Isolated as endophyte in *Erythroxylum monogynum*–Stem; *Glycosmis pentaphylla*–Leaf midrib, Petiole and Stem, Nov. 1988, collected by S. Masilamani, MUBL No: 566; MTCC No: 4276.

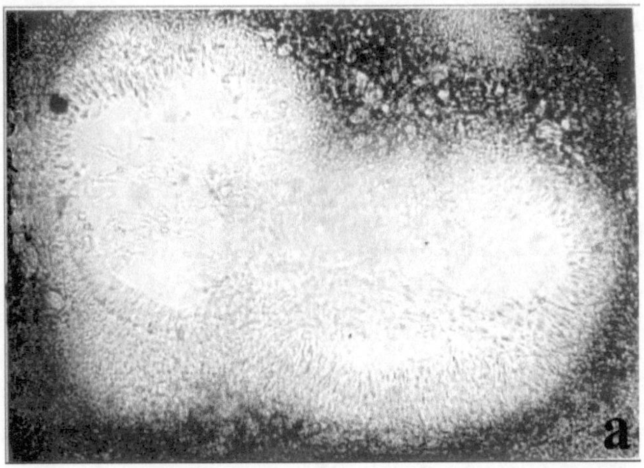

Figure 4.134 *Phomopsis tersa*; V.S. of conidioma (×200).

102. *Phomopsis* sp.

Source

Description

Mycelium bright white to turning grey, conidiomata stromatic, 2–3 mm diameter, dark brown, erumpent or immersed, scattered, confluent, unilocular or multilocular. Ostiole absent. In multilocular forms each locule separated from each other by pseudoparenchymatous tissue. Conidiogenous cells, phialidic, hyaline, cylindrical, 8–11.2 × 1–3 μm, tapering and rounded apices.

Parasite on *Solanum* fruits form pustular growth on the pericarp. Conidiomata eustromatic, dark brown erumpent, unilocular single ostiolate, papillate, the inner wall of conidiomata is lined with hyaline conidiogenous cells. α-conidia hyaline, unicellular 8 × 2 μm. β-conidia absent.

Specimen collected

On litter of unidentified leaves Kodaikanal, Tamil Nadu, Jan 1989, Collected by J. Muthumary. IC. No: 521.

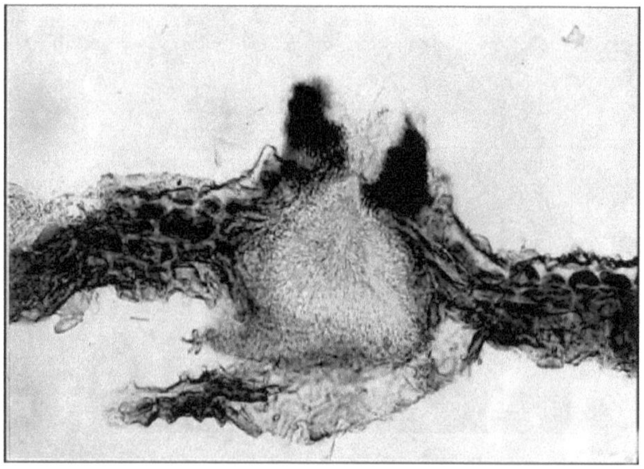

Figure 4.135 *Phomopsis* sp; (b) V.S. of conidioma (×200).

103. *Phyllosticta arecae* Hohnel

Source

Van der Aa, *Stud. Mycol. Baarn,* **5**: p. 33, (1973); *P. arecae* Died., *Ann. Mycol.,* **14**: p. 185 (1916); Sacc., **25**: p. 661 (1931); Vasant Rao, *Mycopath. Mycol. Appl.,* **28**: p. 192 (1966); Seshadri *et al., Indian Phytopath.* **25**: p. 250 (1972); R.S. Mathur, *The Coelomycetes of India,* pp.203–204 (1979); Ali and Saikia, *Inidan Phytopath.,* **50**: p. 200 (1997).

Description

Conidiomata spherical to subspherical, pycnidial, immersed to erumpent, solitary or aggregated, ostiolate. Pycnidial wall composed of many layers of thick-walled, pseudoparenchymatous cells, the outer most layers brown to black, the inner layers hyaline from which the conidiophores arise. Conidiogenous cells short cylindrical or conical, 5.6–11 × 2–3 µm. Conidia one-celled, sometimes remaining in chains, ovoidal, ellipsoidal or short cylindrical, seldom almost globose, when young somewhat clavate with a truncate base, broadly rounded apically and slightly indented, 7.5–14 × 5–8.5 µm (X = 9.36 × 6.78 µm), surrounded by a thick slime layer, containing a dense mass of coarse, greenish guttules and often a large, bright vacuole, with a distinct apical appendage, 3–11 µm long, slightly tapering and disappearing on the older conidia.

Specimen collected

On the leaves of *Caryota urens* (Arecaceae) Mettupalayam, Ooty, Tamil Nadu, April 2002, collected by R. Senthil Kumaran. MUBL No: 593 and MTCC No: 3514.

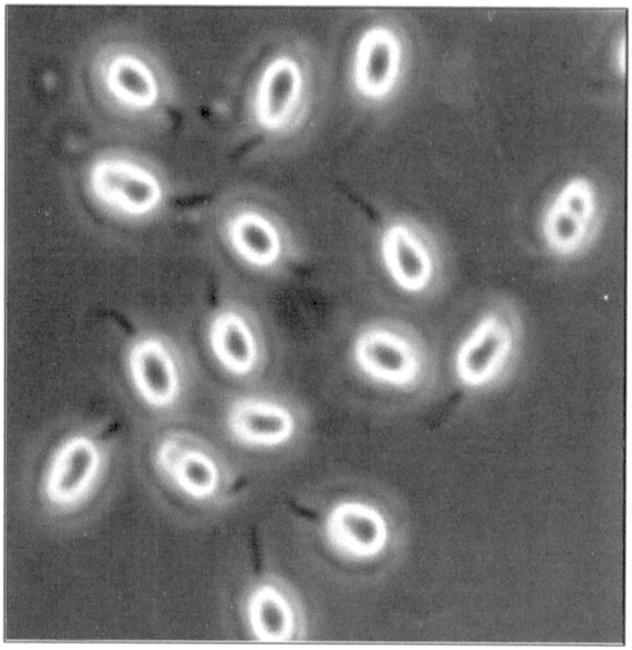

Figure 4.136 *Phyllosticta arecae*; Mature conidia (×900).

104. *Phyllosticta amaryllidicola* Van der Aa

Source

Van der Aa, *Stud. Mycol. Baarn,* 5: p. 27 (1973).

Description

Foliicolous. Infectious near the tip of the leaf, irregular, later coalescing to form large necrotic patches, reddish brown later becoming brownish grey, covered by a brownish yellow border. Pycnidia epiphyllous, sub-epidermal, unilocular, 100–160 µm in diameter, globose or pyriform, with usually well developed stroma in the upper region, composed of subhyaline to brownish pseudoparenchyma, ostiole circular, 7–12 µm wide. Pycnidial wall 2–4 cells (about 14 µm) thick, composed of isodiametrical or rounded, thick-walled, brownish cells, 8–14 µm in diameter and flattened towards the outer surface but hyaline towards the conidiogenous region and isodiametrical or flattened. Conidia 1-celled, hyaline, spherical to subspherical, 8.4–1.5 × 5–7 µm., enveloped by mucilage.

Specimen examined

On the diseased leaves of *Amaryllis belladonna* (Amaryllidaceae) Chennai, Tamil Nadu, collected by R. Senthil Kumaran. MUBL No: 594; IC. No: 522.

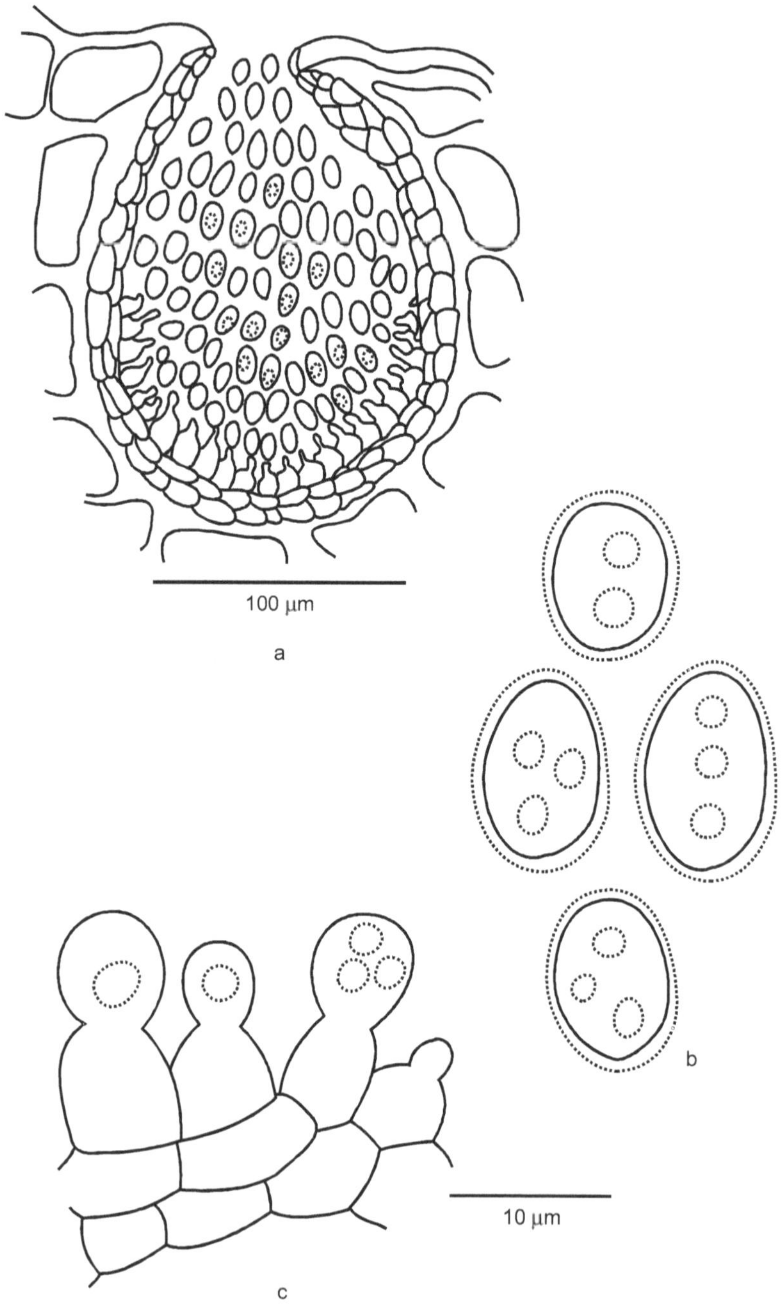

Figure 4.137 *Phyllosticta amaryllidicola*, (a) V.S. of conidioma; (b) Mature conidia; (c) Developing conidia.

105. *Phyllosticta artocarpina* Sydow & Butler

Source

Sydow & Butler, *Annls. Mycol.*, **14**: p. 180 (1916); Syd., *Feddes Repert. Beih.*, **42**: p. 188 (1927); Van der Aa, *Mycol. Baarn,* **5**: p. 35 (1973); R.S. Mathur, *The Coelomycetes of India*, p. 223 (1979).

Description

Foliicolous. Leaf spots subcircular to linear up to 3mm in diameter, later coalescing to form large necrotic area, grey to greyish brown with dark brown margin. Pycnidia epiphyllous, subepidermal, single, unilocular, 130–190 µm in diameter, without distinct stroma, flat or slightly papillate, with a 9–14 µm wide pore. Pycnidial wall 1–4 cells (13–23 µm) thick, composed of irregularly shaped, dark brown cells, darker and thick-walled around the pore, firmly grown together with the epidermis in the upper region, hyaline, isodiametrical or flattened towards the conidiogenous region. Conidiogenous cells lining the pycnidial cavity, cylindrical to doliiform, 3–4 × 2–8 µm. Conidia ovoid to obvoid, hyaline, single celled, 5–8 × 4–5 µm, enveloped in mucilage.

Specimen collected

On diseased leaves of *Artocarpus heterophyllus* (Moraceae) collected by R. Senthil Kumaran from Kolli Hills on May 2002. MUBL No: 596 and MTCC No: 3536.

Figure 4.138 *Phyllosticta amaryllidicola;* (a) infected leaf of *Atrocarpus heterophyllus;* (b) culture; (c) mature conidia (× 900)

106. *Phyllosticta citricarpa* (Mc Alp.) Van der Aa

Source

Van der Aa, *Stud. Mycol. Baarn,* **5**: p. 40 (1973); Path., Fungi & Bact., No. 85 (1996); Punithalingam, *Nova Hedwigia,* **XXXI**: p. 145 (1979); Punithalingam & Woodhams, *Nova Hedwigia,* **XXXVI**: p. 156 (1982).

Description

Pycnidia scattered over infected leaves, infected leaf turning pale whitish. Pycnidia 80–320 µm in diameter, globose or depressed on the leaves, flat or conspicuously papillate with a circular pore of 11–14 µm diameter. Pycnidial wall 3–8 cells thick, composed of isodiametrical or flattened cells, subhyaline to brownish on leaves; cell walls of variable thickness, hyaline towards the conidiogenous region, isodiametrical or flattened and thin-walled. Conidia, 1-celled, hyaline, subspherical to subcylindrical, 9–11 × 6–7 µm, enveloped in mucilage.

Specimen collected

On leaves of *Citrus medica* (Rutaceae) collected by R. Senthil Kumaran from Chennai May 2000. MUBL No: 598

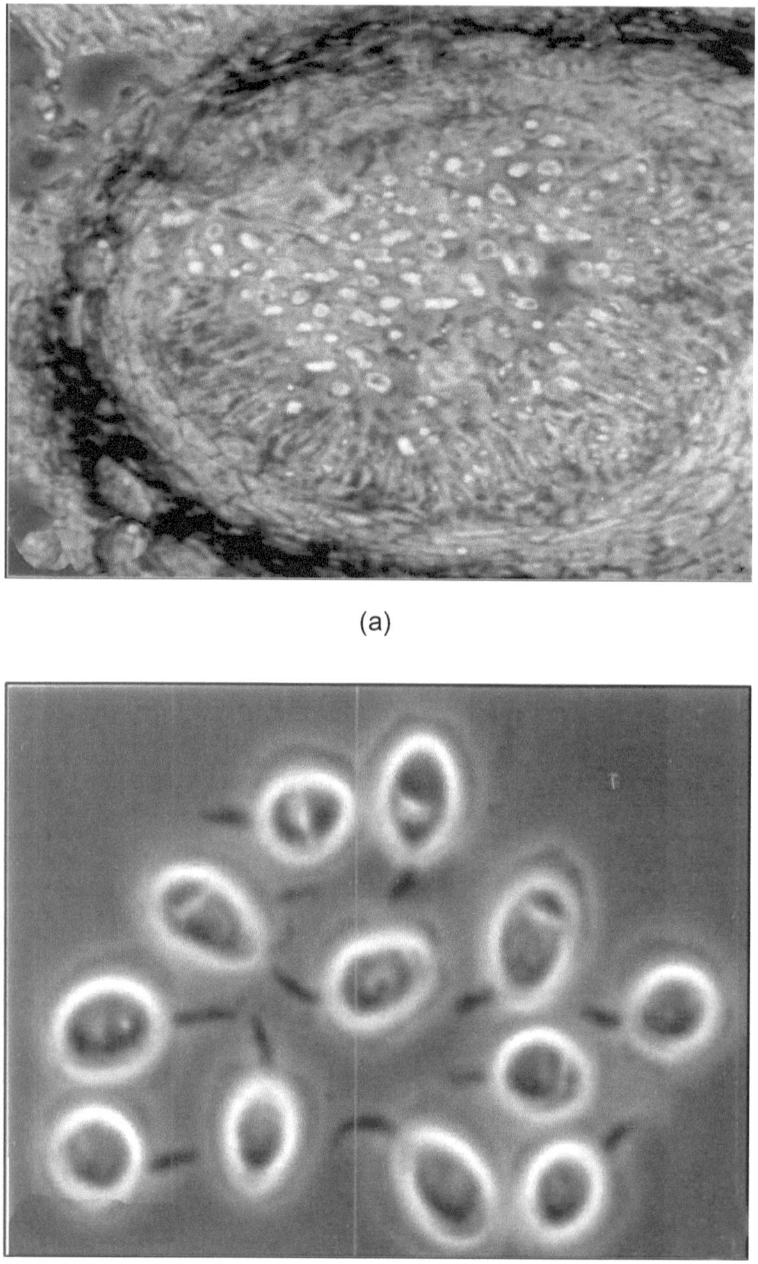

(a)

(b)

Figure 4.139 *Phyllosticta citricarpa;* (a) V.S. of Conidioma (×200); (b) mature conidia (× 900)

107. *Phyllosticta codiaeicola* Died.

Source

Da Costa & Mund., *Proc. Nat. Acad. Sci. India,* **14**: p. 57 (1948); Syd. & *Butl. Ann Mycol.*, **14:** p. 184 (1916); Sacc., **XXV**: p. 35 (1931); Sohi H.S., *Diseases of ornamental plants in India,* IARI, p. 79 (1992).

Description

Foliicolous. Leaf spots initially small, greyish brown with a brownish margin, later coalescing to form necrotic area around the symptom of the infected living leaf. Pycnidia epiphyllous, later amphigenous, unilocular, 70–130 µm in diameter, flat or inconspicuously papillate, with a 10–16 µm wide pore. Pycnidial wall 1–3 (5–13 µm) thick, composed of thin to thick-walled, usually compressed, brownish cells, more rounded, with thicker cellwalls and very dark brown around the pore, hyaline and flattened towards the conidiogenous region. Conidia 1-celled, hyaline, subspherical to obovoidal, 8–10 × 4–5 µm., enveloped in mucilage, with a mucoid appendage.

Specimen collected

On the diseased leaves of *Dieffenbachia* sp. (Araceae) collected by R. Senthil Kumaran from Ooty April 2002. IC No: 523; MUBL No: 599.

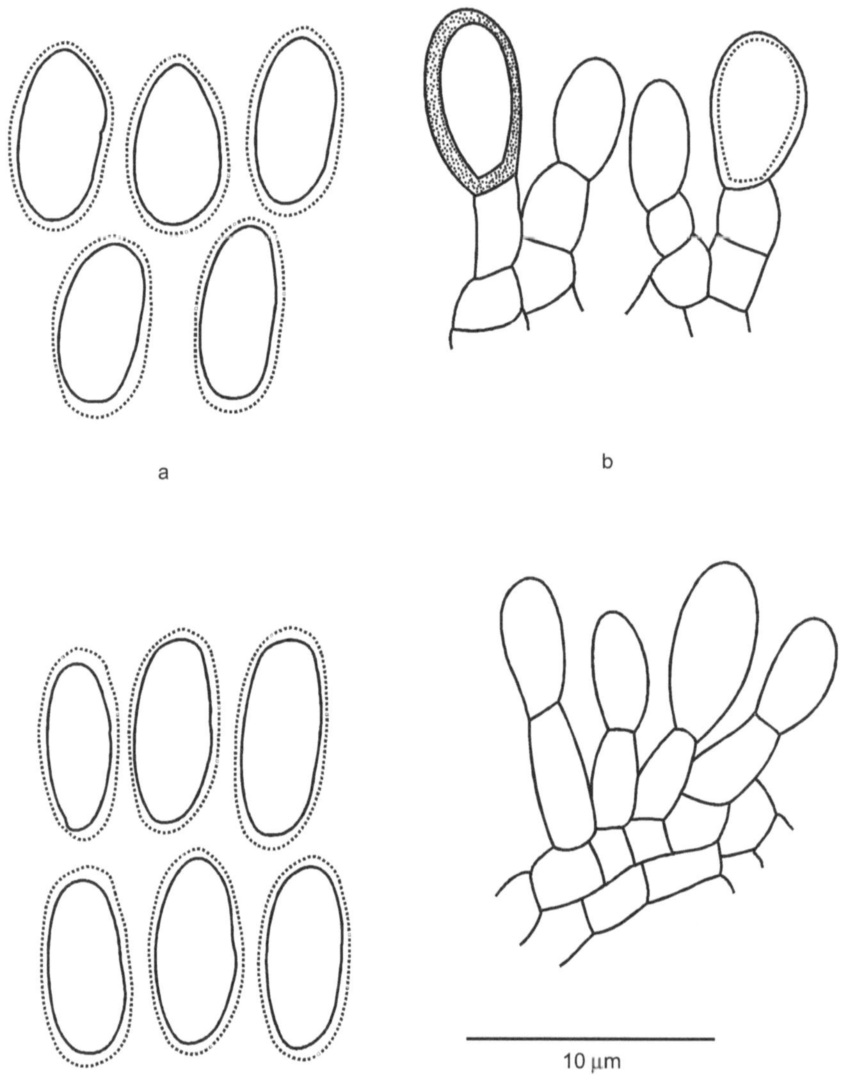

Figure 4.140 *Phyllosticta codiaeicola*, (a) Mature conidia; (b) Developing conidia.

108. *Phyllosticta colocasiicola* Hohnel

Source

Van der Aa, *Stud. Mycol. Baarn,* 5: p. 42 (1973); Hohnel, Sber. K. Akad. Wiss., Wien, Math. Nat. KI. 116, Abt. 1:60 (Fragm. Mykol. 3: 140) (1907).

Description

Foliicolous. Leaf spot initially subcircular, later coalescing to form large marginal necrotic patches, greyish brown to reddish brown surrounded by a brownish yellow margin. Pycnidia numerous, growing densely together, 90–130 μm in diam, globose and distinctly papillate. Pycnidial wall composed of large, dark brown cells. Conidiogenous cells very short. Conidia hyaline, 1-celled, ovoid, containing coarse guttules, measuring 10.5 × 5.5–6 μm. with mucilaginous envelope.

Specimen collected

On the leaves of *Colocasia* sp. (Araceae) collected by R. Senthil Kumaran from Kolli Hills May 2002. MUBL No: 600.

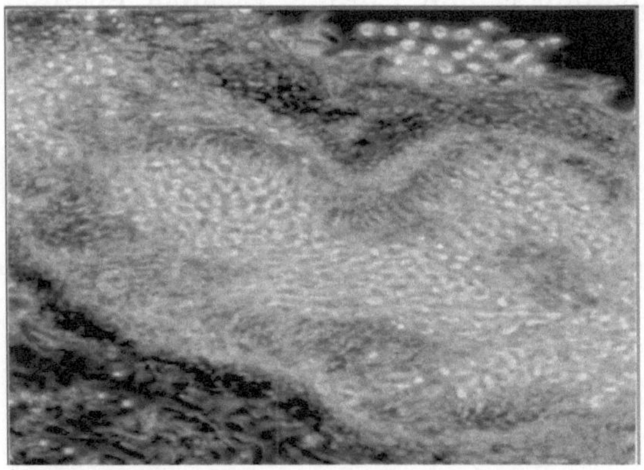

(a)

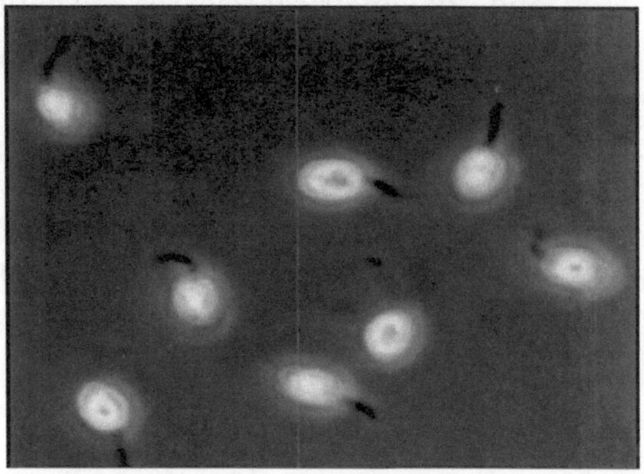

(b)

Figure 4.141 *Phyllosticta colocasiicola;* (a) V.S. of Conidioma (×260); (b) mature conidia (×900)

109. *Phyllosticta cruenta* (Kunze ex Fr.) Kickx.

Source

Van der Aa, Stud. *Mycol. Baarn,* **5**: p. 23 1973; Sacc., **III**: 58 (1884); Syd. and Butl., *Ann. Mycol.,* **14**: p. 185 (1916); R.S. Mathur, *The Coelomycetes of India,* p. 208 (1979); Agarwal, *Indian Phytopathology,* **51**: p. 104 (1998).

Description

Foliicolous, Leaf spot subcircular or ellipsoidal, coalescing to form large necrotic pathches, 0.5–2.5 cm in diameter, at first greyish brown, later pale cream or whitish gray, surrounded by a yellowish brown to almost dark brown margin. Pycnidia epiphyllous, single, unilocular, 90–210, mostly 120–150 µm in diameter, globose, seldom somewhat depressed, the upper region firmly grown together with the epidermis, flat or papillate, with a single circular or somewhat angular pore, of 13–28 µm diameter. Pycnidial wall 1–4 cells (8–20 µm) thick, composed of partly thick-walled, partly thin-walled, brownish cells, darker around the pore, hyaline and isodiametrical towards the conidiogenous region. Conidia 1–celled, hyaline, supspherical, 10–17 × 7–10 µm., enveloped in mucilage.

Specimen collected

On leaves of *Nephrolepis* sp. (Fern–Pteridophyte) collected by R. Senthil Kumaran from Yercaud September 1999. MUBL No: 603.

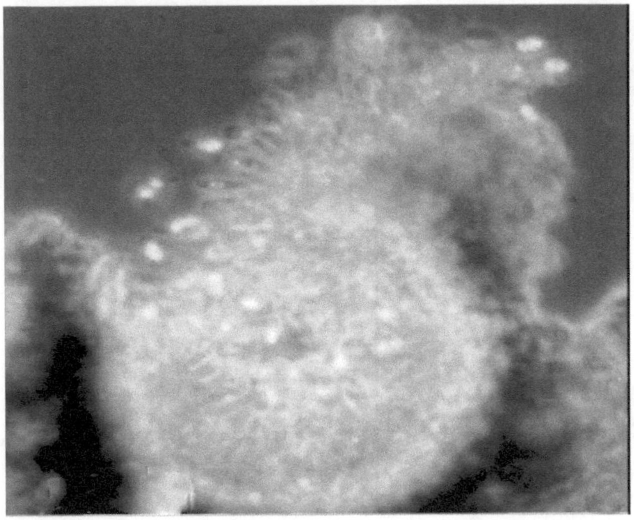

(a)

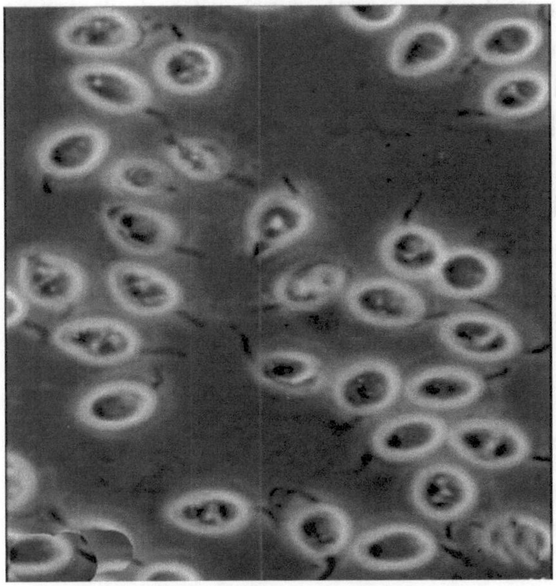

(b)

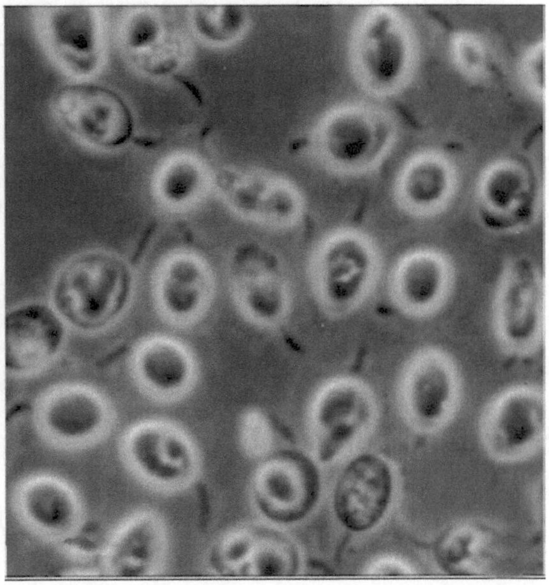

(c)

Figure 4.142 *Phyllosticta cruenta;* (a) V.S. of Conidioma (×200); (b–c) mature conidia (×900)

110. *Phyllosticta ficicola* Pat.

Source

Pat., *Cat. Rais Spl. Cell. Tunisie,* p. 115 (1917); Vasant Rao, J. Poona Univ., **28**: p. 110 (1964); Sacc., **XIV**: p. 861 (1899); Sawada, *Coll. Agric. Nat. Taiwan Univ. Spec. Publ.*, **8**: p. 136 (1959).

Description

Foliicolous. Leaf spot circular, occasionally coalescent, reaching a diameter of 1–4 cm, yellowish grey and later greenish brown, surrounded by a pale yellow brown to dark reddish brown margin. Pycnidia epiphyllous, seldom amphigenous, single, unilocular, 80–170 µm, usually about 110 µm in diameter, globose or depressed, flat or inconspicuously papillate, with a circular pore, 6–18 µm in diam. Pycnidial wall seldom up to 3 cells (4–8 µm) thick composed of isodiametrical, thin-walled, brownish cells, 6–12 µm in diam, smaller around the pore and with thicker cell walls, hyaline towards the conidiogenous region and isodiametrical. Conidia, 1-celled, hyaline, subspherical to ovoidal, 6–10 × 4–5 µm., enveloped in mucilage.

Specimen collected

On the leaves of *Ficus* sp. (Moraceae) collected by R. Senthil Kumaran from Ooty (Ootacamund) March 2000. MUBL No: 608; MTCC No: 4251.

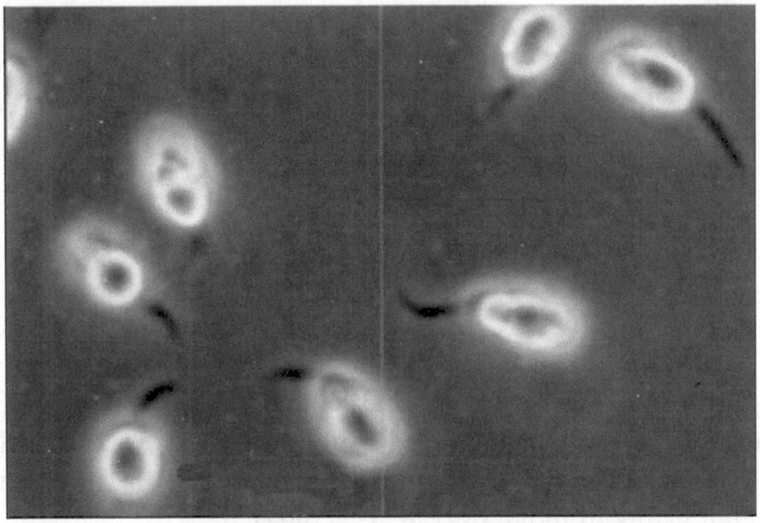

(a)

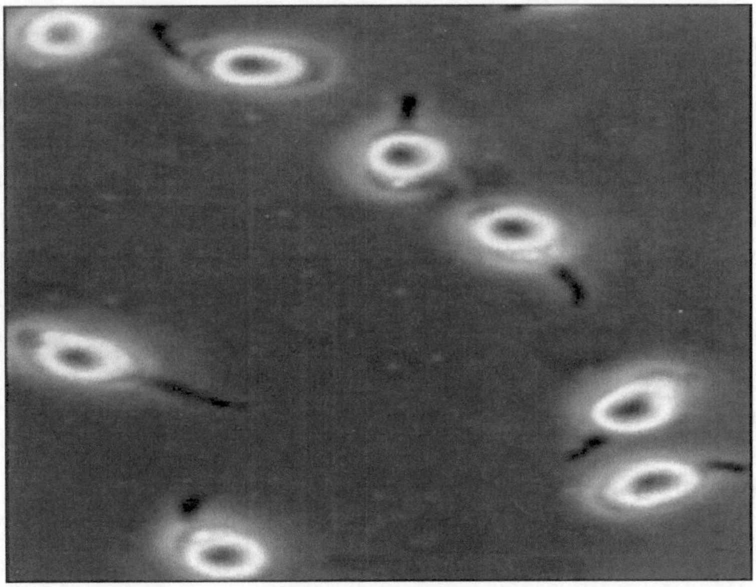

(b)

Figure 4.143 *Phyllosticta ficicola;* (a–b) mature conidia (×900)

111. *Phyllosticta kigeliae* Died.

Source

Syd. & Butl., *Ann. Mycol.* **14**: p. 62 (1916); Sacc., **XXV**: p. 25 (1931); Tandon & Bilgrami, *Curr. Sci.* **30**: p. 348 (1961); Bilgrami, *Proc. Nat. Acad. Sci. India*, **33B**: p. 436 (1963); Mathur, *The Coelomycetes of India*, p. 213 (1979); Sohi, *Diseases of ornamental plants in India*, IARI, p. 33, (1992).

Description

Foliicolous. Leaf spots initially subcircular, later coalescing to form necrotic patches especially to form near the apical tip, dark greyish brown, surrounded by a reddish brown rim. Pycnidia epiphyllous, single or clustered, unilocular, 60–90 μm in diam, globose or somewhat depressed, distinctly papillate, with a subcircular or irregular, 10–16 μm wide pore. Pycnidial wall 1–3 cells (3–12 μm), but in the upper part up to 9 cells thick, composed of rounded, sometimes compressed, brown to dark-brown, thin-walled cells, very dark and with thicker cell-walls around the pore, hyaline and very flat towards the conidiogenous region. Conidia single–celled, hyaline, subglobose to obovoid, guttulate, 7–9 × 5–6 μm., enveloped in mucilage.

Specimen collected

On the diseased leaves of *Kigelia pinnata*, *Crescentia* sp. (Bignoniaceae) collected by R. Senthil Kumaran from Kodaidkanal, October 2001. MUBL No: 612.

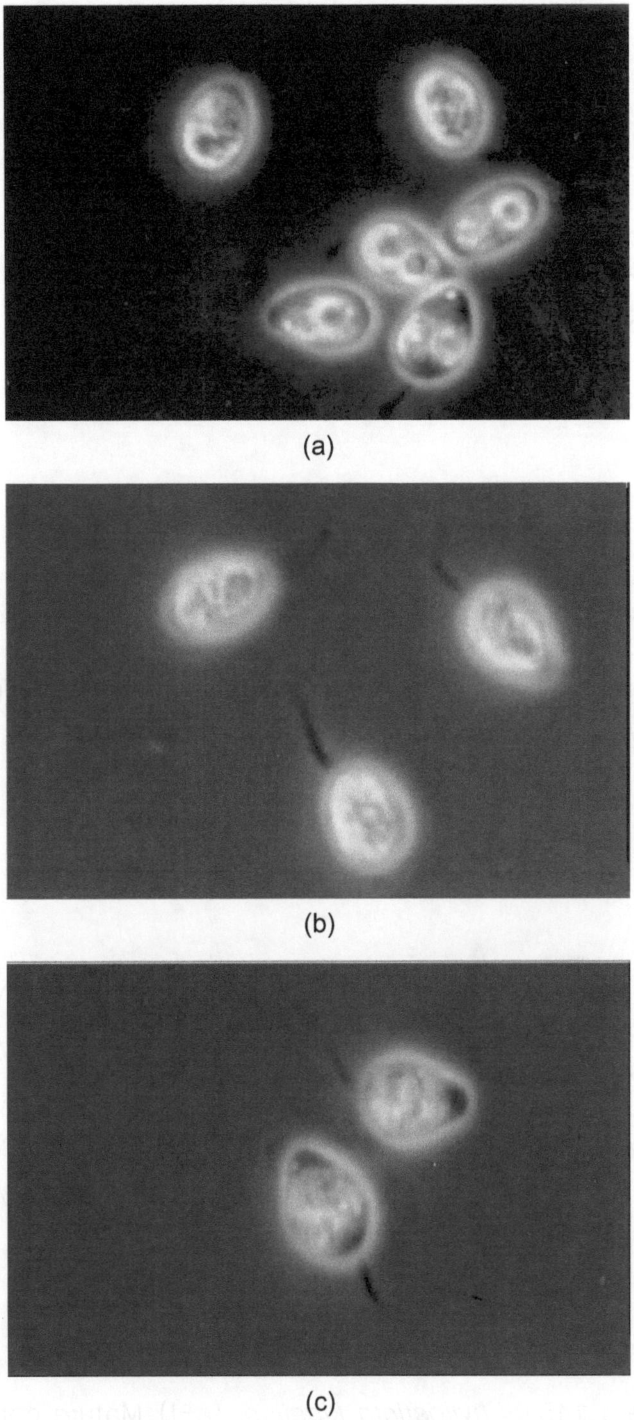

Figure 4.144 *Phyllosticta kigeliae;* (a–c) mature conidia (×900)

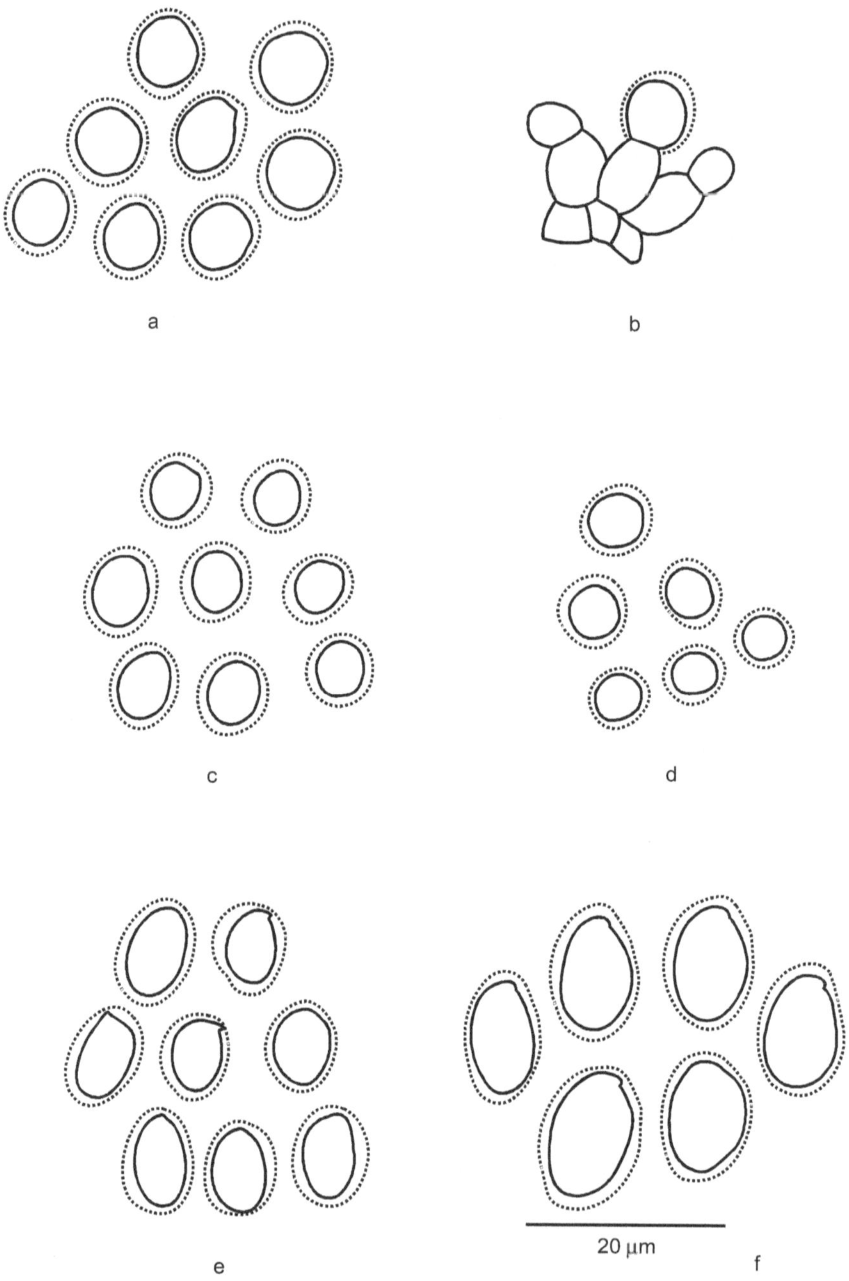

Figure 4.145 *Phyllosticta kigeliae*, (a–f) Mature conidia.

112. *Phyllosticta* state of *Guignardia mangiferae* Roy

Source

Roy, *Indian Phytopath.*, **20**: p. 348 (1968); Muthumary, J. *et al.*, *Mycotaxon,* **XLVII**: pp. 147–155 (1993).

Description

Conidiomata pycnidial, immersed, separate, globose, 100–200 µm diam. Ostiole circular, papillate. In vertical section, pycnidium unilocular, outer layer thick-walled, brown cells and inner layer of thin–walled, hyaline cells giving rise to conidiogenous cells. Conidiogenous cells lining the pycnidial cavity, holoblastic, hyaline, smooth, ampulliform, 5–10 × 2.5 µm. Conidia spherical to subspherical, hyaline, smooth, 10–15 × 5–8 µm, provided with an apical, mucilaginous, appendage and the conidium is surrounded by a sheath revealed by modified Leifson's flagella staining.

Specimen collected

On dead leaves of *Mangifera indica* L., Madras, Tamilnadu, May 1988, collected by J. Muthumary. IC. No: 613.

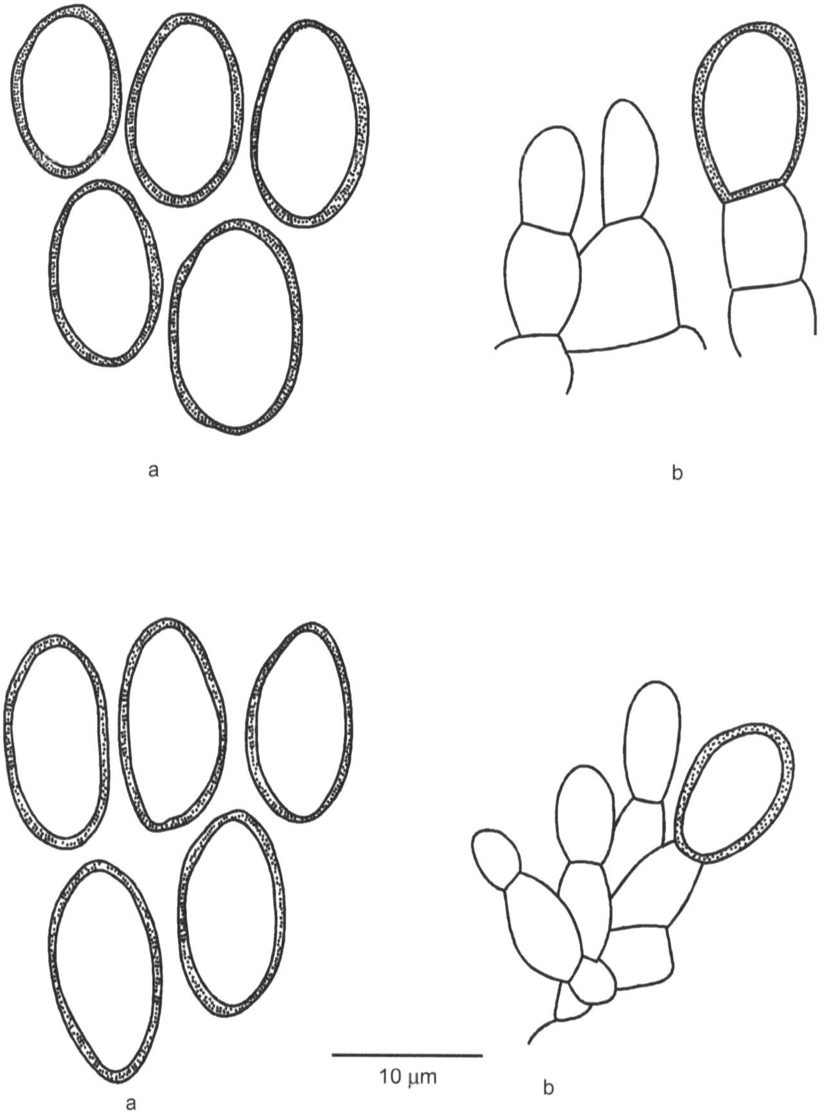

Figure 4.146 *Phyllosticta* state of *Guignardia mangiferae*, (a) Mature conidia; (b) Developing conidia.

113. *Phyllosticta mimusopsis-elengi* Vasant Rao

Source

Vasant Rao, *Mycopath. Mycol. Appl.*, **19**: p. 226 (1963); R.S Mathur, *The Coelomycetes of India*, p. 215, 1979.

Description

Foliicolous. Black pycnidial fruit bodies on the necrotic areas of the living leaves. Pycnidia epiphyllous, unilocular, 85–125 µm in diam, globose or somewhat depressed, with a flattened upper region which is firmly grown together with the epidermis, with a single, circular or somewhat angular ostiole of 8–12 µm diam. Pycnidial wall 2–3 cells (12–15 µm) thick, composed of thin-walled, brownish, isodiametrical or flattened cells, sometimes missing in the lower region of the pycnidium, darker around the pore, hyaline and isodiametrical towards the conidiogenous region. Conidia 1-celled, hyaline, 7–9 × 5–6 µm. Enveloped in slime.

Specimen collected

On the leaves of *Mimusops elengi* (Sapotaceae) collected by R. Senthil Kumaran from Kodaikanal, March 2001. MUBL No: 615.

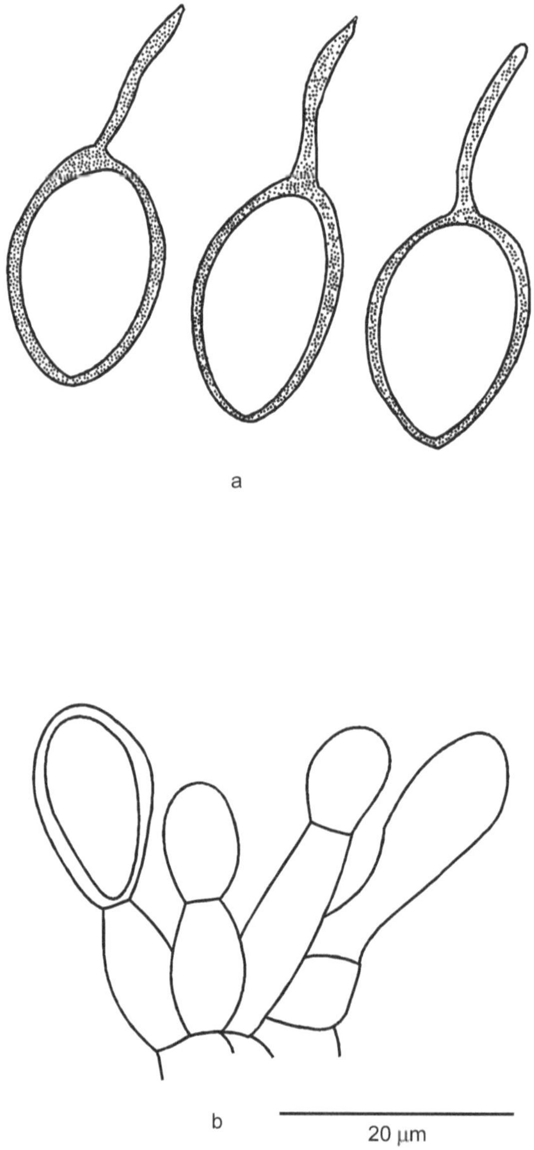

Figure 4.147 *Phyllosticta mimusopsis-elengi*, (a) Mature conidia; (b) Developing conidia

114. *Phyllosticta murnadensis* (Ponnappa) Van der Aa

Source

Van der Aa, *Stud. Mycol. Baarn*, **5**: p. 70 (1973); Mathur, *The Coelomycetes of India*, p. 216 (1979).

Description

Foliicolous. Leaf spots subcircular, up to 8 mm in diameter, dark creamy to pale yellowish white, surrounded by dark reddish brown margin. Pycnidia epiphyllous, single or clustered, unilocular, up to 180 µm in diam, globose or somewhat depressed, conspicuously papillate with a 10–14 µm wide pore. Pycnidial wall 10–22 µm, darker around the pore, hyaline and flattened towards the conidiogenous region. Conidia 1-celled, hyaline, spherical to obovoidal, 8–10 × 4–5 µm., surrounded by slime.

Specimen collected

On living leaves of *Glycosmis cochinchinensis* (Rutaceae) collected by R. Senthil Kumaran from Kodaikanal, March 2001. MUBL No: 617.

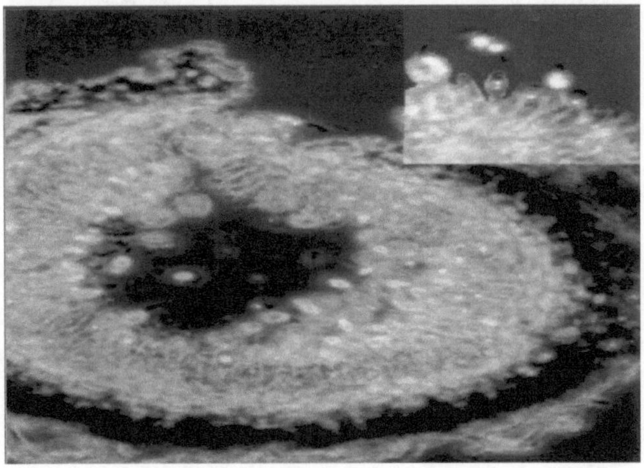

Figure 4.148 *Phyllosticta murnadensis;* V.S. of Conidioma (×200)

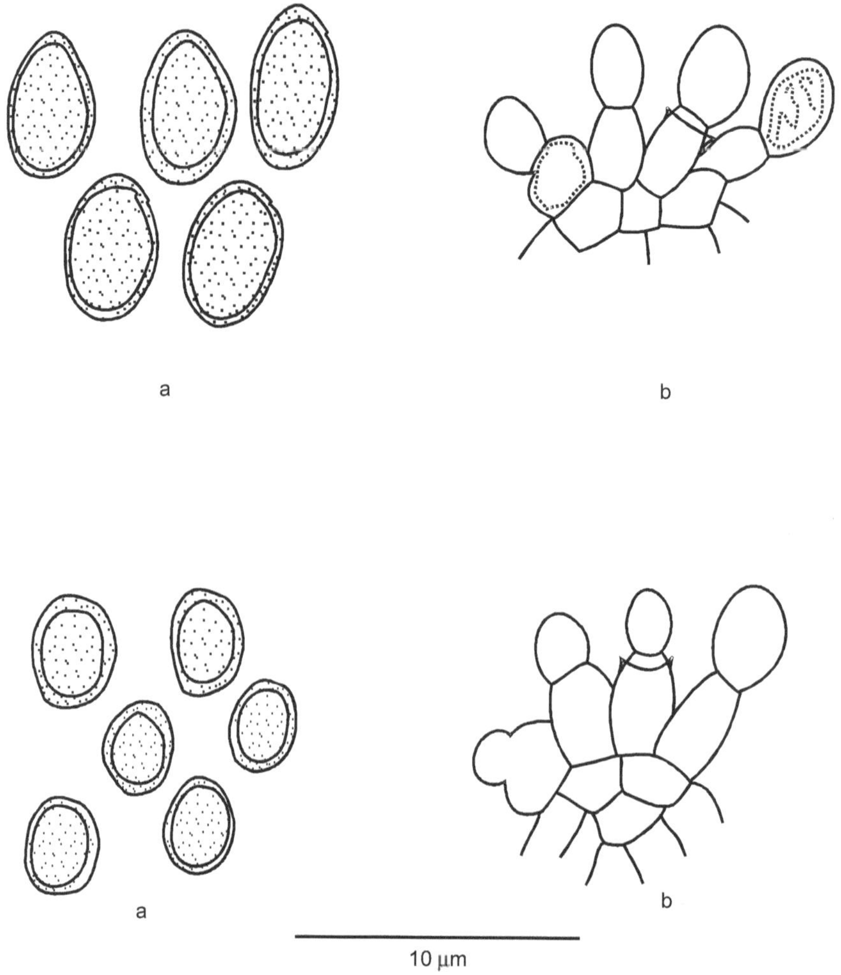

Figure 4.149 *Phyllosticta murnadensis*, (a) Mature conidia; (b) Developing conidia

115. *Phyllosticta murrayicola* Van der Aa

Source

Van der Aa, *Stud. Mycol. Baarn*, **5**: p. 71 (1973); Frag. And Cif., *Boln. R. Soc. Esp. Hist. Nat.*, **27**: p. 328 (1927); Chandra & Tandon, *Curr. Sci.*, **33**: p. 409 (1964); Punithalingam, *Mycol. Pap.*, No. **136**: p. 7 (1974) (as *Phyllostictina murrayae* Syd.); Punithalingam and Woodhams, *Nova Hedwigia*, **XXXVI**: p. 164 (1982); Sohi, *Disease of ornamental plants in India*, IARI, p.166 (1992); Ali and Saikia, *Indian Phytopath.*, **20**: p. 50 (1997).

Description

Foliicolous. Leaf spots small, up to 8 mm in diameter, subcircular, ochreous, later almost brown with distinct purple brownish margin. Pycnidia epiphyllous, single, unilocular, 80–130, sometimes, 180 µm in diam, globose. Somewhat depressed or tympaniform, apically flattened and firmly grown together with the epidermis, inconspicuously papillate, with a pore of about 14 µm diam. Pycnidial wall 1–3 cells (6–16 µm) thick, composed of brownish, rounded or irregular, thin-walled cells, 5–15 µm in size, darker around the pore with thick-walled cells, hyaline towards the conidiogenous region, and with isodiametrical or flattened cells. Conidia single-celled, hyaline, subspherical to obovoid, broadly rounded apically, with a truncate base, 9–10 × 7–9 µm, surrounded by slime layer and with an apical appendage.

Specimen collected

On the leaves of *Murraya koenigii* (Rutaceae) collected by R. Senthil Kumaran from Chennai, July 1999. MUBL NO: 618 and MTCC No: 3900.

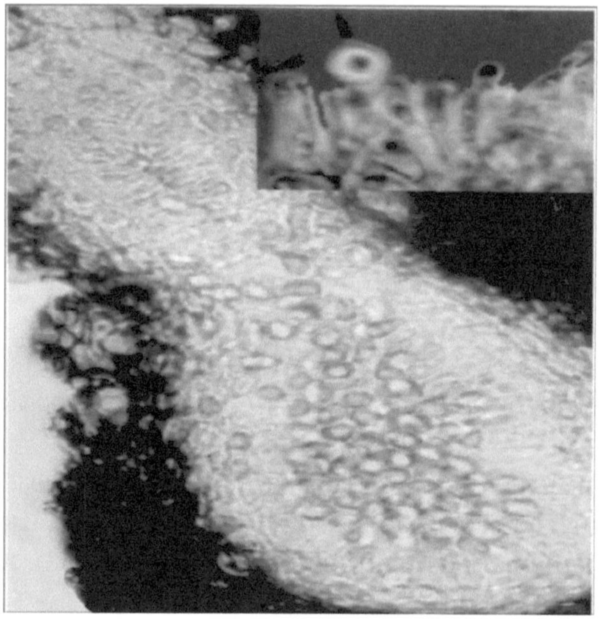

(a)

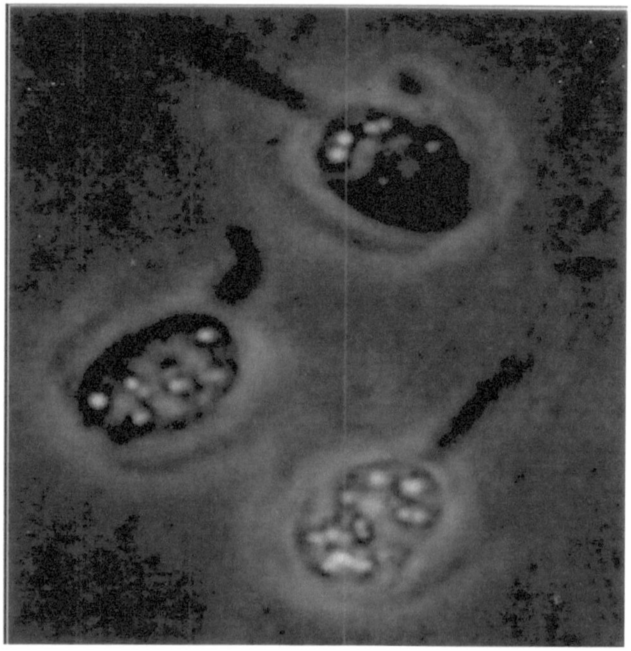

(b)

Figure 1.150 *Phyllosticta murrayicola;* (a) V.S. of Conidioma (×260); (b) mature conidia (×900)

116. *Phyllosticta musarum* (Cooke) Van der Aa

Source

Van der Aa, *Stud. Mycol. Baarn*, **5**: p. 72 (1973); *Sphaeropsis musarum* Cooke, *Grevillea,* **8**: p. 93 (1880) (Basionym); Punithalingam, *Mycol. Pap.*, No. **136**: pp. 33-39 (1974); Punithalingam & Holiday, CMI, *Desc. Path. Fungi & Bact.* No. 467 (1975) (as *Guignardia musae*); *P. musarum* (Petr.) Khune, comb. nov. in Ph.d. thesis, IARI, New Delhi, India (1977); R.S. Mathur, *The Coelomycetes of India*, p. 216 (1979); Punithalingam & Woodhams, *Nova Hedwigia*, **XXXVI**: pp. 167–168 (1982); Ali and Saikia, *Indian Phytopath.*, **50**: p. 200 (1997).

Description

Foliicolous. spots on leaf, stem (sheaths), fruit and almost all parts of the plant, greyish yellow to dark brown. Fruit bodies also developed on necrotic areas of the leaves, later coalescing to form a necrotic region to a great extent. Pycnidia prominent under the raised epidermis, mostly epiphyllous, sometimes amphigenous, single or clustered, unilocular, 70–180, mostly about 135 µm in diameter, globose, pyriform or tympaniform, the upper region firmly grown together with the epidermis and the subepidermal stroma, conspicuously papillate, with a 20–30 µm wide pore. Stroma prosenchymatous, clypeus-like developed around the upper part of the pycnidia, subepidermal, composed of very dark hyphae. Wall in the lower and lateral region hardly developed, only composed of some compressed brownish hyphae or brownish, thin-walled cells, but 1–4 cells thick in the upper part, composed of flattened, dark brown cells, darkest around the pore and hyaline towards the conidiogenous region. Conidia single-celled, obovoid to spherical with a truncate base, broadly rounded apically, 10–14 × 7–9 µm, surrounded by slime layer, bearing an apical appendage.

Specimen collected

On leaves and sheaths of *Musa paradisiaca* (Musaceae) collected by R. Senthil Kumaran from Yercaud, May 2002. MUBL No: 619.

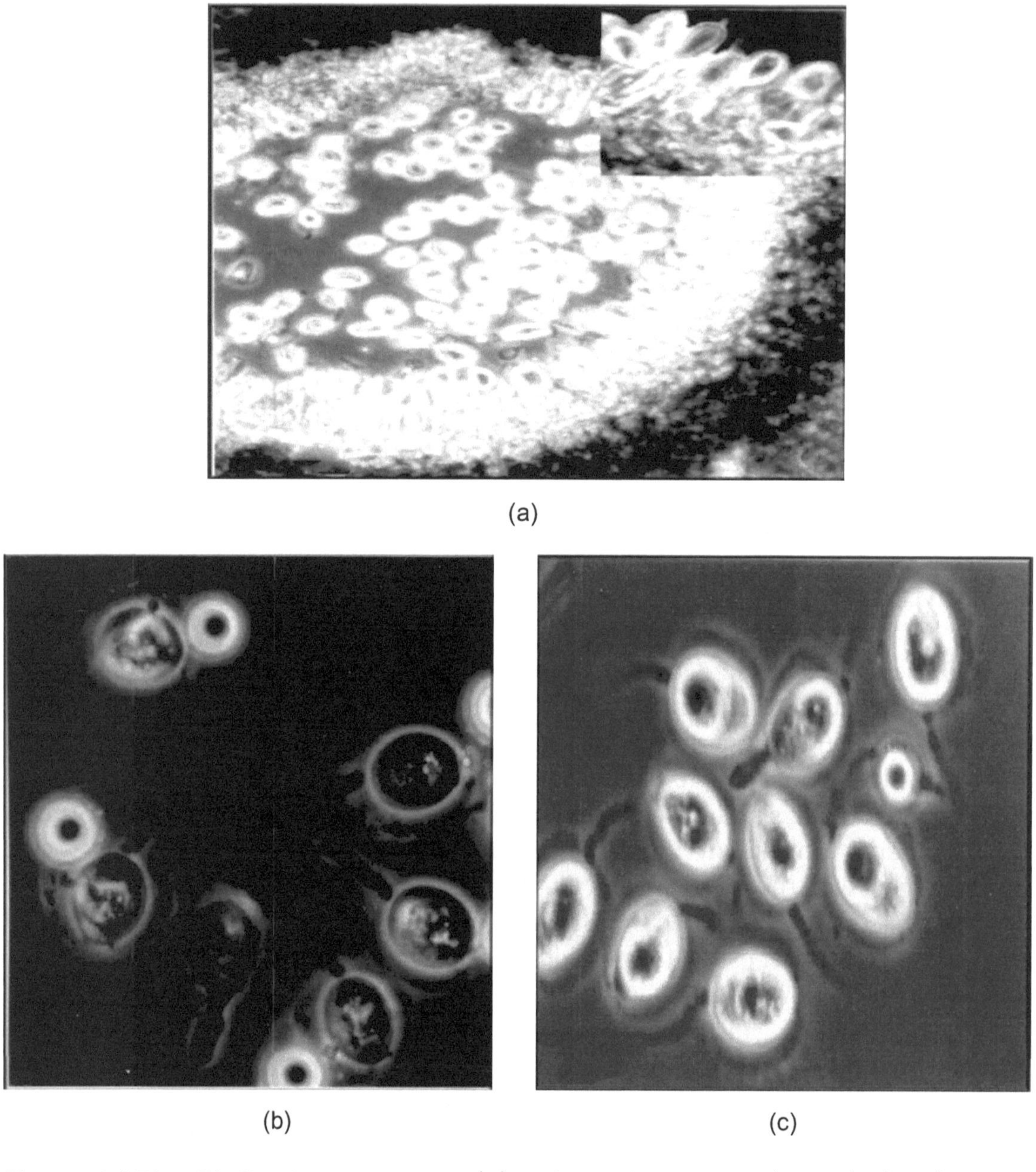

(a)

(b) (c)

Figure 4.151 *Phyllosticta musarum;* (a) V.S. of Conidioma (×260); (b–c) mature conidia (×900)

117. *Phyllosticta pervincae* Bissett & Darbyshire

Source

Bissett & Darbyshire, *Fungi canadenses* #282 (1984); Nag Raj T.R., *Coelomycetous anamorphs with appendage bearing conidia*, Mycologue Publication, Canada, p. 682 (1993).

Description

Foliicolous. leaf spots brown diffuse, covering large parts of the leaf, especially near the tip with yellowish brown margin. Conidiomata pycnidial, mostly epiphyllous, scattered to gregarious, subepidermal in origin, immersed to partly erumpent, globose to ellipsoid, 95–220 µm deep, unilocular, glabrous, brown, ostiolate; ostiole, circular to oval. Conidia 1-celled, ellipsoid to ovoid, 8–10 × 5–6 µm with mucoid sheath.

Specimen examined

On living leaves of *Catharanthus roseus, Plumeria obtusa* (Apocyanaceae) collected by R. Senthil Kumaran from Chennai, July 2000. MUBL No: 621.

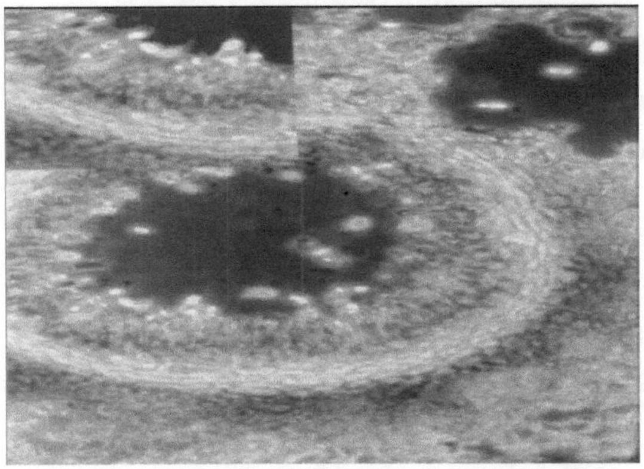

(a)

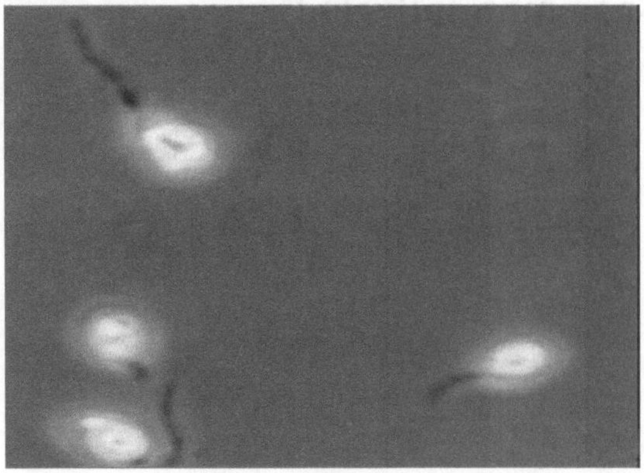

(b)

Figure 4.152 *Phyllosticta pervincae;* (a) V.S. of Conidioma (×260); (b) mature conidia (×900)

118. *Phyllosticta plumbaginicola* Vasanth Rao

Source

Vasanth Rao, *Publ. Inst. Micol. Recife*, **6**: p. 383 (1963); *Phyllostictina plumbaginis* Ponnappa, *Mycologia*, **62**: p. 395 (1970); Van der Aa, *Stud. Mycol.*, **5**: p. 94 (1973); R.S. Mathur, *The Coelomycetes of India*, p. 217 (1979); Nag Raj, T.R. *Coelomycetous anamorphs with appendage-bearing conidia*, Mycologue Publication, p. 682 (1993).

Description

Foliicolous. leaf spots brown, covering larger parts of the leaf especially near the tip with reddish brown margin. Conidioma pycnidial, amphigenous, but predominantly epiphyllus, scattered to gregarious, subepidermal in origin, immersed to semi-immersed, globose to depressed globose, 75–120 µm wide, 60–90 µm deep, unilocular, glabrous, brown, ostiolate. Conidia globose to ellipsoid, hyaline, thin-walled, 7–8 × 4–5 µm., enclosed in a mucoid envelope, appendage occasionally seen.

Specimen collected

On the leaves of *Plumbago* sp. (Plumbaginaceae) and *Guazuma tomentosa* (Sterculiaceae) collected by R. Senthil Kumaran from Kodaikanal, September 2001. MUBL NO: 623.

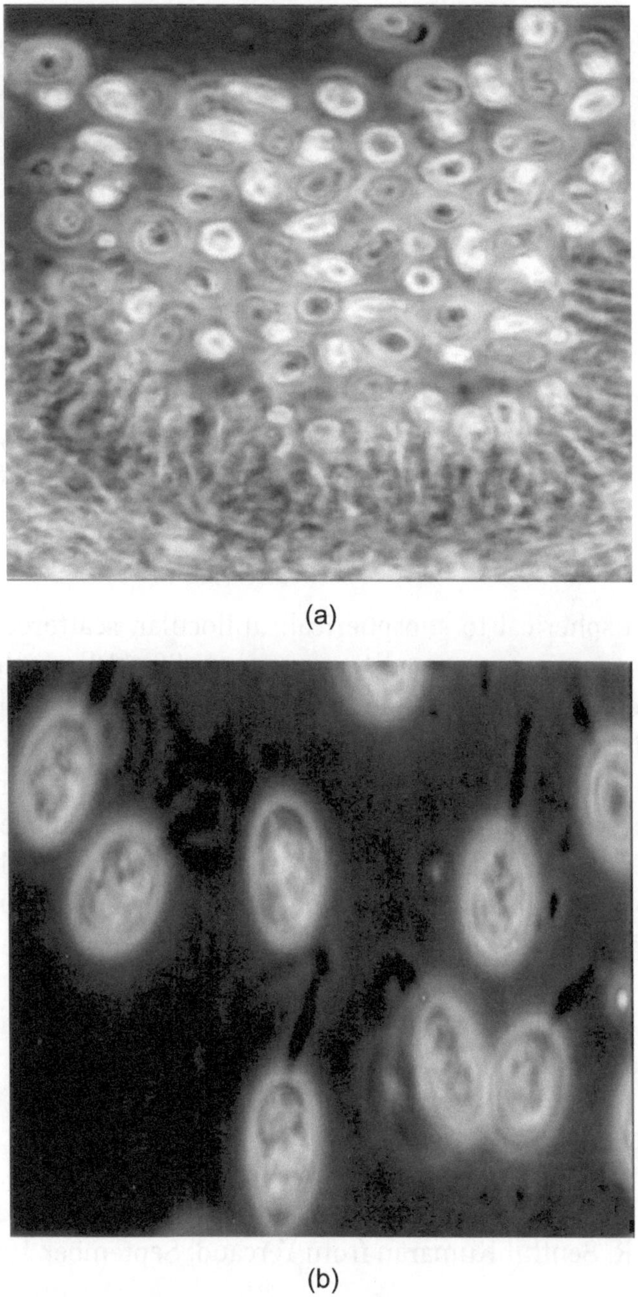

Figure 4.153 *Phyllosticta plumbaginicola;* (a) V.S. of Conidioma (×200); (b) mature conidia (×900)

119. *Phyllosticta spinarum* (Diedicke) Nag Raj & Morelet

Source

Morelet, *Bull. Soc. Sci. nat. Archeol., Toulon var.* **219**:p. 12 (1978); Nag Raj & Morelet, *Can. J. Bot.,* **57**: p. 1297 (1979); Nag Raj, *Coelomycetous anamorphs with appendage bearing conidia,* Mycologue Publication, p. 685 (1993).

Description

Conidiomata spherical to subspherical, unilocular, scattered to gregarious and confluent, pycnidia, immersed to erumpent, 80–110 µm diam, 120–170 µm, globose to depressed globose, glabrous, ostiolate. Pycnidial wall composed of many layers of thick-walled pseudoparenchymatous cells, outer most layers brown to pale brown the inner layers hyaline from which the conidiophores arise. Conidiophores arising all round the cavity, reduced to conidiogenous cells, invested in mucus. Conidiogenous cells cylindrical to subcylindrical, hyaline, 10–15 µm. Conidia obovoid with a broadly rounded apex and narrow, truncate base bearing minute marginal frills, unicellular, hyaline, smooth, guttulate, 8–13.7 × 5–7.5 µ m (X=9.59 × 6.02 µm), enclosed in a mucilaginous sheath and bearing an attenuated, mucoid, apical appendage, 5–14 µm long.

Specimen collected

Isolated as endophyte from fresh leaves of *Juniperus* sp., *Cupressus* sp. (Pinaceae) collected by R. Senthil Kumaran from Yercaud, September 1999. MUBL No: 625.

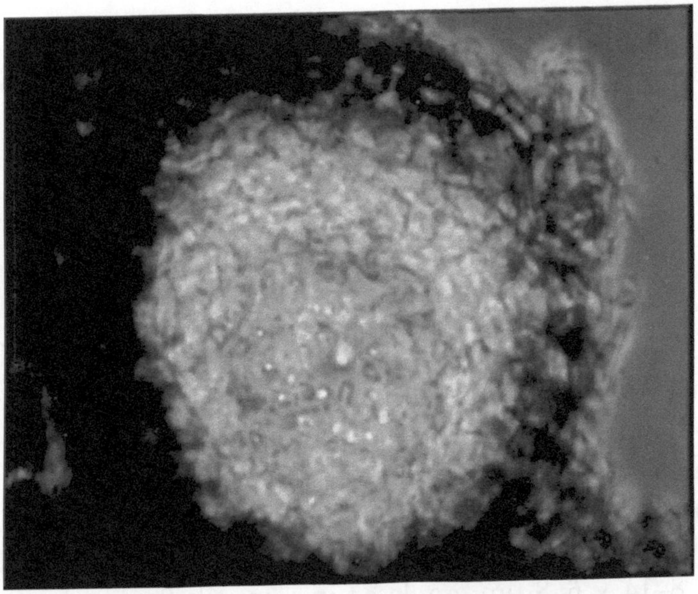

(a)

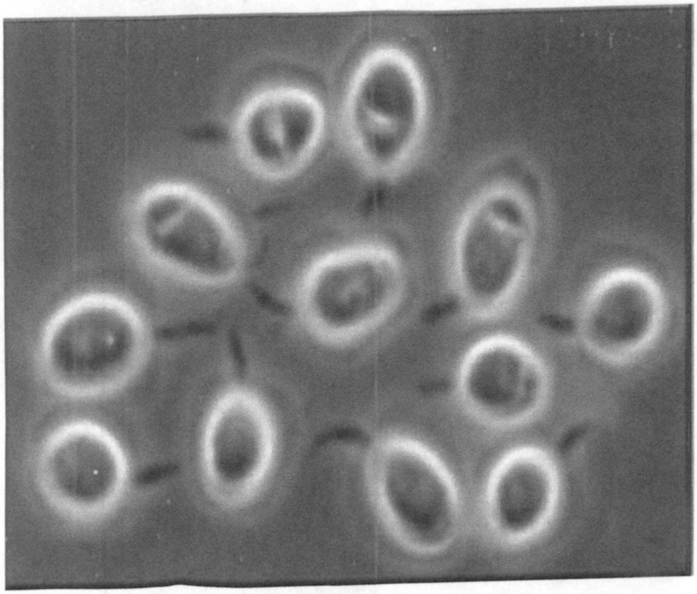

(b)

Figure 4.154 *Phyllosticta spinarum*; (a) V.S. of Conidioma (×200); (b) mature conidia (×900)

120. *Phyllosticta* sp.

Source

Description

Conidiomata pycnidial, 90–120 µm diam, unilocular, glabrous, dark brown, ostiolate. Conidiophores reduced to conidiogenous cells lining the cavity of the conidiomata, invested in mucus. Conidiogenous cells narrowly conical to subcylindrical, colourless. Conidia obovoid with a broadly rounded apex and narrow, truncate base bearing minute marginal frills, unicellular, colourless, smooth guttulate, 8–14 × 6–8 µm, enclosed in a mucilaginous sheath and bearing an attenuated, mucoid, apical appendage, 7–17 µm long.

Recorded as endophyte in

Isolated as endophyte in *Aegle marmelos*, *Andrographis paniculata*, *Cardiospermum helicacabum*, *Justicia gendarussa*, *Mimusops elengi* from Botany Field Research and Herbal Science Centre, Maduravoyal.

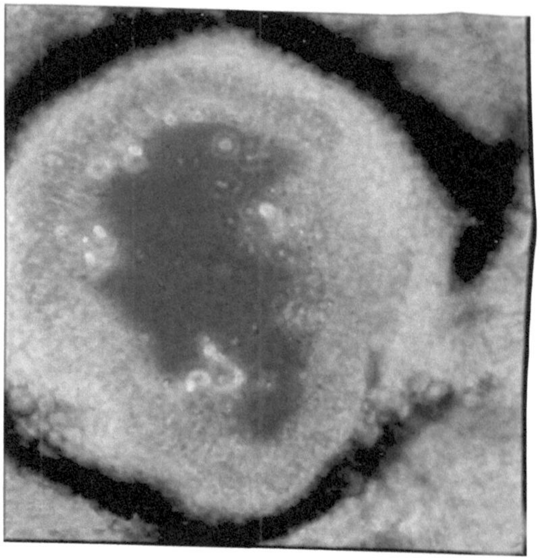

(a)

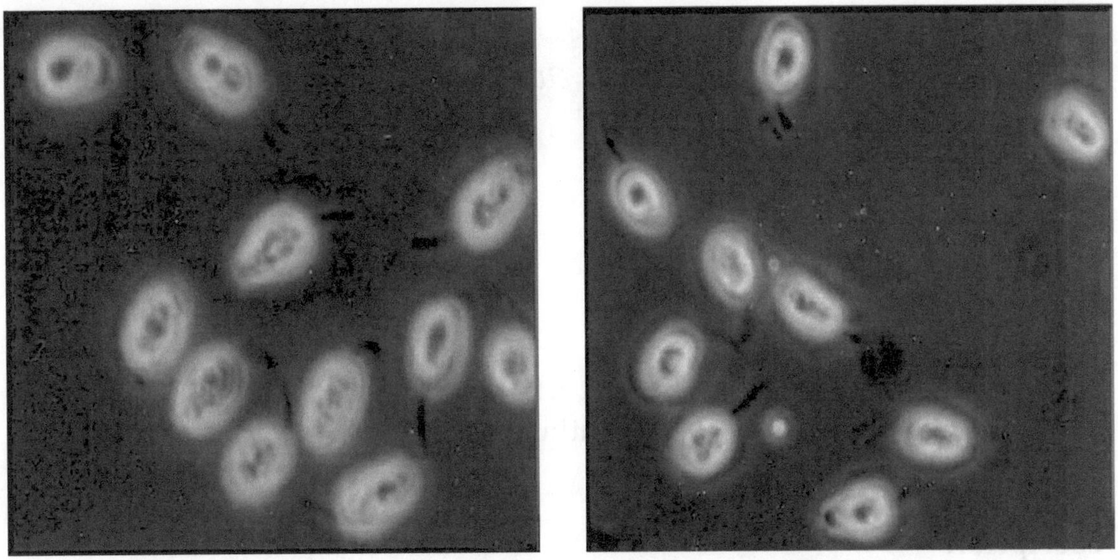

(b) (c)

Figure 4.155 *Phyllosticta* sp; (a) V.S. of Conidioma (×260); (b–c) mature conidia (×900)

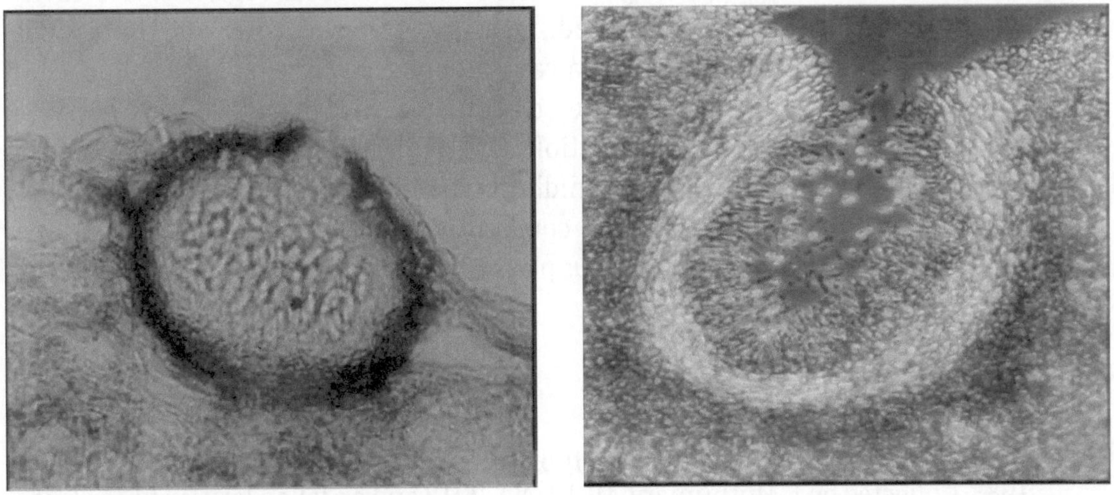

(a) (b)

Figure 4.156 *Phyllosticta* sp; (a–b) V.S. of Conidiomata (×260).

121. *Pilidium acerinum* Kunze

Source

Kunze and Schmidt, *Mykol. Hefte*, **2**: p. 92 (1823); Sutton, B.C., *The Coelomycetes*, CMI, England, p. 564 (1980); Muthumary, J., *Current Science*, **55**(16): pp. 791–792 (1986)

Description

Conidiomata lenticular and stromatic pycnidia. The Indian collection differs from other collections in having much larger pycnidia measuring up to 1.5 mm in diameter whereas in collections from other countries the pycnidia are only 200 µm in diam. The conidiogenous cells line the basal region of the cavity, lageniform, phialidic, integrated, hyaline, tapered towards the apices, 18–25 × 2–3 µm. The conidia are falcate, aseptate, hyaline, slightly colorued in mass, apex acute, base acute to obtuse, thin-walled, eguttulate, 15–18 × 2–2.5 µm. There is no ostiole and dehiscence is by irregular, stellate rupture of the upper wall of the pycnidium. The fungus was collected on leaves of *Calophyllum inophyllum* for three consecutive years in the same season. This shows that this Coelomycetous fungus plays an important role in the decomposition of tropical leaf litter.

Specimen collected

On fallen leaves of *Calophyllum inophyllum* L., Y.M.C.A. Hostel, Madras, October 1983, Collected by J. Muthumary, MUBL No. 2912 and on fallen leaves of *Eucalyptus* sp., Kodaikanal, on the way to Berijam, October 1984, collected by J. Muthumary.

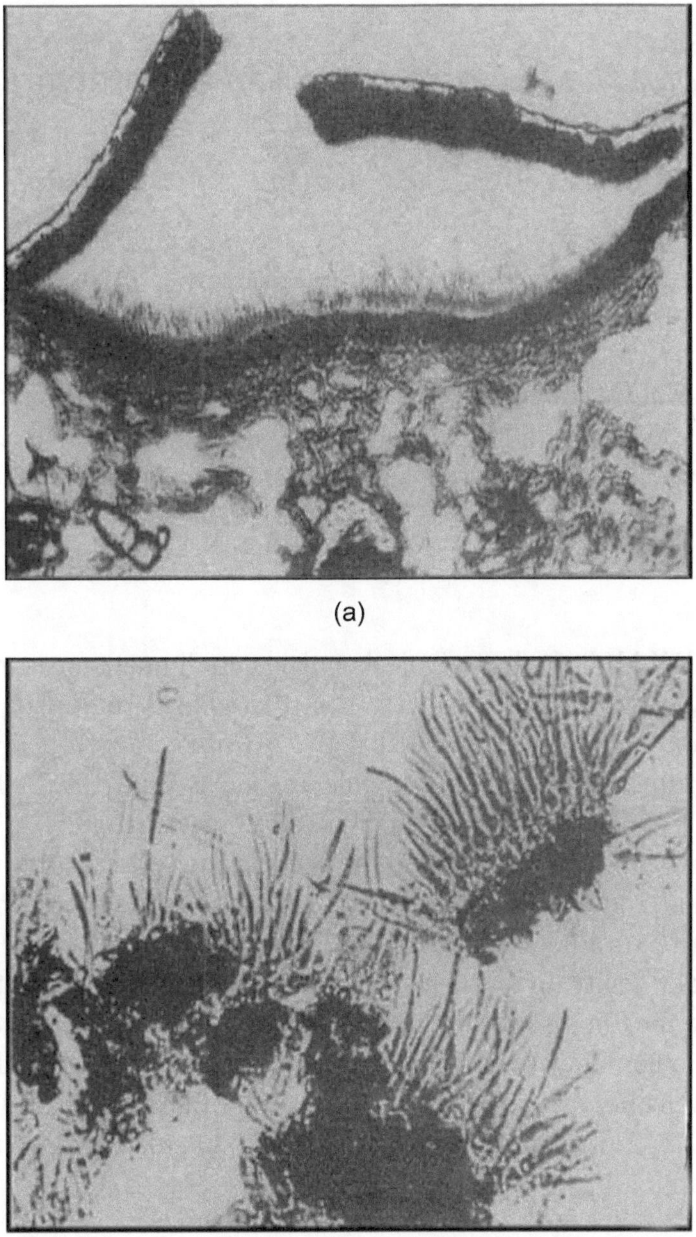

(a)

(b)

Figure 4.157 *Pilidium acerinum;* (a) V.S. of Conidioma (×200); (b) Condiogenous cells and young conidia (×900)

122. *Pseudolachnea longiciliata* (Hino and Katumotto) Sutton

Source

Sutton, B.C., *The Coelomycetes,* CMI, England, p. 461 (1980); Muthumary, J., *Curr. Sci.,* **56** (7): pp. 306-307 (1987). Nag Raj, T.R., *Coelomycetous anamorphs with appendage bearing conidia,* Mycologue Publications, Canada, pp. 725–729 (1993).

Description

The type (IMI 110665) shows typical conical conidiomata measuring 350–500 μm in diam with the conidiophores lining the basal region of the conidioma. The conidiomata black, cylindrical to cornuted, tapering towards the apex, 600–1000 μm high, 300–500 μm wide at the base and 100–200 μm wide at the apex. Thick-walled, dark brown, septate setae measuring 200–250 × 3–5 μm are seen covering the conidioma up to the neck. In the vertical section, the conidioma is unilocular, composed of dark brown to pale brown pseudoparenchymatous cells forming a textura porrecta at the periphery. The conidiophores line the entire conidiomatal cavity up to the neck. Despite these differences in the nature of the conidioma the fungus is accommodated in *P. longiciliata* because of the presence of phialidic conidiogenous cells and the typical cylindrical and septate conidia with an apical and basal appendage. Conidia filiform, curved, 10–15 septate, 60–70 × 1.5–2 μm apical appendage single, 30–40 μm long, basal appendage 5–8 μm long.

Specimen collected

On *Sasa kurilensis* Makino et Shibata, Mt. Nyutozon, Prov. Ugo. collected by H. Muroi, August 1957, IMI 110665 and On culms of *Bambusa* sp.Munnar, Kerala, India, collected by J. Muthumary, March 1978. MUBL No: 2920.

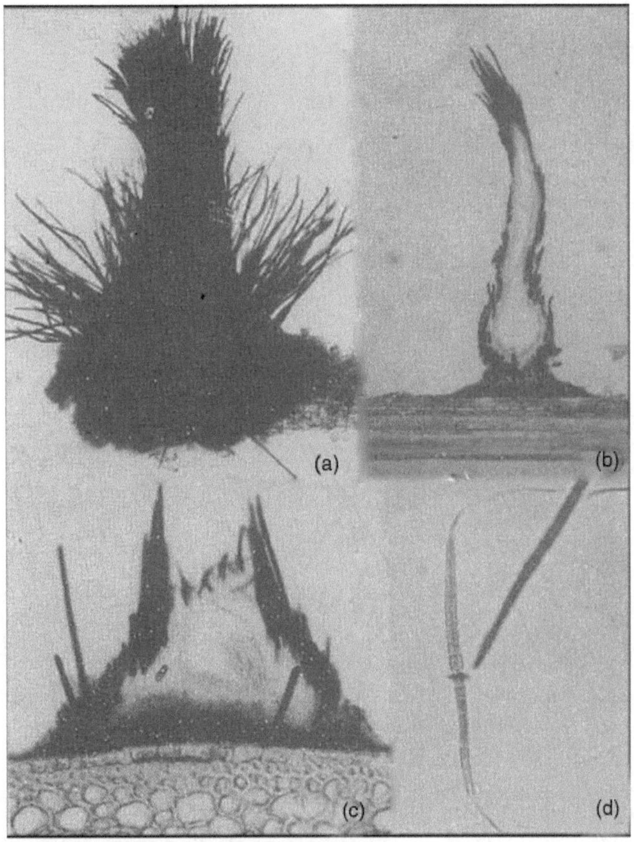

Figure 4.158 *Pseudolachnea longiciliata;* (a) mature conidioma (×200); (b–c) V.S. of Conidiomata (×200); (d) mature conidia (×900).

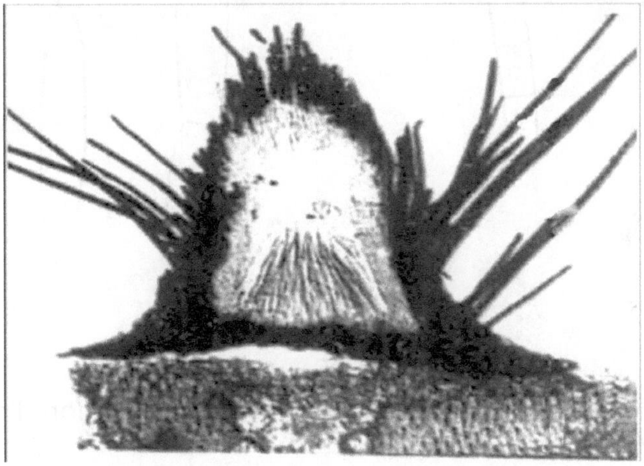

Figure 4.159 *Pseudolachnea longiciliata;* V.S. of Conidioma (×200).

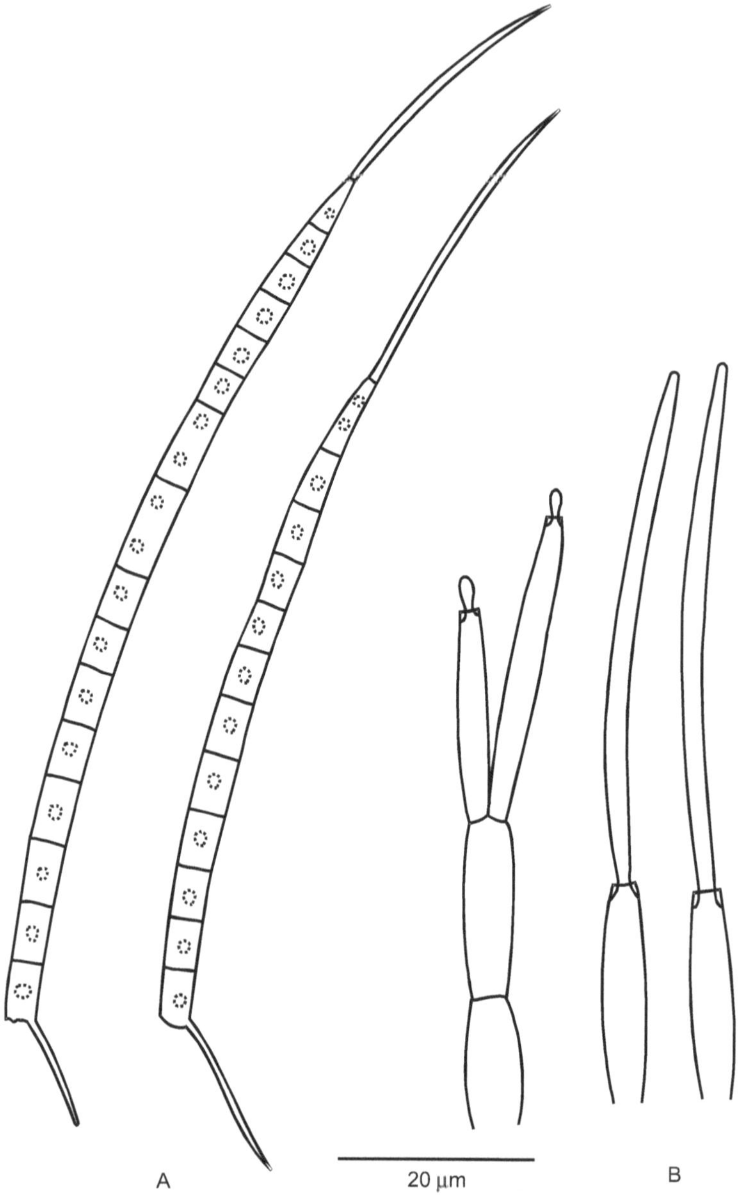

Figure 4.160 *Pseudolachmea longiciliata*, (a) Mature conidia; (b) Developing conidia.

123. *Pyrenochaeta ilicis* Wilson

Source

Wilson, *Scot. Bot. Rev.,* **1**: p. 161 (1912); Sutton, B.C., *The Coelomycetes,* CMI, England, 395 (1980).

Description

Conidiomata pycnidial, separate, globose, superificial, brown, unilocular, up to 750 µm diam. Pycnidial wall made up of thick-walled, brown cells at the outside and thin-walled, hyaline cells at the innerside giving rise to conidiogenous cells. Ostiole single, central, circular. Setae abundant around the ostiole, erect, dark brown, septate. Conidiogenous cells lining the pycnidial cavity enteroblastic, phialidic, integrated, arising as very short lateral branches immediately below transverse septa, 40-60 µm long. Conidia hyaline, aseptate, cylindrical to ellipsoid, 7-8.5 x 1.5 µm.

Specimen collected

On dead leaves of *Glycosmis cohinchinensis* Pierre., Courtallam, Tamilnadu, July 1989, collected by J. Muthumary. IC NO: 189.

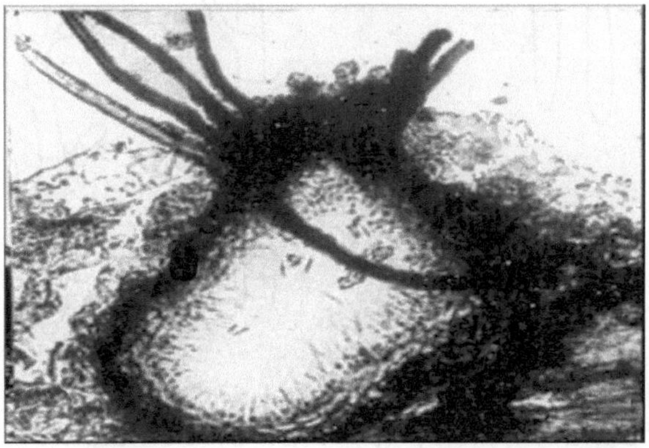

Figure 4.161 *Pyrenochaeta ilicis;* V.S. of Conidioma (×200).

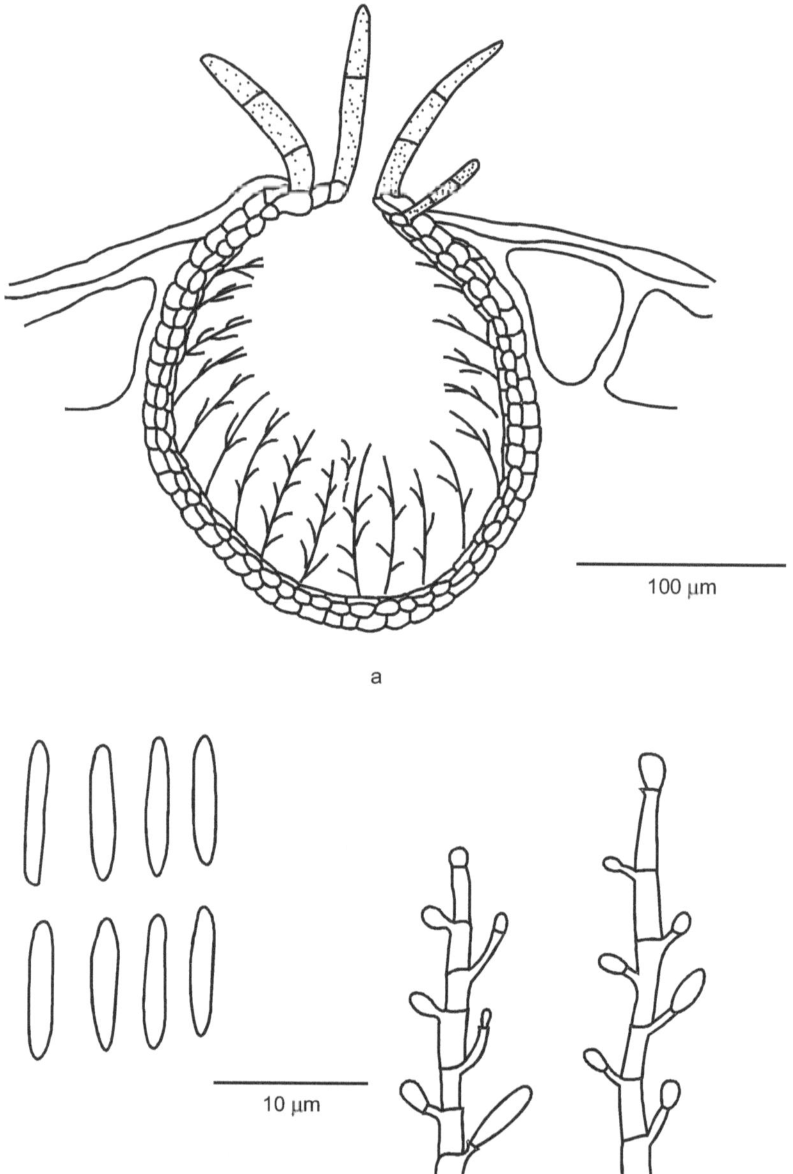

Figure 4.162 *Pyrenochaeta ilicis*, (a) V.S. of conidioma; (b) Mature conidia; (c) Developing conidia.

124. *Robillarda depazeoides* (W. & C.) Sacc.

Source

Sacc., *Syll. Fung.* (1884); Ramakrishnan, K., Ph.D. Thesis, Madras Univ. (1953); R.S. Mathur, *The Coelomycetes of India*, p. 235 (1979).

Description

Conidiomata pycnidial, immersed, dark brown, globose, 130–300 × 120–250 μm. Pycnidial wall made up of thick-walled, brown cells at the outside and thin-walled, hyaline cells at the innerside giving rise to conidiogenous cells. Conidiogenous cells lining the pycnidial cavity, holoblastic, lageniform, hyaline, smooth, 10–12 × 2–3 μm. Conidia hyaline, 0–1 septate, cylindrical, base truncate, apex obtuse, 14–20 × 2–4.5 μm; apical cell modified into 3–appendages, filiform, hyaline, 8–20 μm long.

Specimen collected

On dead leaves of *Commellina nudiflora* L., Madras, Tamilnadu, Sep. 1989, collected by J. Muthumary. IC. No: 185.

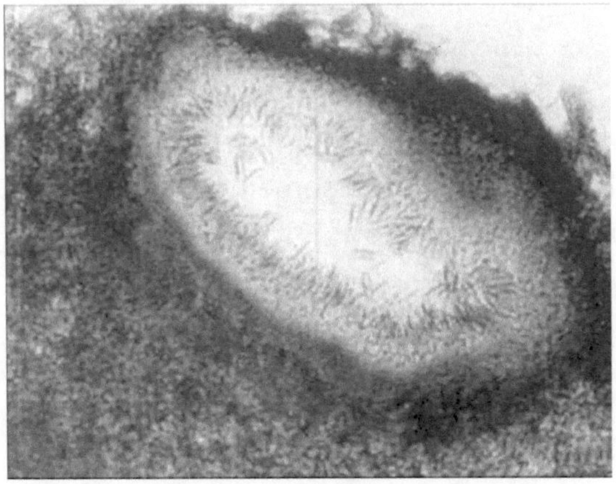

(a)

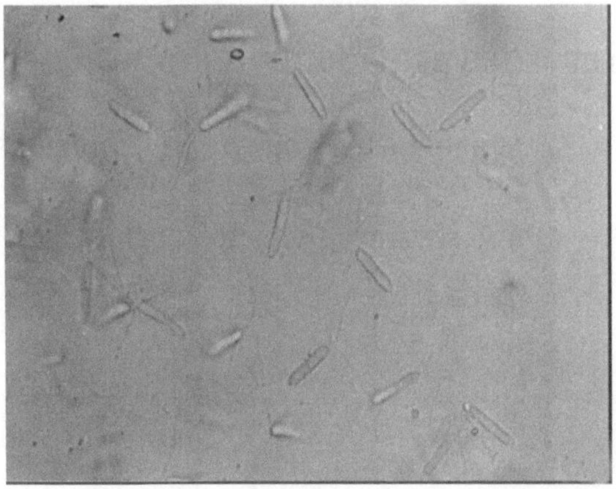

(b)

Figure 4.163 *Robillarda depazeoides;* (a) V.S. of Conidioma (×200); (b) Mature conidia (×900)

125. *Robillarda sessilis (Saccardo) Saccardo*

Source

Saccardo, *Michelia*, **2**(6): 8 (1880); R.S. Mathur, *The Coelomycetes of India*, p. 235 (1979); Nag Raj, T.R., *Coelomycetous anamorphs with appendage bearing conidia*, Mycologue Publications, Canada, pp. 763–767 (1993).

Description

Isolated as an endophyte from leaves of *Mimusops elengi,* colony white to grey on PDA. Pycnidial conidiomata appearing often 7–10 days. Pycnidium black, scattered to gregarious, unilocular, 200–250 μm diam. Wall cells brown to black in the outer layer and hyaline to thin-walled towards the inner conidiogenous region. Conidiogenous cells lining the pycnidial cavity, holoblastic, hyaline, smooth, 4.5–8.5 × 2–3 μm. conidia fusiform, straight to slightly curved, 1-septate, hyaline, smooth, 8–10 × 2–3 μm apical appendage cylindrical at the base, and diverging into 3–branches, each 15–20 μm long, flexuous, attenuated towards the tip.

Recorded as endophyte in

Isolated as endophyte in *Mimusops elengi, Rauvolfia tetraphylla,* V. Gangadevi, 2005, Ph.D. Thesis, "Studies on Fungal Endophytes from some medicinal plants and screening selected coelomycetes for taxol production". MUBL No: 716.

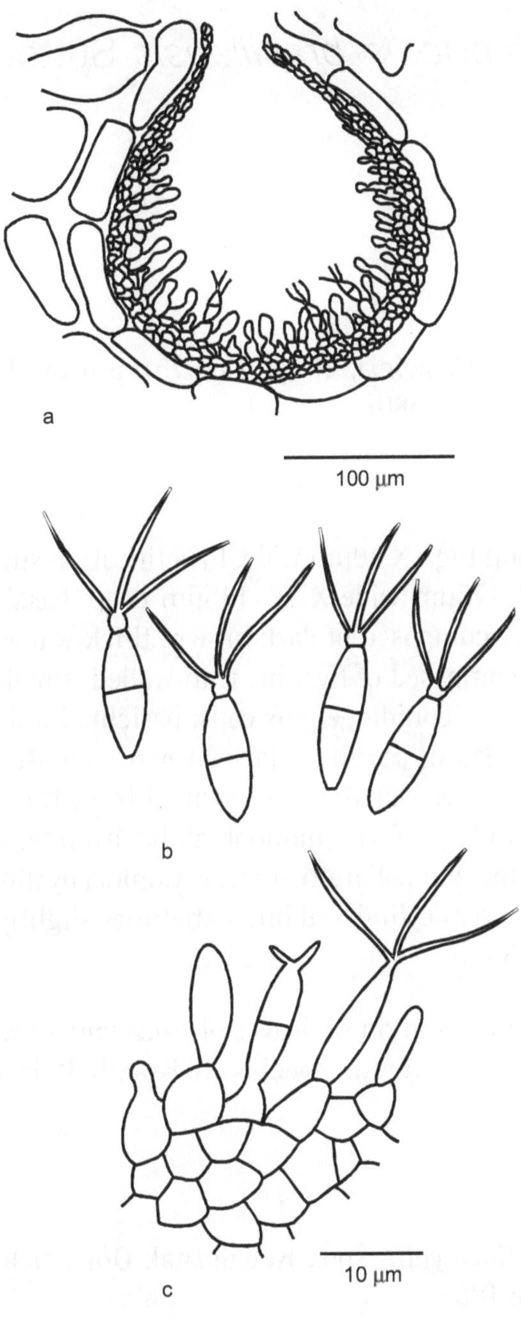

Figure 4.164 *Robillarda sessilis*, (a) V.S. of conidioma; (b) Mature conidia; (c) Developing conidia.

126. *Satchmopsis brasiliensis* Sutton & Hodges

Source

Sutton, B.C., *Nova Hedwigia*, **26**: p. 3 (1975); Sutton, B.C., *The Coelomycetes*, CMI, England, p. 483 (1980)

Description

Stromata up to 60μ high X 40μm wide. Fructification superficial, black, attached centrally, at first 175μm wide X 70–135μm deep; basal wall consisting of two regions, the lower composed of dark brown, thick-walled, larger cells, 3–5 cells thick, the upper composed of hyaline, thin-walled, smaller cells, 5–7 cells thick, giving rise directly to conidiogenous cells; periclinal wall 1-cell thick, composed of vertically elongated, parallel, dark brown, septate hyphae 5–6 μm wide, conidiogenous cells restricted to and formed from the upper cells of the basal wall, papilliform to lageniform, monophialidic, hyaline, smooth, 4–7μm × 1.5–3 μm, aperture distinct but collarette minute. Conidia hyaline, aseptate, occasionally guttulate, more or less cylindrical but sometimes slightly bent both ends obtuse 11.5–15.5 × 1–1.5 μm.

The species is quite common on leaves of *Eucalyptus paniculata, E. grandis* and several unnamed *Eucalyptus* species in Brazil. It is saprophytic occurring consistently in leaf litter.

Specimen collected

On leaf litter of *Eucalyptus* spp., Kodaikanal, Ooty, October 1989, collected by K. Sudha. IC. No: 186.

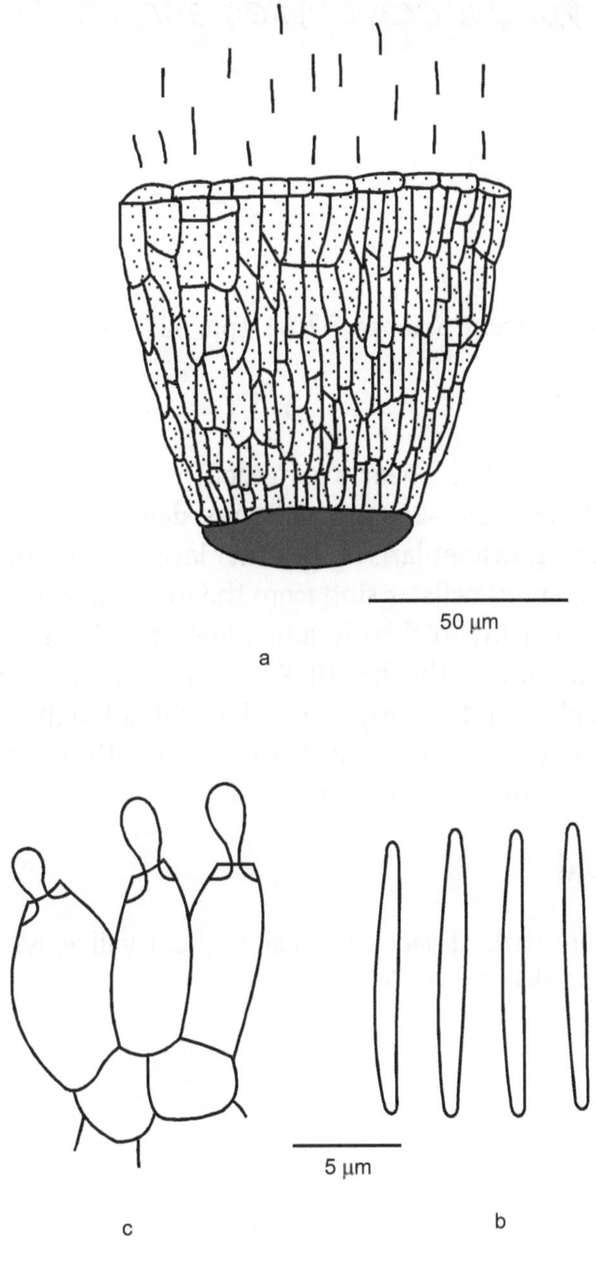

Figure 4.165 *Satchmopsis brasiliensis*, (a) V.S. of conidioma; (b) Mature conidia; (c) Developing conidia.

127. *Scyphospora phyllostachydis* Muthumary

Source

Muthumary, J., *Current Science*, **57**(11): pp. 621–622 (1988).

Description

Conidiomata stromatic, linearly arranged, black, sub-epidermal, unilocular. Flattened, 800–950 × 200–250 µm, glabrous, dehiscing irregularly, wall made up of pale brown, textura angularis in the outer layers, becoming hyaline in the inner layers. Conidiogenous cells arising from the inner most layers of the cells lining the bottom of the cavity, holoblastic, ampuliform to cylindrical, pale brown, smooth to slightly verrucose near the tips. 10–20 × 5–6 µm. Conidia amerosporous, brown, thick-walled, spherical to nearly triangular, with a longitudinal hyaline band on the flat surface, 25.5–37.5 µm in diam., often with an appendage at the base measuring 25–30 µm long and 3–5 µm wide.

Specimen collected

On dead *Bambusa arundinacea*, Dam site, Silent Valley, Kerala, December 1978, collected by G. Sekar. MUBL No: 2921.

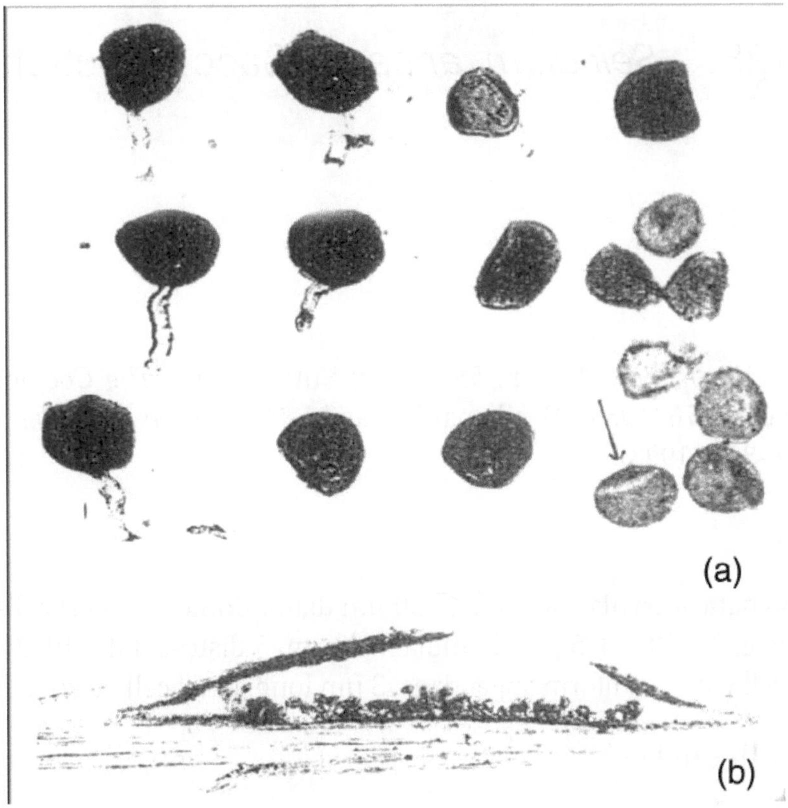

Figure 4.166 *Scyphospora phyllostachydis;* (a) Mature conidia (×900); (b) V.S. of Conidioma (×200).

128. *Seiridium anceps* (Sacc.). Sutton

Source

Sutton, *Mycol. Pap.*, **138**: p. 58 (1975); Sutton, B.C., *The Coelomycetes*, CMI, England, p. 276 (1980); Masilamani, S. and J. Muthumary, *J. Indian Bot. Soc.*, **72**: pp. 311–312 (1993).

Description

Conidiomata acervular, separate, 150 µm diam. Conidiogenous cells holoblastic, annellidic, 25–40 × 3–5 µm. Conidia fusiform, 5-distoseptate, 19–24 × 8–8.5 µm; apical cell with corniform appendage, 3 µm long, basal cell lacking an appendage.

Specimen collected

On dead leaves of *Thuja plicata* D., Kodaikanal, Tamilnadu, Nov. 1987, collected by J. Muthumary. IC. No: 135.

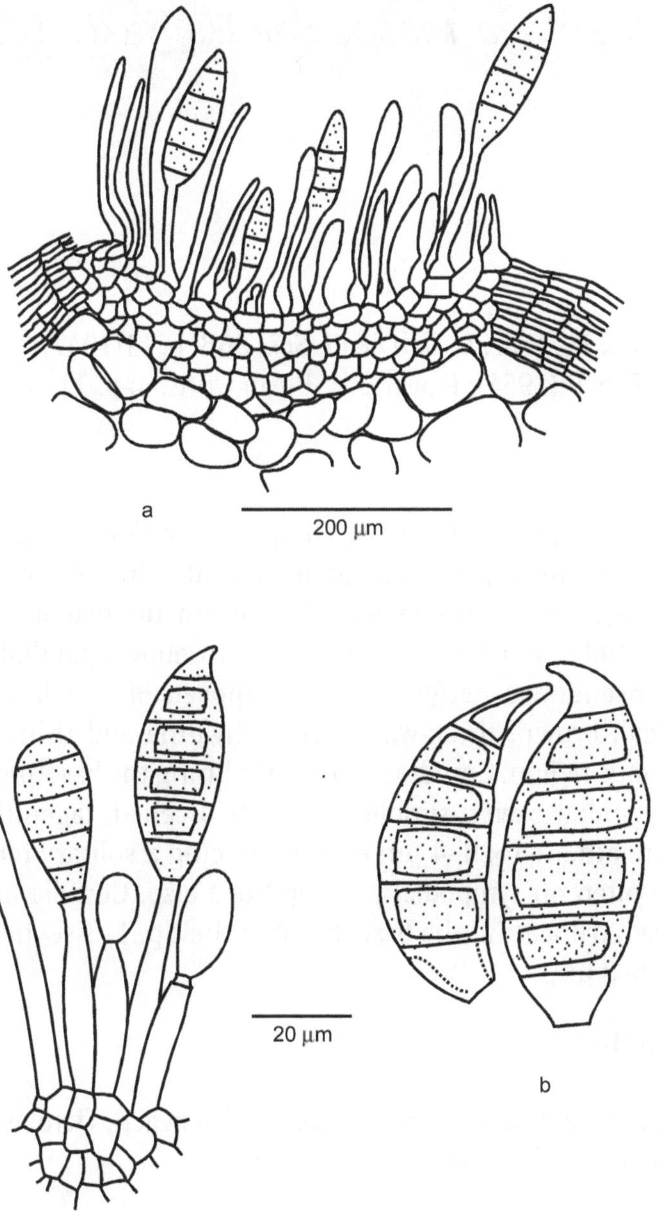

Figure 4.167 *Seiridium anceps*, (a) V.S. of conidioma; (b) Mature conidia; (c) Developing conidia.

129. *Septoria thespesiae* Ramakr. T.S. & K.

Source

Ramakr. T.S. & K., *Proc. Indian Acad. Sci.*, **26B**: p. 11(1947); Sukapure & Thirum., *Sydowia*, **17**: p. 9 (1964); Pawar, J., *Shivaji Univ.*, 4: p. 11 (1971).

Description

Fungus causing spots on leaves of *Thespesia papulnea,* spots amphigenous, dark brown to black, circular to irregular, often coalescing, 3–8 mm in diam. Pycnidia amphigenous, scattered, subepidermal, pyriform, unilocular, 100–200 µm in diam, ostiolate; ostiole circular when viewed from above, papillate, 50 µm in diam. Peridium pseudoparenchymatous, composed of 2–3 layers of spherical to subspherical, brown, thick-walled cells, hyaline and thin-walled towards the conidiogenous region. Conidiogenous cells lining the base and to some extent the sides of pycnidial cavity, simple, one to two-celled, lageniform to cylindrical, hyaline, smooth, 5–8 × 3–4 µm each producing a solitary terminal conidium at its tip and often on sympodial proliferations also. Conidia fusiform, straight to often curved, truncate at the base, blunt at the tip, 1–3-septate, hyaline, smooth, 20–30 × 2.5–3 µm.

Specimen collected

On dead leaves of *Thespesia populnea*, Marina beach, October 1983, collected by J. Muthumary. IC. No: 104.

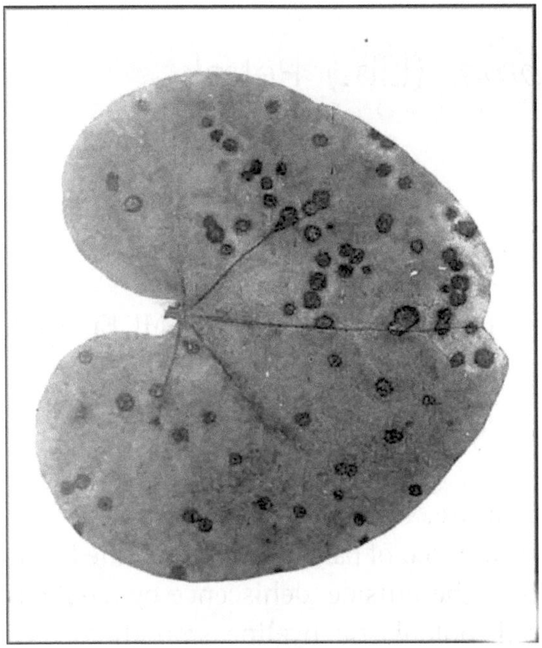

(a)

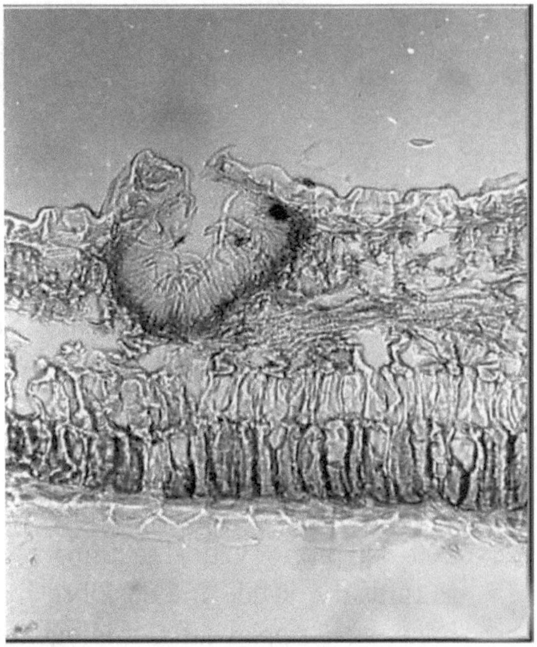

(b)

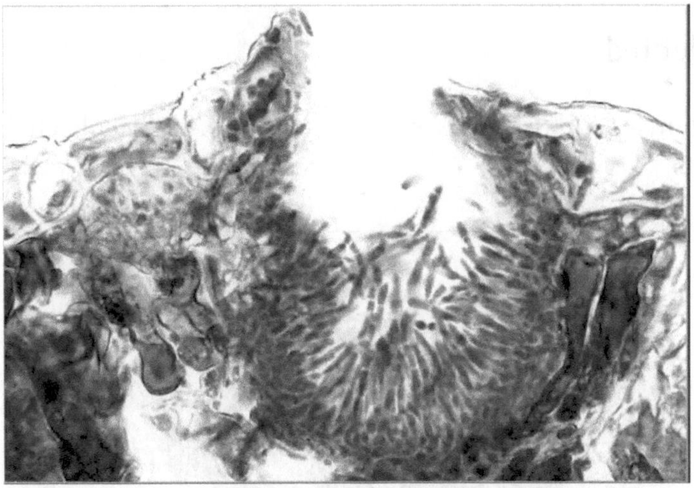

(c)

Figure 4.168 *Septoria thespesiae;* (a) infected leaf of *Thespesia populnea*; (b–c) V.S. of Conidiomata (×200)

130. *Septocyta ruborum* (Lib.) Petrak

Source

Petrak, *Sydowia,* **21**: p. 245 (1968); Sutton, B.C., *The Coelomycetes,* CMI, England, pp. 250–251(1980).

Description

Conidiomata stromatic, immersed, separate, brown to black, multilocular, 200–300 µm in vertical section, stroma composed of pale brown thin walled cells towards the inside and dark brown cells at the outside, dehiscence by irregular breaking of upperwall. Conidiogenous cells holoblastic, hyaline, smooth, formed from the inner cells of the stroma. Conidia hyaline, smooth, 3–4 septate, straight to slightly curved, filiform, apex obtuse, base truncate 15–25 × 1–1.5 µm.

Specimen collected

On dead stems of *Rubus idaeus,* Kambakkam, Feb. 1983, collected by J. Muthumary. IC. No: 105.

Figure 4.169 *Septocyta ruborum;* V.S. of Conidioma (×260).

131. *phaerellopsis filum* (Bivona-Bornar: : Fries) Sutton

Source

Sutton, B.C., *Mycol. Pap.,* **141**: p. 196 (1977); Sutton, B.C., *The coelomycetes,* CMI, England, p. 471 (1980); Masilamani, S. and J. Muthumary, *J. Indian Bot. Soc.,* **72**: pp. 311–312 (1993).

Description

Conidiomata eustromatic, immersed, unilocular, 200–250 µm diam. wall composed of thin-hyaline to pale brown thick-walled cells. conidiogenous cells enteroblastic, phialidic, cylindrical, hyaline, smooth, 10–25 × 5–7 µm. Conidia hyaline, 1-euseptae, apex obtuse, base truncate, straight, ellipsoid, 17–20 × 5.5–6.5 µm, the apex with gelatinous cap. Nag Raj (1993) had elaborately diccussed about the species with synonyms.

Specimen collected

On dead leaves of *Araucaria bidwillii* hook., Nilgiris, Tamilnadu, Feb. 1988, collected by J. Muthumary. IC. No: 161.

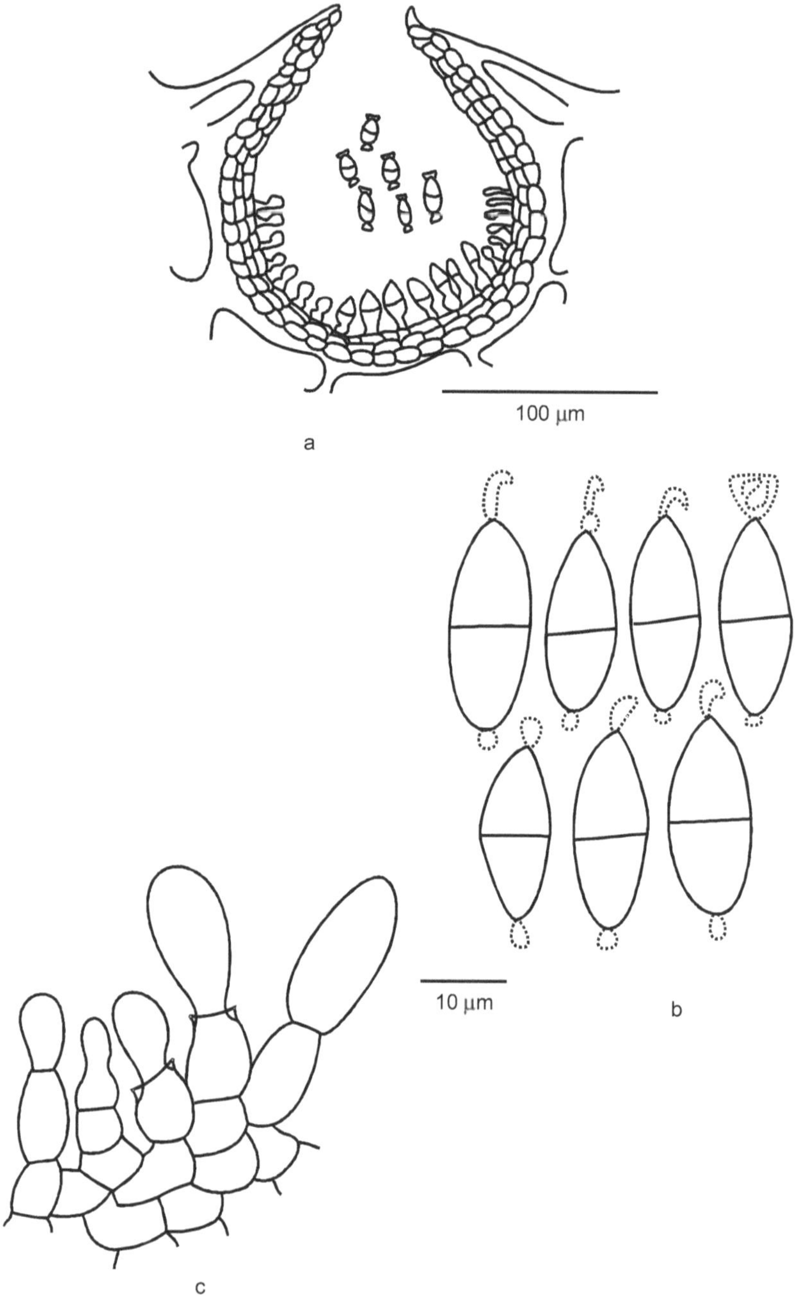

Figure 4.170 *Sphaerellopsis filum*, (a) V.S. of conidioma; (b) Mature conidia; (c) Developing conidia.

132. *Sphaeropsis tumefaciens* Hedges

Source

Hedges, *Phytopathology*, **1**: pp. 63-65 (1911); Prasad, N. & G.C. Bhatnagar, *Curr. Sci.*, **30**: pp. 110-111 (1961); R.S. Mathur, *The Coelomycetes of India*, p. 265 (1979).

Description

Conidiomata pycnidial, sub-globose, dark brown, 160–190 µm diam. Ostiole central, circular. Pycnidial wall made up of thick-walled, brown cells at the outside and thin-walled, hyaline cells at the innerside giving rise to conidiogenous cells. Conidiogenous cells lining the pycnidial cavity cylindrical, hyaline, smooth, annellidic, 5–12 × 3–4 µm. Conidia single celled, pale brown to dark brown, oblong to ovoid, 17.5–24 × 9–12 µm.

Specimen collected

On twigs of *Citrus medica* L., On dead leaves of *Dracaena* sp., Madras, Tamilnadu, May 1988; On twigs of *Azadirachta indica* A. Juss. and on twigs of *Citrus limonum* L. Madras, Tamilnadu, July 1989, collected by J. Muthumary. IC No: 187.

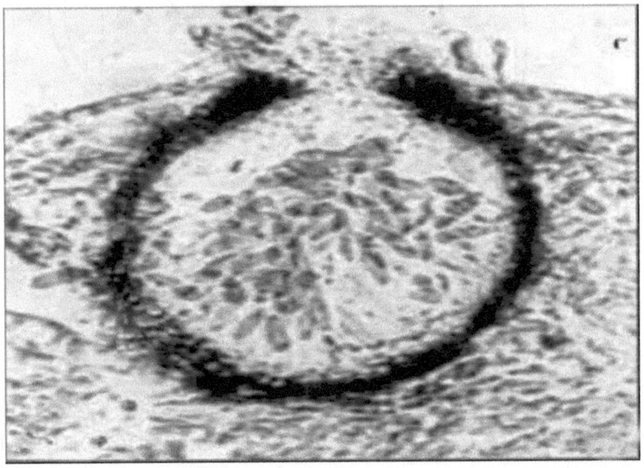

Figure 4.171 *Sphaeropsis tumefaciens*; V.S. of Conidioma (×260).

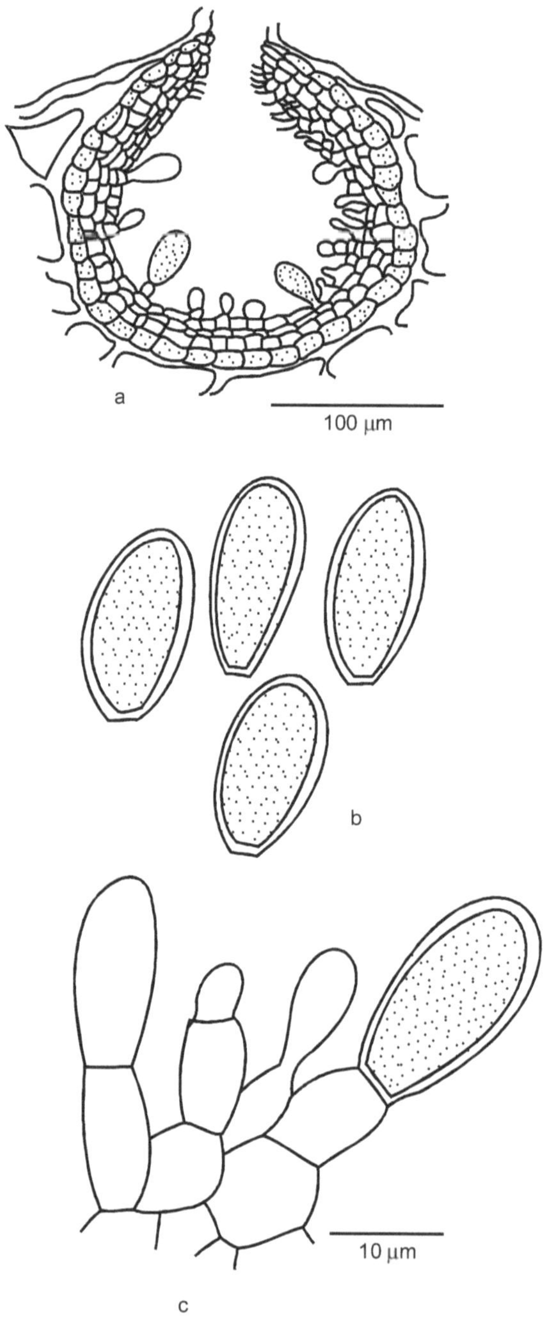

Figure 4.172 *Sphaeropsis tumefaciens*, (a) V.S. of conidioma; (b) Mature conidia; (c) Developing conidia.

133. *Stagonospora brideliae* Thirum. & Narasimhan

Source

Thirum. & Narasimhan, *Sydowia*, **4**: p. 71 (1950); R.S Mathur, *The Coelomycetes of India*, p. 265 (1979).

Description

Conidiomata pycnidial, immersed, separate, globose, unilocular, 200–250 µm diam. Ostiole central, circular. Pycnidial wall made up of thick-walled, brown cells at the outside and thin-walled, hyaline cells at the innerside giving rise to conidiogenous cells. Conidiogenous cells lining the pycnidial cavity holoblastic, discrete, doliiform, 71–10 × 3.5 µm. Conidia hyaline, smooth, with 2–3 transverse eusepta, sometimes constricted at the septa.

Specimen collected

On dead twigs of *Acacia decurrens* Willd, Kodaikanal, Tamilnadu, Nov. 1987, collected by J. Muthumary. IC No: 144.

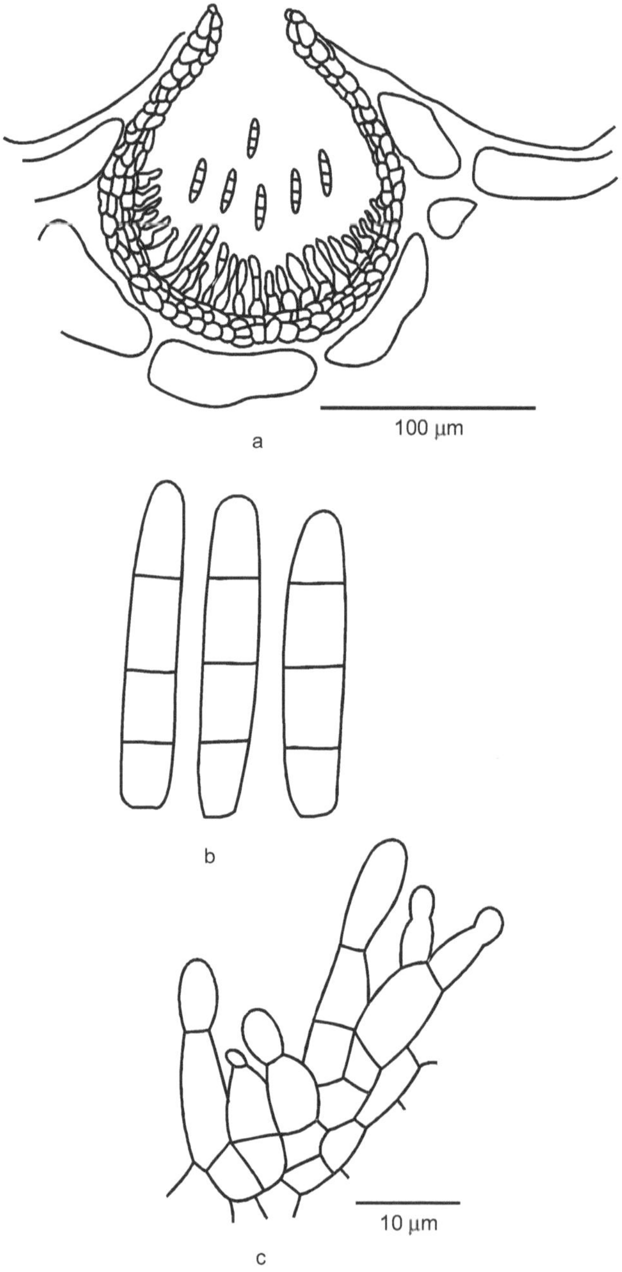

Figure 4.173 *Stagonospora brideliae*, (a) V.S. of conidioma; (b) Mature conidia; (c) Developing conidia.

134. *Stauronema cruciferum* (Ellis) Sydow

Source

Sydow, P. and E.J. Butler, *Ann. Mycol.*, **14**: p. 217 (1916); Nag Raj, T.R., *Coelomycetous anamorphs with appendage bearing conidia,* Mycologue Publications, Canada, pp. 905–906 (1993); Mathur, *The Coelomycetes of India,* p. 267 (1979); Sutton, B.C., *The Coelomycetes,* CMI, England, p. 466 (1980).

Description

Foliicolous conidiomata stromatic, cupulate, scattered to gregarious, superficial, up to 350 μm in diam. black, setose. Basal stroma made of pale brown to brown, thick-walled, pseudoparenchymatous cells at the base, hyaline to pale brown cells towards the conidiogenous region. Conidiomatal setae surrounding the cupulate stroma, 90–150 μm long wider at the base, alternatively towards the tip, thick–walled, smooth, dark brown, conidiogenous cells lining the base and to some extent the sides of the cupulate stroma, septate, hyaline, smooth, cylindrical, with periclinal thickening, phialidic, 10–15 × 1.5–2.5 μm. conidia ellipsoid to fusiform, hyaline, smooth, one-celled, 8–15 × 2–3 μm, apical and basal appendages 9–15μm long, basal appendage excentric, lateral appendages 2–3, all appendages filiform, 0.5–1.0 μm thick.

Specimen collected

Dead leaves of *Zea mays,* Theni and Madurai. Tamilnadu, Oct. 1988, collected by J. Muthumary. IC. No: 145.

Figure 4.174 *Stauronema cruciferum;* V.S. of Conidioma (×260).

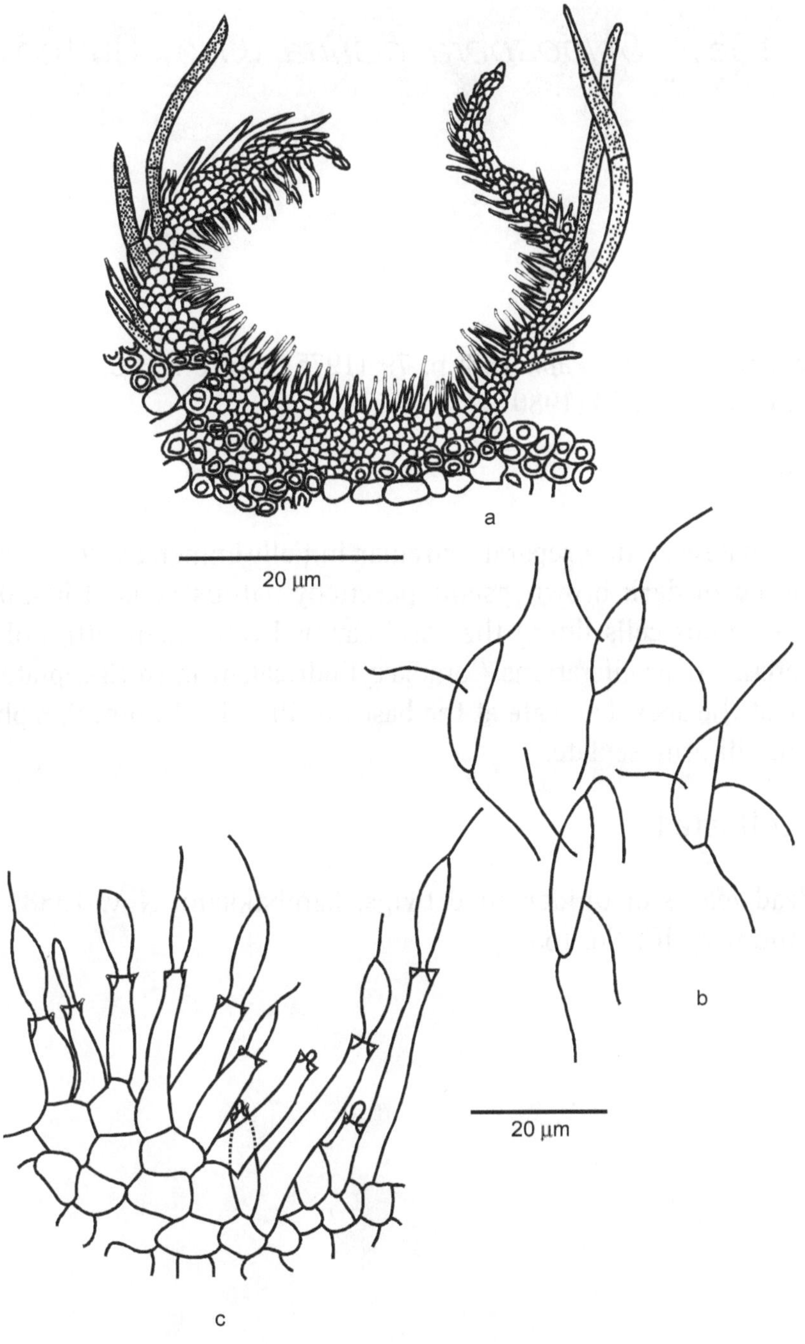

Figure 4.175 *Stauronema cruciferum*, (a) V.S. of conidioma; (b) Mature conidia; (c) Developing conidia.

135. *Stilbospora cistina* (Cke) Sutton

Source

Sutton, B.C., *Mycol. Pap.*, **138**: p. 79 (1975); Sutton, B.C., *The Coelomycetes*, CMI, England, p. 343 (1980).

Description

Conidiomata acervular, separate, circular, initially immersed, dehiscing irregularly, composed of dark brown pseudoparenchymatous cells, 150–200 µm diam. Conidiogenous cells lining the basal cavity, hyaline, smooth, holoblastic with 1–2 percurrent proliferations. Conidia cylindrical, fusiform, 3-septate, dark brown, obtuse at the apex, truncate at the base 30–40 × 10–15 µm. Paraphysis hyaline, smooth, filiform, septate.

Specimen collected

On dead *leaves* of unidentified twigs, kambakkam, Nov. 1988, collected by J. Muthumary. IC. No: 163.

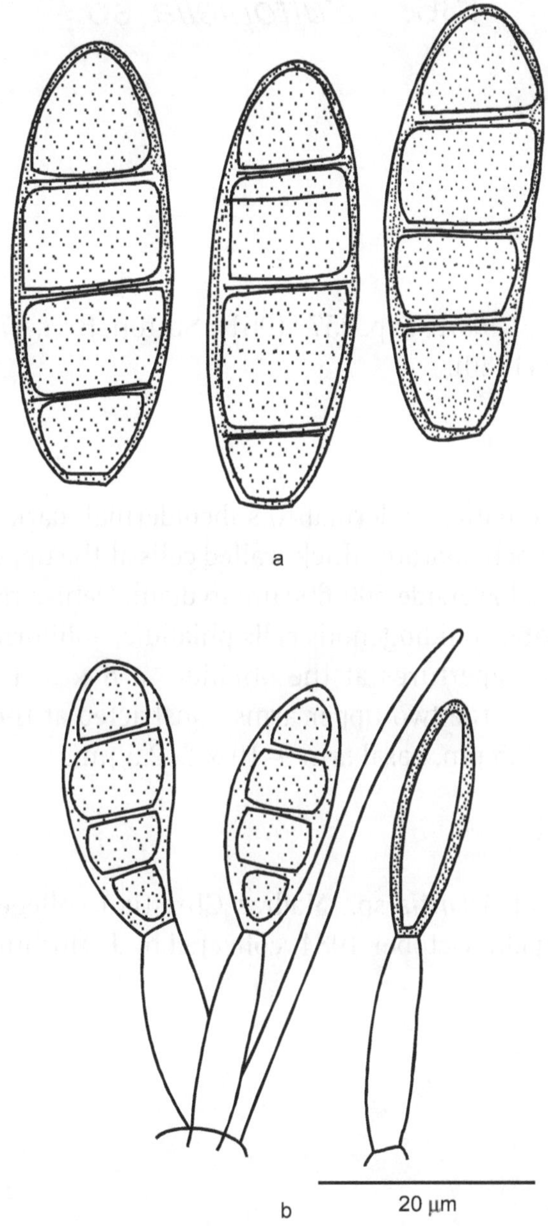

Figure 4.176 *Stilbospora cistina*, (a) Mature conidia; (b) Developing conidia.

136. *Suttoniella* sp.

Source

Ahmad, *Biologia, Labore,* **6**: p. 257 (1961); Sutton, B.C., *The Coelomycetes,* CMI, England, p. 450 (1980).

Description

Conidiomata stromatic, epidermal to subepidermal, dark brown to black, made up of pseudoparenchymatous, thick-walled cells at the upperside, hyaline to thin-walled cells at the lowerside 500-600 µm in diam. Dehiscing irregularly at the top of the conidiomata, conidiogenous cells phialidic, doliiform to irregular, hyaline, smooth, with 1-2 apertures at the phialide, 7–8 × 3–4 µm. Conidia hyaline, 1-celled, Y–shaped, the two upper arms constricted at the base, smooth, upper arms 10–12 × 2–2.5 µm, basal arm 9–10 × 2–2.5 µm.

Specimen collected

On dead leaves of *Atlantia* sp., Madras Christian college campus, Tambaram, Chennai, Tamilnadu, October 1984, collected by J. Muthumary. IC. No:108.

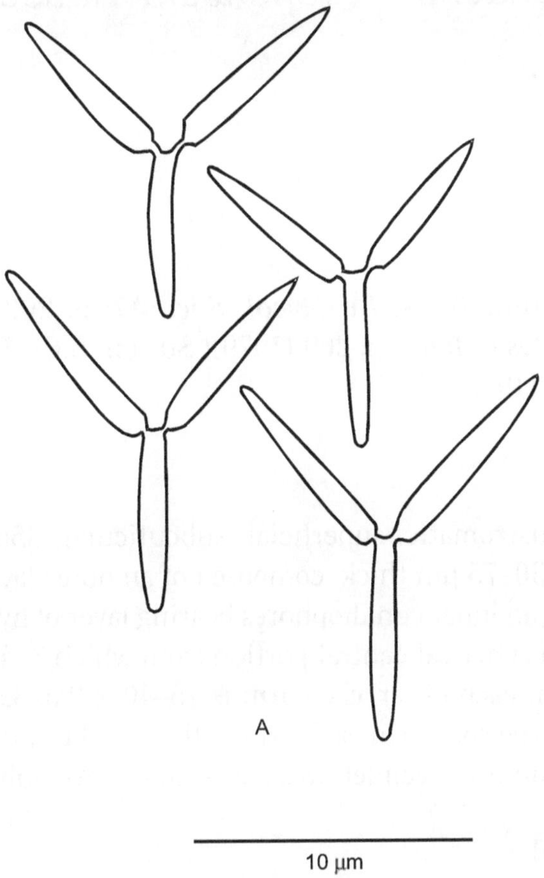

Figure 4.177 *Suttoniella* sp., Mature conidia.

137. *Tetranacrium gramineum* Hudson & Sutton

Source

Hudson & Sutton, *Trans. Br. Mycol. Soc.,* **47**: p. 202 (1964); R.S. Mathur, *The Coelomycetes of India*, p. 269 (1979); Sutton, B.C., *The coelomycetes,* CMI, England, p.142 (1980).

Description

Conidiomata eustromatic, superficial, subcuticular, 350–550 × 120–200 µm. Stromatic wall 30–75 µm thick, composed of an outer layer of dark brown thin-walled cells and an inner conidiophores bearing layer of hyaline thin-walled cells. Conidium has a spherical central portion from which 3–4, divergent arms arise, equidistant from each other, each arm is 15–40 × 2.5–3.5 µm and up to 1 to 2 septate. In the type specimen each arm of the conidium has 2–4 septa but in the present collection the specimen was observed to have only 2-septa.

Specimen collected

On dead leaf of *Bambusa arundinacea* Willd., Kodaikanal, Tamilnadu, Nov. 1987, Collected by J. Muthumary. IC. No: 146.

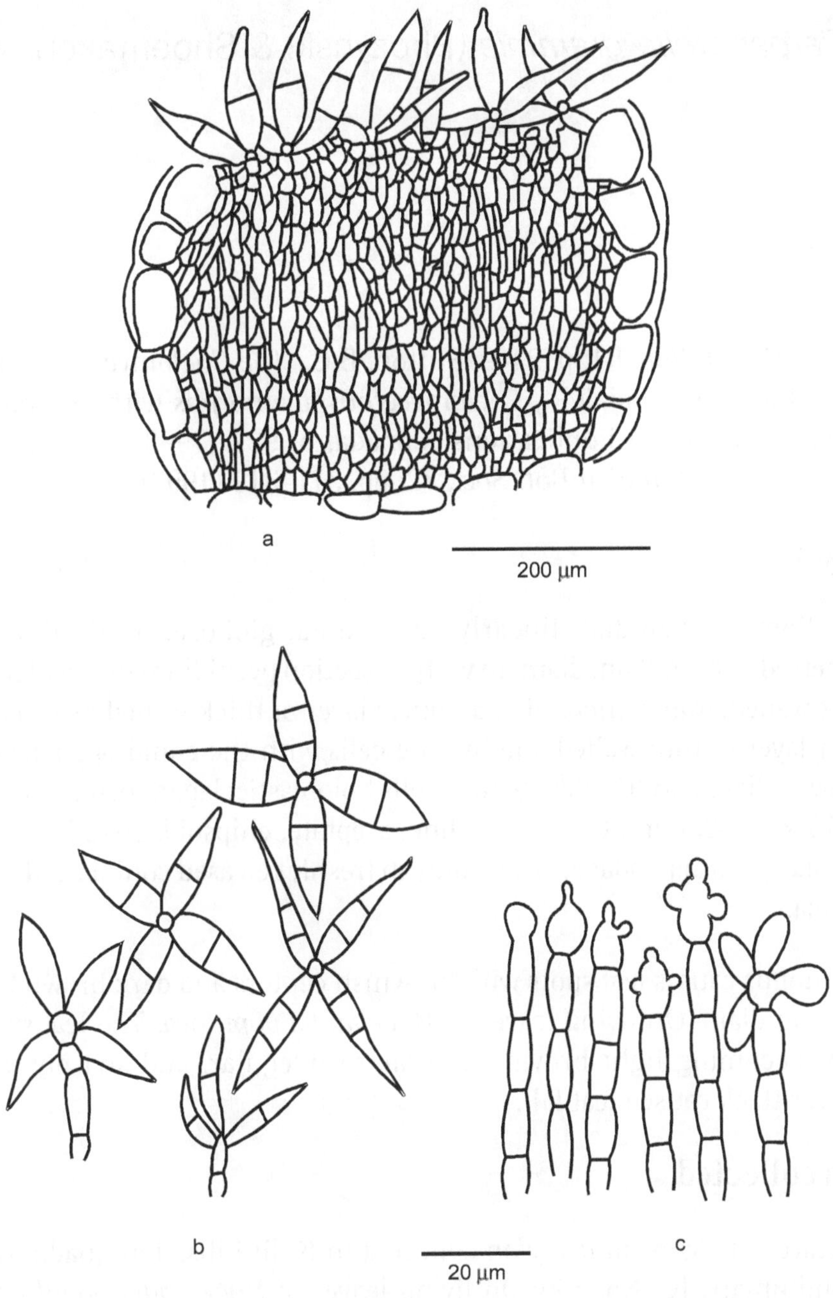

Figure 4.178 *Tetranacrium gramineum*, (a) V.S. of conidioma; (b) Mature conidia; (c) Developing conidia.

138. *Tiarosporella graminis* (Pirozynski & Shoemaker) Nag Raj

Source

Nagraj, *Can. J. Bot.*, **51**: p. 2469: Sutton, B.C., *The Coelomycetes*, CMI, England, p. 58 (1980); Nag Raj, T.R., *Coelomycetous anamorphs with appendage bearing conidia*, Mycologue Publications, Canada, p. 927 (1993); Masilamani, S. and J. Muthumary, *J. Indian Bot. Soc.*, **72**: pp. 311–312 (1993).

Description

Conidiomata pycnidial, linearly aggregated, globose, dark brown, glabrous immersed, 200–250 µm diam. In vertical section pycnidium unilocular, immersed, thick-walled, wall composed of an outer layer of thick-walled brown cells and an inner layer of thin-walled and hyaline cells with the conidiogenous cells lining the pycnidial cavity. Conidiogenous cells holoblastic, lageniform, hyaline, smooth, 10–15 × 1.5–2.0 µm. Conidia hyaline, aseptate, ellipsoid, 20–25 × 5–6 µm, with mucoid apical appendage, visible only in freshly released conidia and absent in old conidia.

The fungus causes leaf spots with brownish centers and dark brown border up to 8 mm in diam. On living leaves of *Bucklandia populnea*. The leaf spots initially grew, becoming light brown, circular to irregular, and sharply demarcated. Severe attack caused leaf fall.

Specimen collected

On leaves of *Cynodon dactylon* collected in Kolli hills, Tamilnadu, collected by J. Muthumary, IC. No. 556; On living leaves of *Bucklandia populnea* Br., Ooty, Tamilnadu, Nov. 1987, collected by J. Muthumary. IC. No: 99.

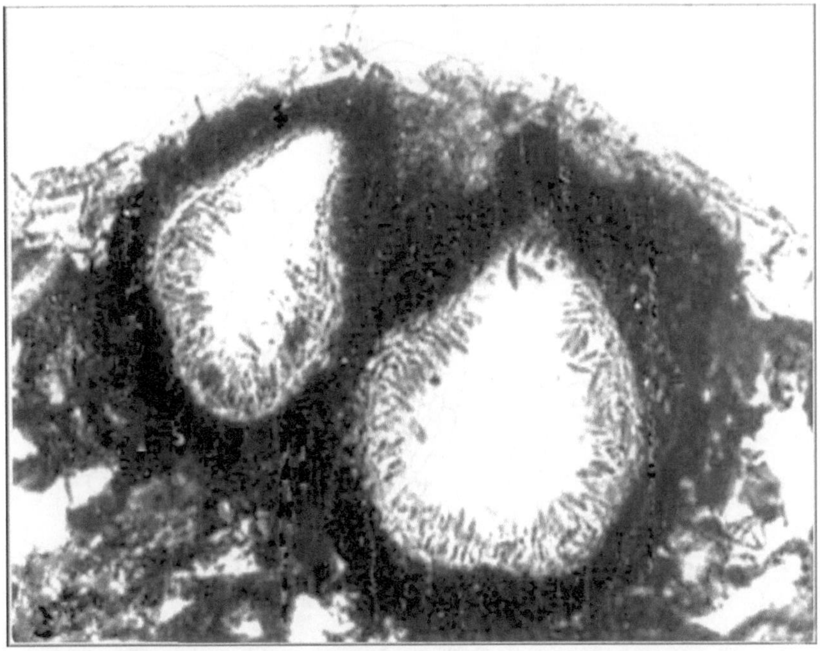

Figure 4.179 *Tiarosporella graminis;* V.S. of Conidioma (×260).

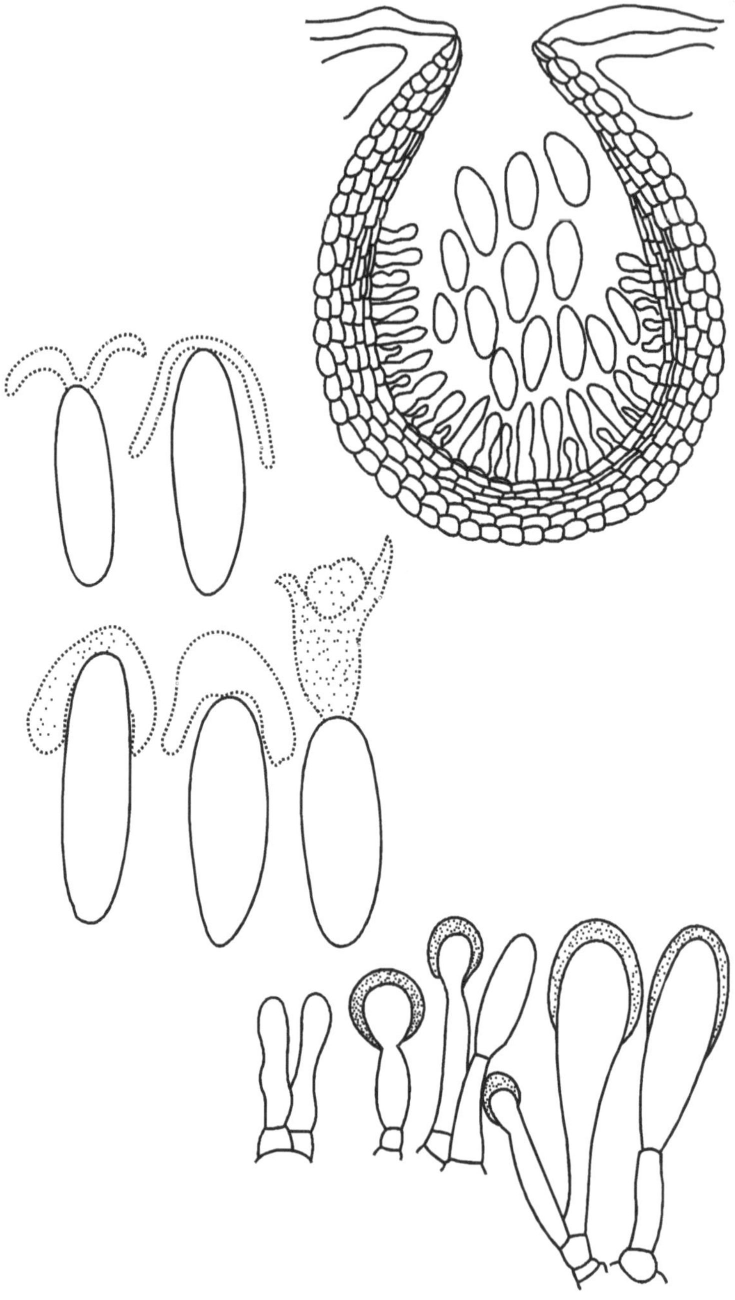

Figure 4.180 *Tiarosporella graminis*, (a) V.S. of conidioma; (b) Mature conidia; (c) Developing conidia.

139. *Tiarosporella madreeya* (Subram. & Ramakr.) Nag Raj

Source

Nag Raj, *Can. J. Bot.*, **51**: p. 2470 (1973) = *Sakireeta madreeya* Subram. & Ramakr., *J. Indian Bot. Soc.*, **36**: p. 84 (1957); R.S. Mathur, *The Coelomycetes of India,* p. 270 (1979); Nag Raj, T.R., *Coelomycetous anamorphs with appendage bearing conidia,* Mycologue Publications, Canada, p. 927 (1993).

Description

Conidiomata pycnidial, brown, immersed, unilocular, 200–220 µm diam. Ostiole central, circular. Pycnidial wall made up of thick-walled, brown cells at the outside and thin-walled, hyaline cells at the innerside giving rise to conidiogenous cells. Conidiogenous cells lining the pycnidial cavity lageniform, 7–10 × 2–4.5 µm. Conidia cylindrical, straight, with an apical conical, gelatinous appendage, revealed by modified Leifson's flagella staining, 17–25 × 4–6 µm.

Known distribution, India mainly from grasses.

Specimen collected

On dead leaves of *Rhododendron hodgsoni* Hook., Shimla, Himachala Pradesh, Oct. 1989, collected by J. Muthumary. IC No: 188.

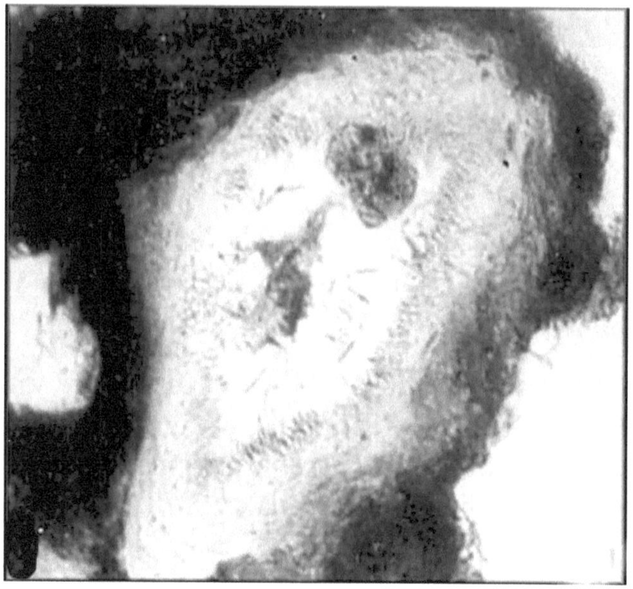

(a)

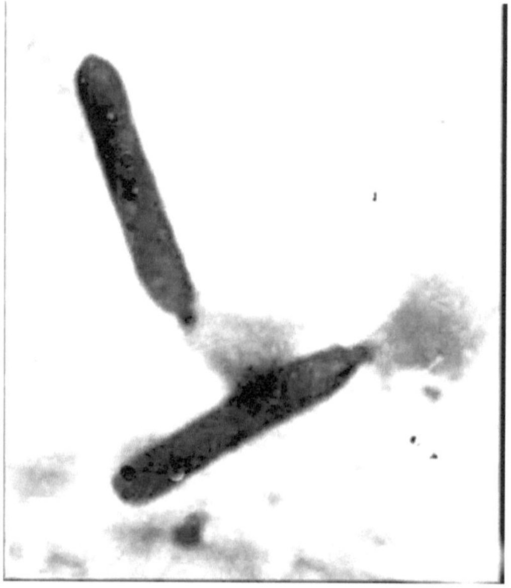

(b)

Figure 4.181 *Tiarosporella madreeya;* (a) V.S. of Conidioma (×260); (b) Mature conidia (×900).

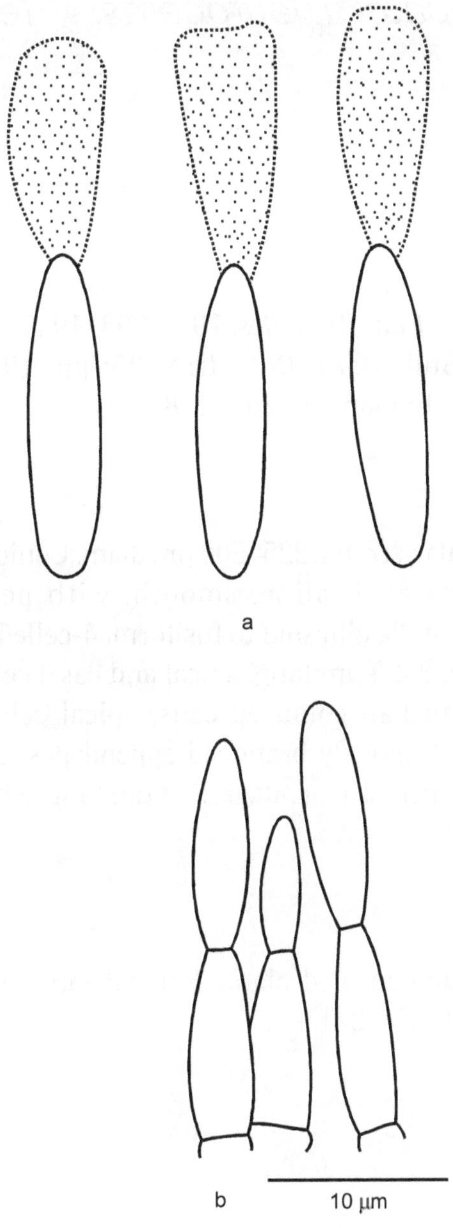

Figure 4.182 *Tiarosporella madreeya*, (a) Mature conidia; (b) Developing conidia.

140. *Truncatella laurocerasi* (Westd.) Stey.

Source

Stey., *Bull. Jard. Bot. Etat. Bruxelles*, **19**: p. 298 (1949); Bull., *Acad. Roy. Bolg.Ser.* II, **2**:556 (1857); Bull., *Jard. Bot. Brux*, **25**: pp. 191-199; Sutton, B.C., *The Coelomycetes*, CMI, England, p. 265 (1980).

Description

Conidiomata acervular, stroma 225–500 µm diam. Conidiogenous cells holoblastic, annellidic, cylindrical, hyaline, smooth, with percurrent proliferations, 10–20 × 2–3 µm. Conidia ellipsoid to fusiform, 4-celled, 18–25 × 5.5–7.5 µm, two median cells brown, 8–20 µm long, apical and basal cells hyaline, hemispherical, smaller than the median coloured cells, apical cell with 1–4 simple or 2–4 irregularly or dichotomously branched appendages, 10–30 µm long, basal cell with a simple, unbranched appendage, 2–3 µm long, which disappears at maturity of conidium.

Specimen collected

On dead twigs of unidentified plant, Kambakkam, October 1990, collected by J. A. Jayachandra. IC. No: 211.

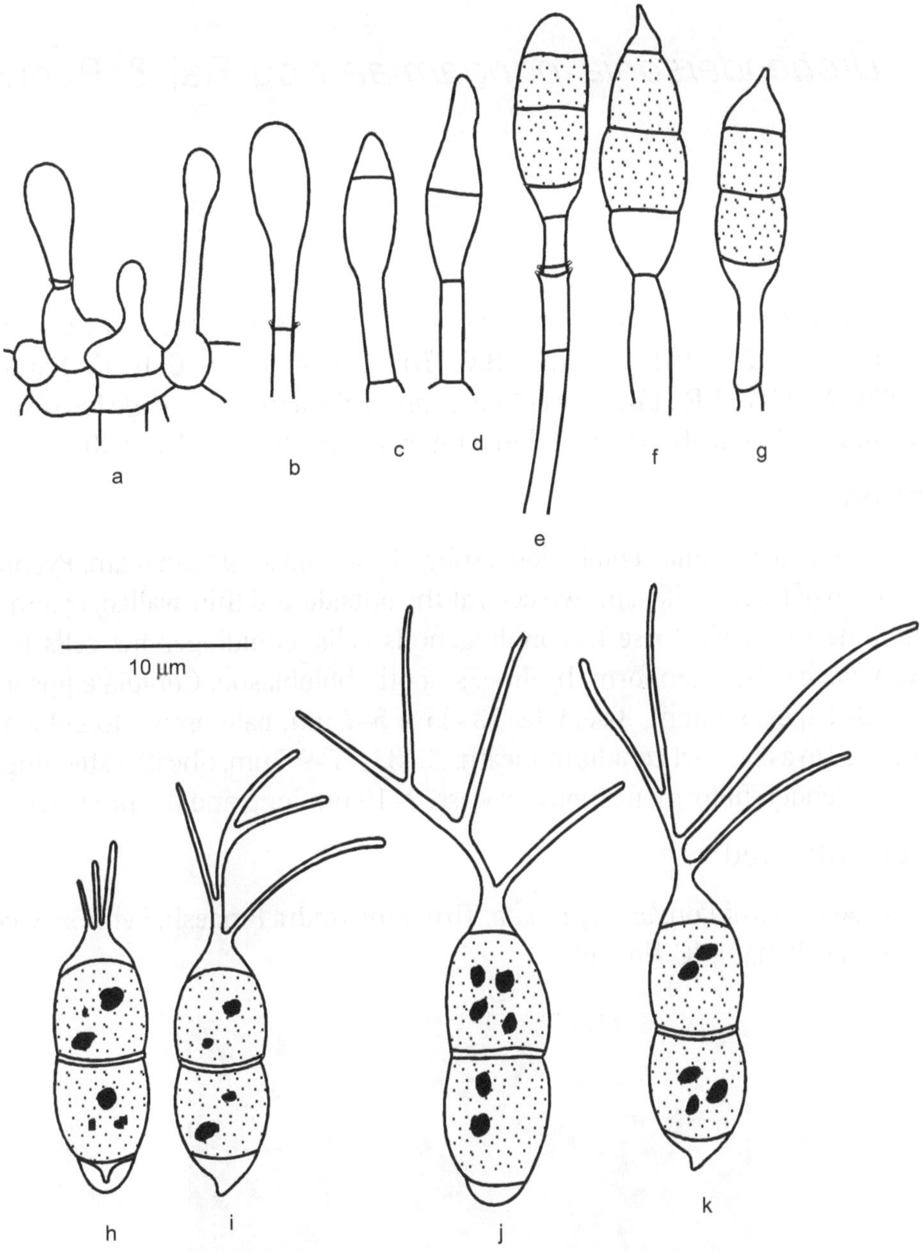

Figure 4.183 *Truncatella laurocerasi*, (a–b) Conidiogenous cell with young conidia; (c–d) Young conidia showing the first formed septum; (e) Annelledes; (f–g) Young conidia showing initiation of apical appendage and two median coloured cells; (h–i) Apical appendage and minute endogenous appendage in 'h'; (j–k) Mature conidia.

141. *Urohendersonia pongamiae* Nag Raj & Ponnappa

Source

Nag Raj & Ponnappa., *Curr. Sci.*, **37**: p. 417 (1968); R.S. Mathur, *The Coelomycetes of India*, p. 271 (1979); Sutton, B.C. *The Coelomycetes* CMI, England, p. 112 (1980); Nag Raj, T.R., *Coelomycetous anamorphs with appendage bearing conidia*, Mycologue Publications, Waterloo, Ontario, Canada, p. 951 (1993).

Description

Conidiomata pycnidial, epiphyllous, subglobose, 160 × 190 µm diam. Pycnidial wall made up of thick-walled, brown cells at the outside and thin-walled, hyaline cells at the innerside giving rise to conidiogenous cells. conidiogenous cells lining the pycnidial cavity, lageniform, hyaline, smooth, holoblastic. Conidia ellipsoidal with rounded apex, straight, 3-septate, 13–25 × 5–7 µm, pale brown to golden brown, enclosed in a compact conidium sheath, 23–33 × 7–9.5 µm. Sheath extending apically into a slender, filiform, flexuous process, 2–10 µm long and 0.5 µm wide.

Specimen collected

On dead leaves of *Pongamia pinnata*, Tirupathi, Andra Pradesh, Feb. 1988, collected by S. Masilamani. IC No: 162.

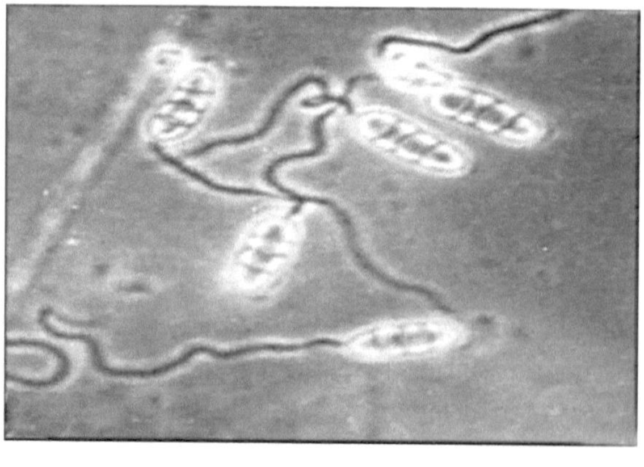

Figure 4.184 *Urohendersonia pongamiae;* Mature conidia (×900)

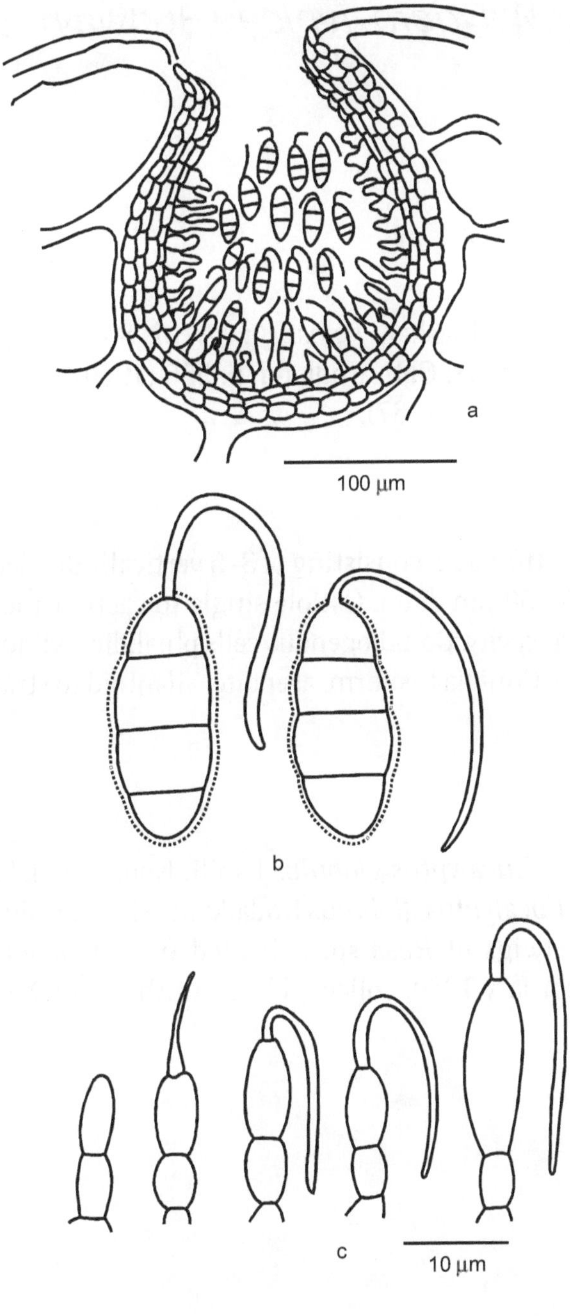

Figure 4.185 *Urohendersonia pongamiae*, (a) V.S. of conidioma; (b) Mature conidia; (c) Developing conidia.

142. *Waydora typica* (Rodway) Sutton

Source

Sutton, V.G. Rao & Mhaskar, *Trans. Br. Mycol. Soc.*, **67**: p. 248 (1976); Sutton, B.C., *The Coelomycetes*, CMI, England, p. 556 (1980); Muthumary, J., *Curr. Sci.*, **56** (23): pp. 1245–1247 (1987).

Description

Conidiomata eustromatic, consisting of 3–5 vertically divided segments, pulvinate, 600–750 × 200–250 µm diam. Ostiole single to each locule. Conidiophore lining the entire locular cavity. Conidiogenous cells phialidic, cylindrical, hyaline, smooth, 10–20 × 5–8 µm. Conidia fusiform, aseptate, allantoid to straight, hyaline, smooth, 5–8 × 1.5–2 µm.

Specimen collected

On dead fruits of *Eucalyptus globulus* Labill, Kodaikanal, Tamilnadu, Nov. 1987; On capsules of *Eucalyptus globulus* Kodaikanal, TamilNadu, August 1980, MUBL: 2918; on dead twigs of *Rosa* sp. collected from Lalbagh garden, Bangalore, Kodaikanal, February 1980, collected by J. Muthumary. MUBL No: 2919.

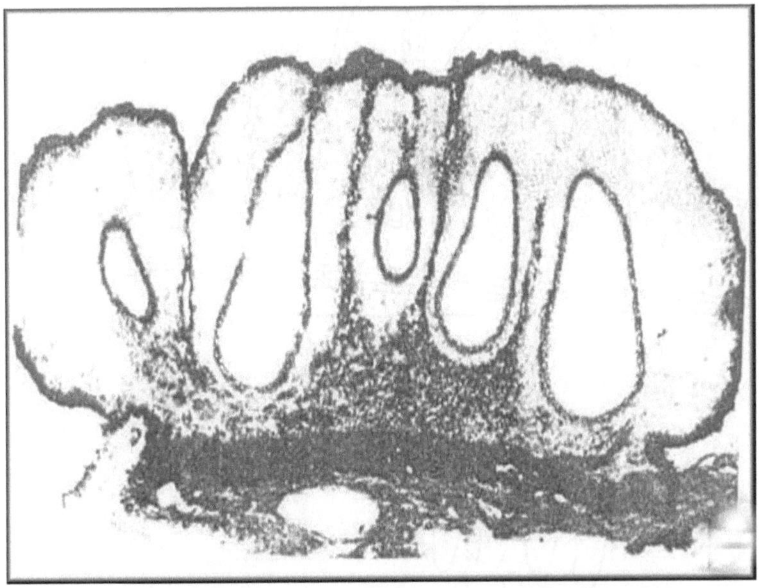

(a)

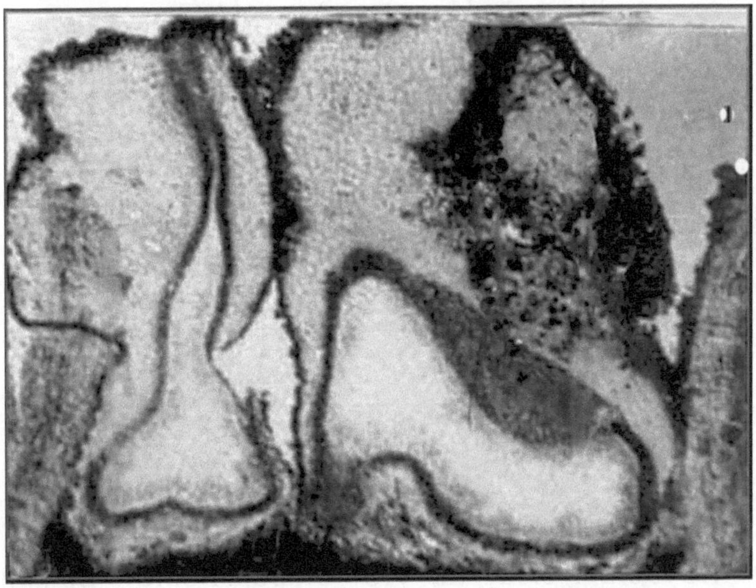

(b)

Figure 4.186 *Waydora typica;* (a–b) V.S. of conidiomata (×200)

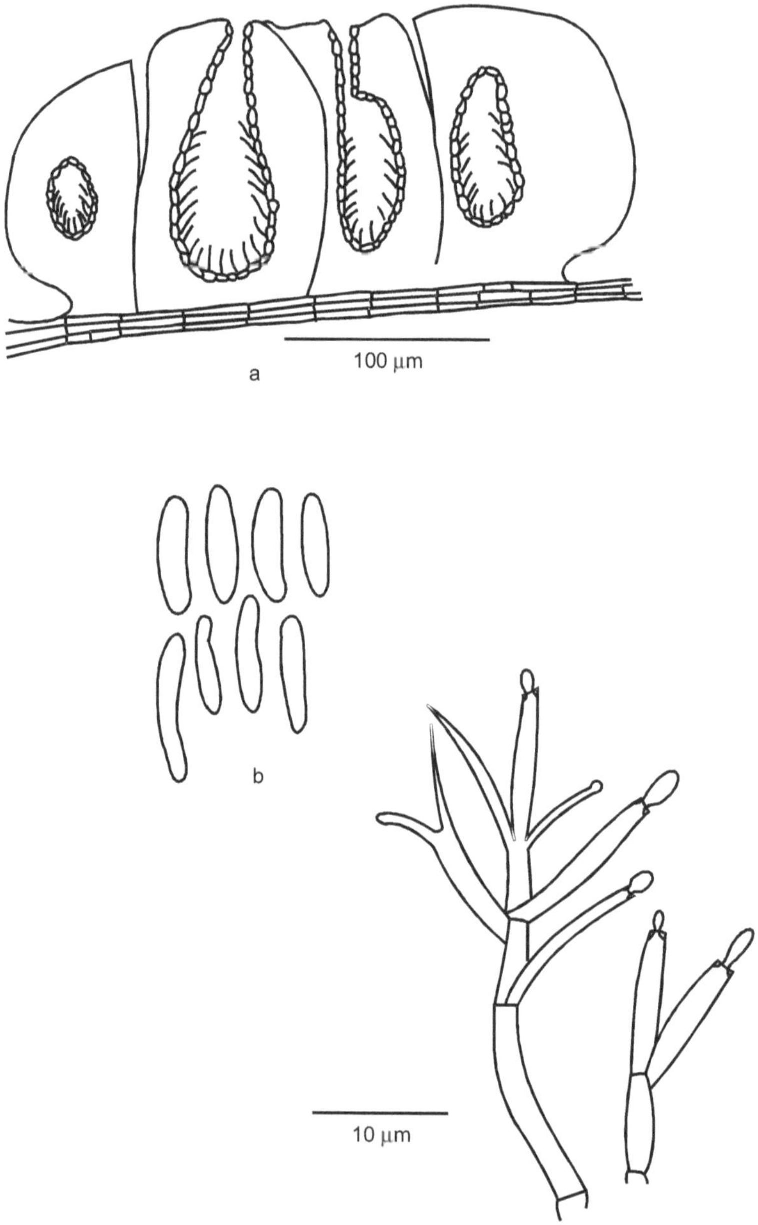

Figure 4.189 *Waydora typica*, (a) V.S. of conidioma; (b) Mature conidia; (c) Developing conidia.

5

TRANSMISSION ELECTRON MICROSCOPIC STUDIES

STILBOSPORA CISTINA—ELECTRON MICROSCOPIC STUDIES

The conidia are fusiform, brown and 4-celled; obtuse at the apex, truncate at the base with a basal frill. The base of each conidium bears a delicate, but distinct marginal frill which is easily observed upon detailed examination under phase-contrast illumination. The part of wall encircled by the frill, which was originally the septum that delimited the conidia, is somewhat thinner than the remainder.

In transmission electron microscopy conidium showed a single wall layer that was differentiated into an outer electron dense zone and an inner electron transparent zone. The pigmentation within conidia was located in somewhat as a continuous layer in the outer zone of the cell wall. Immature conidia lacked pigmentation as well as an electron dense layer within the conidial cell wall. The deposition of pigmented substances occurred throughout the entire circumstance of the outer cell wall zone. The conidium in *S. cistina* has an inner electron transparent zone.

In a longitudinal section through a mature conidium, the periclinal wall of the conidium is distinctly two-layered and the conidium is sheathed externally in a thick cuticle (Figure 5.1, Figure 5.2).

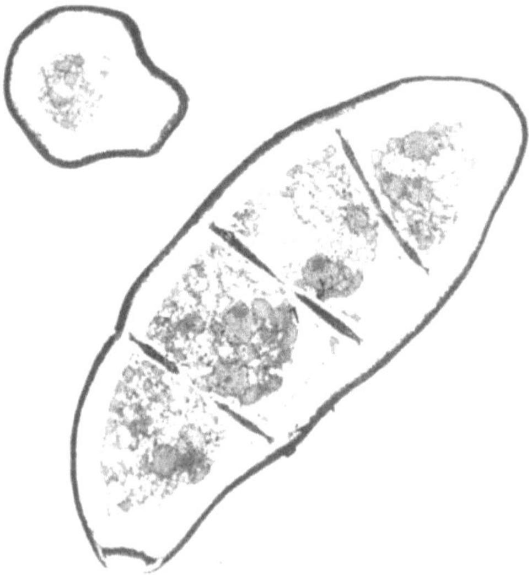

Figure 5.1 *Stilbospora cistina;* Electron micrograph of longitudinal section of mature conidium (×8000).

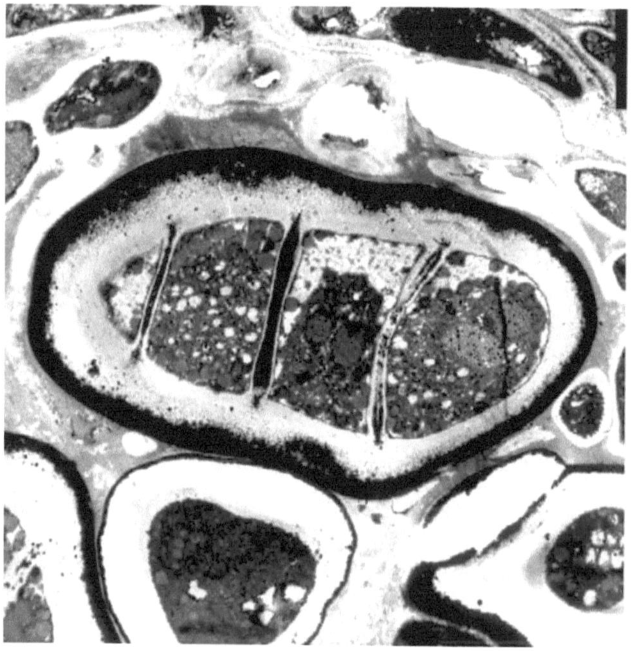

Figure 5.2 *Stilbospora cistina;* Electron micrograph of longitudinal section through a young conidium showing septum formation (×8000).

The outer layer of the wall is thicker than the inner layer and is distinctly granular with electron dense particles embedded towards the inner zone. It is this inner zone of the outer wall layer which extends into and across the septa. The inner layer of the wall immediately bounding the plasmalemma is thinner than the outer layer and takes part in septum formation (Figure 5.3).

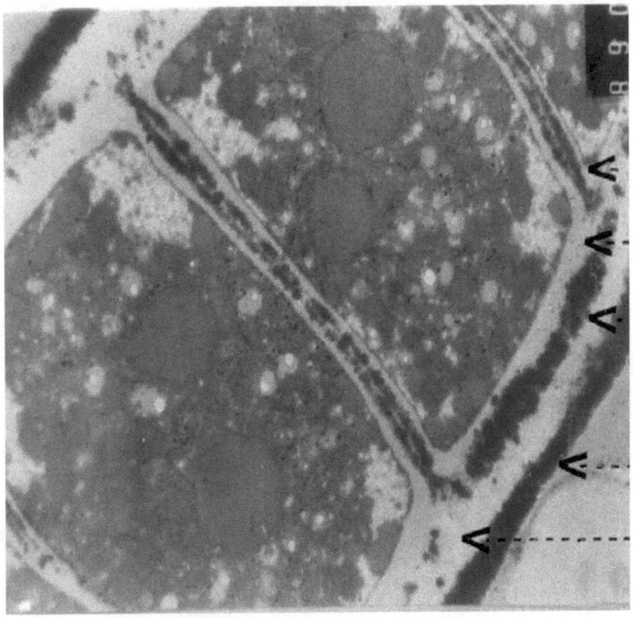

Figure 5.3 *Stilbospora cistina;* Electron micrograph of longitudinal section through a young conidium showing septum formation (×8000).

Conidiogenesis

The young conidium initial is delimited from the tip of the conidiogenous cell by the formation of a transverse wall at the base of the initial. Septal pore was noticed between the tip of the conidiogenous cell and the base of the developing conidium. The transverse wall is laid down by invagination of the lateral wall layer.

The cuticle of the conidium extends down to some extent on to the outer wall of the conidiogenous cell. Rupture of the conidium occurs where the transverse wall layer abuts on to the periclinal wall layer.

The geometry of the apex of the conidiogenous cell and the base of the conidium suggests that the secession is effected by circumscissile rupture of the outer periclinal wall.

The transmission electron microscopic studies show that the wall of the conidiophore is continuous with the wall of the conidium. But after the formation of the cuticle in the conidium, the basal frill becomes visible at the base of the conidium initial. Usually one flared annellation is visible in the present fungus (Figure 5.4).

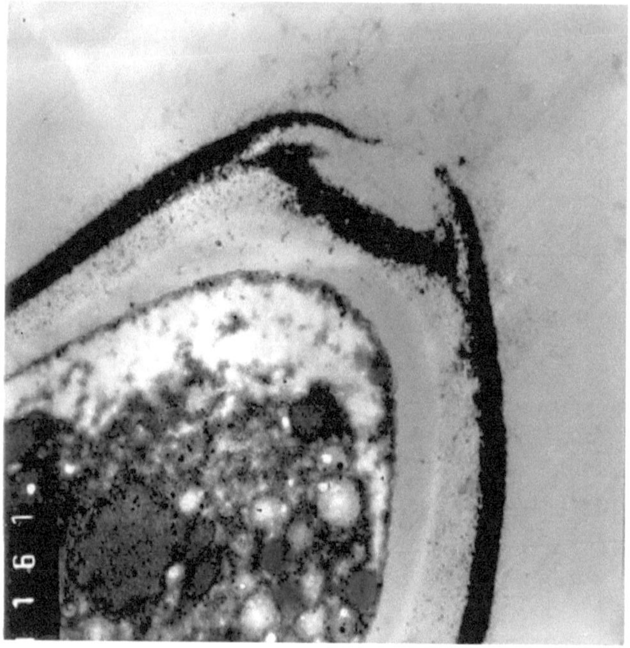

Figure 5.4 *Stilbospora cistina;* Electron micrograph of longitudinal section of basal cell showing frill (×8000).

Another interesting feature found in *S. cistina* is the presence of the second type of conidiogenous cell. These are very rarely found in pycnidial sections and they are pigmented, whereas the normal conidiogenous cells are hyaline. These are 1–2 celled, dark brown, 8–10 µm long, 5–6 µm wide, with 0–1 distinct, cup-like annellation.

PESTALOTIOPSIS DISSEMINATA — ELECTRON MICROSCOPIC STUDIES

The conidia consist of three thick-walled median cells capped by hyaline end cells. In the distilled water mounts of the conidia the wall of the three median cells appears granular and pale brown. During germination the germ tubes arise exclusively from the lowermost median cells. The hyaline basal cell bears an unbranched appendage. Appendages (4–5) also arise from the hyaline apical cell.

Median Coloured Cells

Transconidial septa arise as outgrowths of the conidial wall and new wall material is deposited external to the invaginations. During the growth of membrane across the conidium, wall material is continuously produced through the cell wall developed by each membrane. The septal pores are present between all the five cells of the conidium. The septum is formed as a result of cessation of wall deposition at the junction where the plasma membranes from either side fuse to form the trans-septal membrane. Simultaneously in the septal formation, wall deposition occurs over the entire inner surface of the conidium. The peripheral walls and the septa of the conidia become distinctly electron dense. The thickness is more in the upper two-median cells when compared with the lower median cell. The lower median cell is structurally different from the other median cells of the conidium in that it showed pronounced wrinkling of the outer pigmented wall (Figure 5.5). Probably because of this difference in the nature of the wall it appears pale brown under the light microscope.

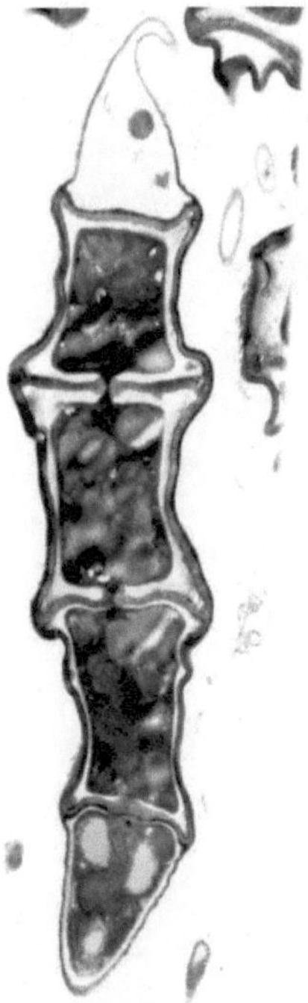

Figure 5.5 *Pestalotiopsis disseminata;* Section of conidium showing the basal cell with the appendage.

Apical and basal cells

The apical and basal cells are morphologically indistinguishable from the median cells during the early stages of development of the conidium. At maturity the end cells showed cytolysis and the cytoplasmic content completely disappears from these cells. The thicknesses of the electron-dense layer of the apical and basal cells gradually decrease.

Basal appendage The conidia were produced from the annellides and more than 5 annellations were observed in some conidiogenous cells (Figure 5.6 a, b). Occasionally the basal appendage of the developing conidium (Figure 5.7 and 5.8) was observed within the annellation, which proves that the basal appendage is endogenously produced.

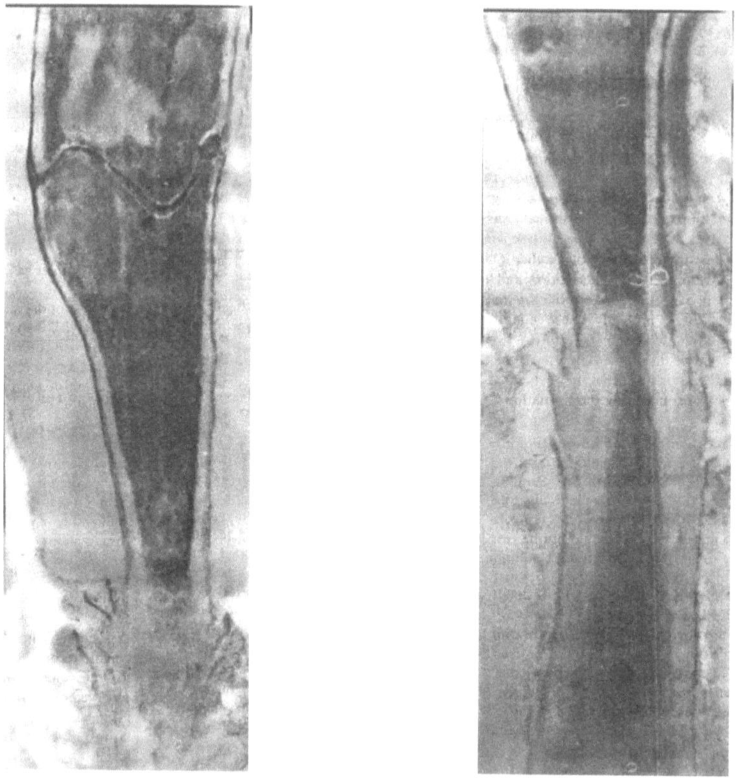

(a) (b)

Figure 5.6 *Pestalotiopsis disseminata;* (a–b) Annelations.

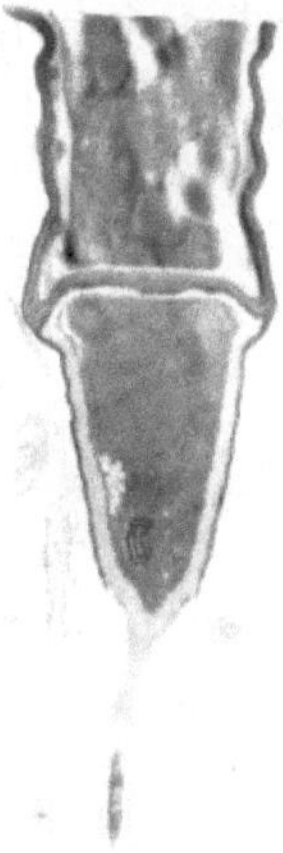

Figure 5.7 *Pestalotiopsis disseminata;* Section of conidiun showing lowermost median cell preparing for germination.

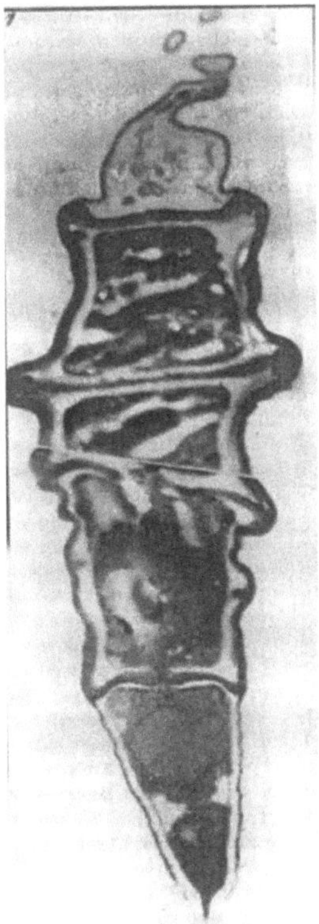

Figure 5.8 *Pestalotiopsis disseminata;* Longitudinal section of mature conidium showing the 4-septate conidium.

Apical appendage The apical appendage is originated from the apex of the conidium as a simple elongation of a small bud produced at an early stage of the development of the conidium.

REFERENCES

Arx. J.A. Von (1963). Revision der zu Gloeosporium gestellten Pilze. Nachtrage und Berichtigungen. Verh. K. Ned. *Akad. Wet. Amst.* **66**: 172-182.

Arx. J.A. Von (1970). A revision of the fungi classified as Gloeosporium. *Bibl. Myc.* **24**: 1-203.

Boerema, G.H. (1963). *Phoma herbarum* Westend., the type-species of the form-genus *Phoma* Sacc. *Persoonia* **3**: 9-15.

Cole, G.T. and Kendrick, W.B. (Eds.) 1981. *Biology of conidial fungi.* 2 vols. Academic Press, NY.

Grove, W.B. (1937). *British Stem- and Leaf-Fungi* **2**: 1-406. C.U.P.

Mordue, J.E.M. (1971). CMI *Descriptions of Pathogenic Fungi and Bacteria* nos. 315,316, 317.

Mordue, J.E.M. (1976). CMI *Descriptions of Pathogenic Fungi and Bacteria* nos. 131, 132, 133.

Morgan-Jones, G. (1964). *Taxonomic and biological studies in the Coelomycetes.* Ph.D. Thesis, University of Nottingham.

Morgan-Jones, G. (1971a), Conidium ontogeny in Coelomycetes. II. Some Melanconiales which possess phialides. *Can. J. Bot.* **49**: 1931-1937.

Morgan-Jones, G. (1971b). Conidium ontogeny in Coelomycetes. III. Meristem thalloconidia. *Can. J. Bot.* **49**: 1939-1940.

Morgan-Jones, G. (1971c). A new species of *Dinemasporium* from Ontario. *Can. J. Bot.* **49**: 1363-1365.

Morgan-Jones, G. and Kendrick, B. (1972). *Icones generum Coelomycetum. III.* Biology Series #3, University of Waterloo, Waterloo. 42 pp. (Revised edn. 1980).

Morgan-Jones, G., Nag Raj, T.R. and Kendrick, B. (1972a). Conidium ontogeny in Coelomycetes. IV. Percurrently proliferating phialides. *Can. J. Bot.* **50**: 2009-2014.

Morgan-Jones, G., Nag Raj, T.R., and Kendrick, B. (1972b). *Icones generum Coelomycetum. II.* Biology series #4, University of Waterloo. 40 pp.

Morgan-Jones, G., Nag Raj, T.R., and Kendrick, B. (1972c). *Icones generum Coelomycetum. IV.* Biology series #6, University of Waterloo. 42 pp.

Morgan-Jones, G., Nag Raj, T.R., and Kendrick, B. (1972d). *Icones generum Coelomycetum. V.* Biology series #6, University of Waterloo. 52 pp.

Morgan-Jones, G., Nag Raj, T.R., and Kendrick, B. (1972e). Genera Coelomycetarum. VI. Kellermania. *Can. J. Bot.* **50**: 1641-1648.

Munjal, R.L. (1966). The genus *Ciliochorella* in India. *Indian Phytopath.* **19**: 198-200.

Nag Raj, T.R. (1974). *Icones generum Coelomycetum VI.* Univ. Waterloo Biol. Ser. **13**: 1-41.

Nag Raj, T.R. (1975). Genera Coelomycetum XI. *Hyalotia, Hyalotiella* and *Hyalotiopsis. Can. J. Bot.* **53**: 1615-1624.

Nag Raj, T.R. (1977a). Miscellaneous microfungi II. *Can. J. Bot.* **55**: 757-765.

Nag Raj, T.R. (1977b). *Ypsilonia, Acanthotheciella,* and *Kazulia* gen.nov. *Can. J. Bot.* **55**: 1599-1622.

Nag Raj, T.R. (1985). Redisposals and redescriptions in the *Monochaetia-Seiridium, Pestalotia-Pestalotiopsis* complexes. I. The correct name for the type species of *Pestalotiopsis. Mycotaxon* **22**: 43-51.

Nag Raj, T.R. (1989). Redisposals and redescriptions in the *Monochaetia-Seiridium, Pestalotia-Pestalotiopsis* complexes. IX. On the status of *Pestalotia eupyrena, P. gastrolobi* and *Pestalotia veneta. Mycotaxon* **35**: 287-296.

Pirozynski, K.A. & Shoemaker, R.A. (1971). Some Coelomycetes with appendaged conidia. *Can. J. Bot.* **49**: 529-541.

Punithalingam, E. (1970). Studies on Sphaeropsidales in culture. *Mycol. Pap.* **119**: 1-24.

Punithalingam, E. (1974). Studies on Sphaeropsidales in culture. II. *Mycol. Pap.* **136**: 1-63.

Subramanian, C.V. and Ramakrishnan, K. (1954). *Alpakesa,* a new genus of Sphaeropsidales. *J. Ind. Bot. Soc.* **33**: 203-205.

Subramanian, C.V. and Ramakrishnan, K. (1956). *Ciliochorella* Sydow, *Plagionema* Subram. & Ramakr., and Shanoria gen. nov. *Trans. Br. Mycol. Soc.* **39**: 314-318.

Subramanian, C.V. and Reddy, K.R.C. (1974). The genus *Discosia.* I. Taxonomy. *Kavaka* **2**: 57-89.

Van der Aa, H.A. (1973). Studies in *Phyllosticta* I. *Studies in Mycology* **5**: 1-110.

BIBLIOGRAPHY

Archer, S.A., 1926. *Phomopsis pittospori* sp. nov., associated with leaf –spotting of *Pittosporum tenuifolium. Transactions of British Mycological society* **61**: 221-225.

Baxter, A.P. 1981. A study of the morphology and taxonomy of certain South African species of *Colletotrichum* Corda. M.Sc. Thesis, University of Pretoria, Pretoria, South Africa.

Baxter, A.P., G.C.A. Van Der Westhuiwen and A. Eicker. 1985. A review of literature on the taxonomy, morphology and biology of the fungal genus *Colletotrichum. Phytophylactia.* **17**: 15-18.

Bhuvaneswari, V. and J. Muthumary. 2005. Diversity and Seasonal influence of fungal endophytes in five medicinal plant species. *Kavaka.* Vol – **33**: 39 – 55.

Booth, C. 1971. Introduction to general methods. In: Methods in Microbiology, Vol. 4, Ed. Booth, C., Academic Press, London and NY. pp. 1-47.

Boerema, G.H. Dorenbosch, M.M.J. and Kesteren, H.A. Van. 1965. Remarks on species of *Phoma* referred to *Peyronellaea. Persoonia* **4**: 47-68.

Boerema, G.H. Dorenbosch, M.M.J. and Kesteren, H.A. Van. 1968. Remarks on species of *Phoma* referred to *Peyronellaea. Persoonia* **5**: 201-205.

Boerema, G.H. Dorenbosch, M.M.J. and Kesteren, H.A. Van. 1971. Remarks on species of *Phoma* referred to *Peyronellaea. Persoonia* **6**: 171-177.

Boerema, G.H. Dorenbosch, M.M.J. and Kesteren, H.A. Van. 1973. Remarks on species of *Phoma* referred to *Peyronellaea* – IV. *Persoonia* **7**: 131-139.

Boerema, G.H. Dorenbosch, M.M.J. and Kesteren, H.A. Van. 1977. Remarks on species of *Phoma* referred to *Peyronellaea* - V. *Kew Bull.* **31**: 533-534.

Brown, K.B., K.D. Hyde and D.I. Guest. 1998. Preliminary studies on endophytic fungal communities of *Musa acuminate* species complex in Hong Kong and Australia. *Fungal Diversity,* **1**: 27-51.

Clay, K. 1989. Clavicipitaceous endophytes of grasses: their potential as biocontrol agents. *Mycol. Res.*, **92**: 1-12.

De Bary, A. 1866. Morphologie und Physiologie der Pilze Flechten und Myxomyceten. Vol. II. Hofmeister's handbook of physiological botany. Leipzig.

Dorenbosch, M.M.J. 1970. Key to nine ubiquitous soil-borne *Phoma*-like fungi. *Persoonia* **6**: 1-14.

Guba, E.F. 1961. Monograph of *Monochaetia* and *Pestalotia*. Harvard University Press. Cambridge, Massachusetts. pp. 342.

Hamer, L., Adachi, K., Montenegro-Chamorro, M.V., Tanzer, M.M., Mahanty, S.K. Lo, C., Tarpey, R.W., Skalchunes, A.R. 2001. Gene discovery and gene function assignment in filamentous fungi. *Proc. Natl. Acad.* U.S.A. **98**: 5110-5115.

Hawksworth, D.L. 1974. Mycologist's Handbook. Commonwealth Mycological Institute, Kew. 231 pp.

Hawksworth, D.L. 2001. The magnitude of fungal diversity: the 1.5 million species estimate revisited. *Mycological Research* **105**: 1422-1432.

Hawksworth, D.L., Sutton, B.C. & Ainsworth, G.C. 1983. *Ainsworth & Bisby's Dictionary of the Fungi.* 7th Edition (CAB, IMI) Kew, Surrey UK. Pp. 445.

Hrushovetz, S.B. 1956. Cytological studies of *Helminthosporium* sativum. Can. J. Bot. **34**: 321-327.

Hughes, S.J. 1953a. Fungi from the Gold Coast. II *Mycol. Pap.* (CAB, IMI) Kew **50**:1-104.

Hughes, S.J. 1953b. Conidiophores, conidia and classification. Can. J. Bot. 31: 577-659.

Jayachandra J.A. and J. Muthumary. 1998. Development of conidiomata in *Gloeosporium musarum* Cke & Massee. *Journal of Indian Botanical Society* **77**: 83-86.

Jayachandra J.A. and J. Muthumary. 2001. Developmental studies in *Colletotrichum capsici* (Syd.) Butler & Bisby and its persistence in soil. *Journal of Indian Botanical Society* **80**: 163-167.

Kirk, P.M., P.F. Cannon, J.C. David and J.A. Stalpers. 2001. *Ainsworth and Bisby's Dictionary of the fungi.* 9th ed., CABI Bioscience, U.K. 655 pp.

Kumar, A., and Snyder, M. 2001. Emerging technologies in yeast genomics. *Nat. Rev. Genet.* **2**: 302-312.

Leach, C.M. 1962. Sporulation of diverse species of fungi under near-ultraviolet radiation. *Can. J. Bot.* **40**: 151-161.

Masilamani, S. and J. Muthumary. 1994. Development of conidiomata in *Ciliochorella mangiferae. Mycological Research* **98(8)**: 857-861.

Masilamani, S. and J. Muthumary. 1995. Pycnidium ontogeny in *Coleophoma cylindrospora. Mycological Research* **99(6)**: 693-696.

Masilamani, S. and J. Muthumary. 1996. Development of Conidiomata in *Botryodiplodia theobromae. Mycological Research* **100(11)**: 1383 – 1387.

Murugan, M. and J. Muthumary. 2001. Developmental morphology and Ultra structure of Pestalotiopsis maculans. *Mycotaxon.,* **LXXIX**: 455-465.

Murugan. M. and J. Muthumary. 2003. Studies on Developmental Morphology of the Conidiomata in *Pestalotiopsis uvicola* with a note on the Ultrastructure of the Conidia and Conidiogenous Cell. *Journal of Mycology and Plant Pathology* **33(2)**: 204-211.

Muthumary, J. 1986a. *Conostroma quercicola* sp. nov. from India. *Current Science* **55(21)**: 1081 - 1082.

Muthumary, J. 1986b. *Pilidium acerinum* Kunze, a new generic record for India. *Current Science* **55(16)**: 791-792.

Muthumary, J. 1986c. *Heteropatella indica* Muthumary sp. nov. from India. *Current Science* **55(5)**: 250 - 251.

Muthumary, J. 1987a. *Pseudolachnea longiciliata* (Hino & Katumato) Sutton. new record for India. *Current Science* **56**: 306 – 307.

Muthumary, J. 1987b. New records of Coelomycetes from India. *Current Science* **56(23)**: 1245 - 1247.

Muthumary, J. 1987c. Observations of conidiogenesis in *Entomosporium mespili* (DC EX DUBY). *Current Science* **56(4)**: 195 - 196.

Muthumary, J. 1987d. Notes on *Cytosphaera mangiferae. Current Science* **56(9)**: 426 - 427.

Muthumary, J. 1988. *Scyphosphora phyllostachydis* Kantschaveli, a new generic record for India. *Current Science* **57**: 621 - 622.

Muthumary, J. 1993. The gelatinous covering of a Phyllosticta condium-SEM study. The mycologist. **7**: 16.

Muthumary, J., and B.C. Sutton. 1986. *Coryneum quercinum* sp. nov. on *Quercus alba* from India. *Transactions of British Mycological society* **86(3)**: 512-515.

Muthumary, J., and R. Vanaja. 1986. Development of conidiomata in *Coniella fragariae. Transacions of British Mycological Society* **87(1)**: 109-113.

Muthumary, J., and S. Masilamani. 1989. Additions to Coelomycetes of India. *Journal of Indian Botanical Society* **68**: 131- 133.

Muthumary, J., S.Q. Abbas and B.C. Sutton. 1986. A reassessment of *Monochaetia terminaliae Transactions of British Mycological Society* **87(1)**: 109-113.

Muthumary, J., A. Jayachandra and B. Preetha. 1993. Development of conidiomata in the *Phyllosticta* state of *Guignardia mangiferae* Roy and observations on the structure of the conidium. *Mycotaxon* XLVII: 147 – 155.

Nagraj. T.R. 1981. Genera Coelomycetarum XIX. *Discosiella,* a licjhenized mycobiont. *Canadian Journal of Botany.* **59**: 2519-1533.

Nagraj. T.R. 1983. Genera Coelomycetarum. XXI. *Strasseria,* and two new anamorph-genera: *Apostrasseria* and *Nothostrasseria. Canadian Journal of Botany.* **61**: 1-30.

Nagraj, T.R. 1993. Coelomycetous anamorphs with appendage-bearing Conidia. Mycologue Publications. Waterloo.

Nagraj, T.R. and W.B. Kendrick. 1971. Genera Coelomycetum. I. *Urohendersonia. Canadian Journal of Botany.* **49** : 1853-1862.

Nagraj, T.R. and W.B. Kendrick 1972a . Genera Coelomycetum II *Doliomyces. Canadian Journal of Botany.* **50**: 45-48.

Nagraj, T.R. and W.B. Kendrick. 1972b. Genera Coelomycetarum III *Pestalozziella. Canadian Journal of Botany.* **50**: 607-617.

Nagraj, T.R. and W.B. Kendrick. 1978. Genera Coelomycetarum XV. *Belaina, Belainopsis* and *Crucellisporium. Canadian Journal of Botany.* **56**: 708-714.

Nag Raj, T.R. and Kendrick, W.B. 1981. *Infundibura,* a new hyphomycete with unique appendages. Can. J. Bot. **59**: 542-546.

Nag Raj, T.R. and Kendrick, B. 1986a. On *Dinemasporium* adeanum Petr. *Mycotaxon* **25**: 15-18.

Nag Raj, T.R. and Kendrick, B. 1986b. Ellurema gen. nov., with notes on *Lepteutypa cisticola* and *Seiridium canariense. Sydowia* **38**: 178-193.

Nicosia, M.G.L., Brocard-Masson, C., Demais, S., Van, H., Daboussi, M.J., and Scazzocchio, C. 2001. Heterologous transposition in *Aspergillus nidulans, Mol. Microbiol.* **39**: 1330-1344.

Phizicky, E.M. and Fields, S. 1995. Protein-protein interactions: Methods for detection and analysis. *Microbiol. Rev.* **59**: 94-123.

Pumpyanskaya, L.V. 1964. *Mikrobiologiya* **33**: 1065-1070.

Punithalingam, E. 1970a. Studies on *Sphaeropsidales* in culture. *Mycol. Pap.* (CAB, IMI) Kew **119**: 1-24.

Punithalingam, E. 1970b. Spore surface ornamentation in *Coniothyrium* species. *Trans. Br. Mycol. Soc.* **55**: 154-156.

Punithalingam, E. 1970c. A new species of *Coniothyrium. Trans. Br. Mycol. Soc.* **58**: 225-230.

Punithalingam, E. 1974a. Cytology of *Septoria leucanthemi. Trans. Br. Mycol. Soc.* **63**: 255-260.

Punithalingam, E. 1974b. Studies on *Sphaeropsidales* in culture II. *Mycol. Pap.* (CAB, IMI) Kew **136**: 1-63.

Punithalingam, E. 1979. Graminicolous *Ascochyta* species. *Mycol. Pap.* (CAB, IMI) Kew, **142**: 1-214.

Punithalingam, E. 1981. Studies on *Sphaeropsidales* in culture III. *Mycol. Pap.* (CAB, IMI) Kew **149**: 1-42 + 18 plates.

Punithalingam, E. 1983. The nuclei of *Macrophomina phaseolina* (Tassi) Goid. *Nova Hedwigia* **38**: 339-367.

Punithalingam, E. 1984a. The nuclei and conidial appendages of *Robillarda sessilis* (Sacc.) Sacc. *Nova Hedwigia* **40**: 45-58.

Punithalingam, E. 1984b. The nuclei and 'conidial appendages' of *Chaetoconis polygoni* (Ell. & Ever.) Clem. *Nova Hedwigia* **40**: 99-112.

Punithalingam, E. 1989a. Techniques for staining fungal nuclei and appendages. *Bot. J. Linn. Soc.* **99**: 19-32.

Punithalingam, E. 1989b. The significance of nucleate conidial appendages in Coelomycetes. *Studies in Mycology* **31**: 113-121.

Punithalingam, E. 1989c. The nuclei and conidial arms in *Crucellisporium* species. *Nova Hedwigia* **48**: 290-307+8 plates.

Punithalingam, E. & J.M. Waller, 1973. CMI Descriptions of pathogenic fungi and bacteria No. 394.

Punithalingam, E. & Woodhams, J.E. 1982. The conidial appendage in *Phyllosticta* spp. *Nova Hedwigia* **36**: 151-198.

Punithalingam, E. & Woodhams, J.E. 1984. A modified Leifson's flagella staining technique for revealing appendages in fungi. *Nova Hedwigia* **40:** 31-43.

Reddell, R.W. 1950. Permanent stained mycological preparations obtained by slide culture. *Mycologia.* **42**: 265.

Royer, J.C., Christanson, L.M., Yoder, W.T., Gambetta, G.A., Klotz, A.V., Morris, C.L., Brody, H., and Otani, S. 1999. Deletion of the trichodiene synthase gene of *Fusarium venenatum*: Two systems for repeated gene deletions. *Fungal Genet. Biol.* **28**: 68-78.

Stierle, A. G.A. Strobel, and D.C. Stierle, 1995. The search for a taxol producing microorganism among the endophytic fungi of the Pacific Yew, *Taxus brevifolia. J. Nat. Prod.* **58**: 1315-1324.

Strobel, G.A. 2003. Endophytes as source of bioactive products. *Microbes and Infection* **5**: 535-544.

Strobel, G.A., A. Stierle, D. Stierle, W.M. Hess, 1993. *Taxomyces andreanae* a proposed new taxon for a bulbilliferous hyphomycete associated with Pacific Yew, *Mycotaxon* **47**:71-78.

Subramanian, C.V. 1961. On some species of *Cryptostictis*. *Proceedings of the National Institute of Scoences of India*. **27**(4): 236-245.

Subramanian, C.V. and K. Ramakrishnan. 1953(a). *Plagionema*, a new genus of the Sphaeropsidales. *J. Ind. Bot. Soc.* **32**:131-136.

Subramanian, C.V. and K. Ramakrishnan. 1953(b). On the Nature of the spore appendage in *Neottiospora* Desm. *Proc. Indian Academy of Sciences*. Sect. B, **37**(6): 228-231.

Subramanian, C.V. and K. Ramakrishnan. 1954. *Alpakesa*, A new genus of the Spheropsidales. *J. Ind. Bot. Soc.* **33** (4).

Subramanian, C.V. and K. Ramakrishnan. 1958. Notes on some species of *Cryptostictis* – I. *J. Ind. Bot. Soc.* Vol. XL. No.4: 578-591.

Subramanian, C.V. and K.R.C. Reddy. 1974. *Discosiella* Sydow and *Discosiellina* gen. nov. *Proc. Indian Academy of Sciences*. sect. B, **75(3)**: 111-116.

Subramanian, C.V. and K.R.C. Reddy. 1974. *Discosiella* Sydow and *Discosiellina* gen. nov. *Proc. Indian Academy of Sciences*. sect. B, **75(3)**: 111-116.

Suryanarayanan, T.S. and Kumaresan, V. 1998. A simple method of storing and transporting fungal cultures. *Mycologist* **12(4)**: 173.

Suryanarayanan, T.S. T.S. Muralli and G. Venkatesan. 2002. Occurrence and distribution of fungal endophytes in tropical forests across a rainfall gradient. *Can. J.Bot.*, **80**: 818-826.

Sutton, B.C. 1961. Coelomycetes, I., Part A: Developmental studies in *Pestalotiopsis*; Part B: Five species of *Pestalotiopsis*. *Mycol. Pap.* (CAB, IMI) Kew **80**: 1-16.

Sutton, B.C. 1962. *Colletotrichum dematium* (Pers Fr.) Grove and *C. trichellum* (Fr. Ex Fr.) Duke. *Transactions of British Mycological Society*. **45**: 222-232.

Sutton, B.C. 1963. Coelomycetes II. *Neobarclaya, Mycohypallage, Bleptosporium* and *Cryptostictis*. *Mycol. Pap.* (CAB, IMI) Kew 1-50.

Sutton, B.C. 1965. Studies on the taxonomy of *Colletotrichum* Cda with especial reference to *C. graminicola* (Ces.) Wilson, Ph.D. Thesis, University of London.

Sutton, B.C. 1967a Redescription of *Ajrekarella* Kamat & Kalani. *Mycopath. Mycol. appl.* **33**: 76-80.

Sutton, B.C. 1967b, Two new genera of the Sphaeropsidales and their relationships with *Diachorella, Strasseria* and *Plagiorhadbus*. Can. J. Bot. **45**: 1249-1263.

Sutton, B.C. 1968a. *Kellaermania* and its generic segregates. *Can. J. Bot.* **46**: 181-196.

Sutton, B.C. 1968b. The appressoria of *Colletotrichum graminicola* and *C. falcatum*. *Canadian Journal of Botany*. **46**: 8873-876.

Sutton, B.C. 1968c. *Polynema*, an earlier name for *Neobarclaya*. *Mycologia* **60**: 201-203.

Sutton, B.C. 1969. Forest microfungi. III. The heterogeneity *of Pestalotia* de Not. Section Sexloculatae Klebahn sensu Guba. *Can. J. Bot.* **47:** 2083-2094.

Sutton, B.C. 1971. Coelomycetes, IV. The genus *Harknessia* and similar fungi on *Eucalyptus. Mycol. Pap.* (CAB, IMI) Kew **123:** 1-46.

Sutton, B.C. 1972a. *Wakefieldia punctata* Arnaud and Blennoria buxi Fr. Sensu Arnaud. *Trans. Br. Mycol. Soc.* **59**: 285-294.

Sutton, B.C. 1972b. Nomenclature of *Ceuthospora, Pyrenophora* and *Blennoria* (Fungi). *Taxon* . **21**: 319-326.

Sutton, B.C. 1973. Ch.11. Coelomycetes, pp. 513-582 in *The Fungi IVA* (Eds. G.C. Ainsworth, F.K.Sparrow & A.S. Sussman). Academic Press.

Sutton, B.C. 1974. Miscellaneous Coelomycetes on *Eucalyptus. Nova Hedwigia* **25**: 161-172.

Sutton, B.C. 1975. Coelomycetes V. *Coryneum. Mycological Papers.* 138: 1-224.

Sutton, B.C. 1977. Coelomycetes VI. Nomenclature of generic names proposed for Coelomycetes. *Mycological Papers.* 141-1-253.

Sutton, B.C. 1980. *The Coelomycetes Fungi imperfect with Pycnidia and Acervuli and Stromata.* Common Wealth Mucological Institute, England. 696 pp.

Sutton, B.C. & Alcorn, J.L. 1984. Additions to *Hyphopolynema* (Hyphomycetes). *Aust. J. Bot.* **32:** 551-559.

Sweigard, J.A., and Ebbole, D.J. 2001. Functional analysis of pathogenicity genes in a genomics world. *Curr. Opin. Microbiol.* **4**: 387-392.

Talbot, N. 2001. Molecular and Cellular Biology of Filamentous Fungi. A Practical Approach. Oxford University Press, Oxford.

Wani, M.C. H.L. Taylor, M.E. Wall, P. Coggon and A.T. McPhail. 1971. Plant antitumor agents VI. The isolation and structure of Taxol, a novel antileukemic and antitumor agent from *Taxus brevifolia. Journal of American Chemical Society* **93**: 2325-2327.

Yang, Y.H. and Speed, T. 2002. Design issues for cDNA microarray experiments. *Nature Rev. Genet.* **3**: 579-588.

Zhang, Y,-Z., Perry, K., Vinci, V.A., Powell, K., Stemmer, W.P.C., and del Cardayre, S.B. 2002. Genome shuffling leads to rapid phenotypic improvement in bacteria. *Nature* **415**: 644-646.

FUNGUS INDEX

R

S

HOST INDEX

O

P

Q

R

S

T

U

V

W

Y

Z

HOST FUNGUS INDEX

D

E

S

T